TEACHING MATHEMATICS TO CHILDREN WITH SPECIAL NEEDS

Carol A. Thornton
Benny F. Tucker
John A. Dossey
Edna F. Bazik
Mathematics Department
Illinois State University

▲▼ ADDISON–WESLEY PUBLISHING COMPANY
Menlo Park, California • Reading, Massachusetts
London • Amsterdam • Don Mills, Ontario • Sydney

Acknowledgments   The processes involved in the development and eventual publication of a text are numerous. To credit everyone involved in one facet or another of the operation is impossible. We are indebted to our teachers, who prepared the way for us; to our students, who have taught us as we attempted to teach them; to our colleagues, who have served as sounding boards; and to our reviewers, who have provided us with constructive and frank comments. Special credit must go to our department chairman Dr. Albert Otto for his support and assistance throughout the project. Additional credit also goes to Hope Barrowes, Anne Dossey, Marilyn Parmantie, Gayle Pilon, and Jody Tucker for their rapid and efficient assistance in the various stages of the preparation of the manuscript. Above all, we thank the members of our families for their continued support and understanding throughout the writing and production of the text.

QA
135.5
.T43
1983

Text and Cover Designer: Michelle Taverniti
Illustrators: Valerie Randall; Ellen Blonder

This book is published by the Addison-Wesley Innovative Division.

ISBN-0-201-07728-0
ABCDEFGHIJK-DO-898765432

PREFACE

Teaching Mathematics to Children with Special Needs has evolved from our work with elementary, special education, and middle school teachers over the last few years. The text covers a wide variety of topics related both to the content and methods of teaching elementary school mathematics, as well as presenting a program for dealing with the special needs child in the classroom.

In writing this text, the authors interpret the term *special needs child* to mean a child who needs extra instructional attention due to a learning problem, a learning disability, or a physical or sensory handicap. However, the materials and methods presented in the text need not be restricted to use with special needs children alone. Well-planned instruction based on a carefully conceived diagnostic and developmental foundation is appropriate for *all* children.

The first five chapters of the text provide an introduction to the mathematics curriculum, the special needs child, the diagnostic-prescriptive model for instruction, models for learning mathematics, and aspects of classroom management related to classroom instruction of mathematics. The remaining eight chapters focus on teaching the content of elementary school mathematics within this framework. This content is first presented through the interaction of children with *concrete materials* as they model concepts and operations. Later, activities are introduced that develop the *appropriate language* and the *accompanying symbolism*. These types of educational experiences appear throughout the text in a wide variety of settings.

Special features of the text include:

What Do YOU Say? These sections give the reader a chance to reflect on key questions before reading the chapter. They help focus the reader's thought on the major points to be discussed.

Activity File

These sections of the text contain classroom learning activities related to specific mathematical concepts and skills and specific learning objectives. They provide the teacher with a wealth of carefully designed and tested activities for all children.

Case Studies

The case studies relate situations involving actual children to the application of the content and methods discussed in the chapter. They show how the materials and activities have been used in the classroom. These case studies have resulted from the authors' experiences and those of Nancy Bley, a learning disabilities teacher at Century Park School in California.

Troubleshooting

These sections focus on the application of methods for correcting common computation problems. These activities simulate the decision-making processes teachers have to employ in the classroom.

Your Time

These end-of-chapter sections contain activities and questions designed to relate the text to classroom practices. Many of these activities involve work in clinical settings or classrooms. Others involve the analysis of additional case studies by the reader.

The textbook concludes with three appendices. Appendix A includes a typical set of K-8 mathematics objectives. Appendix B lists sources of diagnostic tests and teaching aids. The companies named offer a wide variety of materials that you can examine by writing for catalogs. Appendix C is a mathematics laboratory section containing activities: diagnostic/evaluative, developmental, practice, and application. Topics for each of these activities are correlated with the chapter titles.

TABLE OF CONTENTS

1 THE MATHEMATICS CURRICULUM: CONTENT AND TEACHER EXPECTATIONS

Nurturing the development of mathematical ideas in young children is a challenging but rewarding task. When children have special learning needs, innovative or individualized methods are often necessary. This calls for teacher originality and perseverance grounded in an understanding of the child's learning patterns and a knowledge of both standard and adaptive procedures for teaching mathematics that meet the goals of a balanced curriculum. This book was written to help teachers deal more effectively with the handicapped learner, the slow learner and the special learner. Chapter 1 focuses on the mathematics curriculum for grades K–8 and on adaptations that must be made for children with special needs. We will also consider the role of the teacher in implementing the mathematics curriculum.

FACTORS THAT SHAPE THE K–8 MATHEMATICS CURRICULUM

The term *curriculum* can have many different meanings. We will use it to mean *the total set of general and specific content goals together with suggested means for achieving them.* The mathematics curriculum for a particular school is sometimes developed at the state level. In other cases the curriculum is determined at a district level—or even by the local school itself. Once adopted, the set of mathematics goals and methods provides the framework for teaching mathematics in the school.

> **How Would You Handle These Situations?**
>
> 1. Dear Ms. Jones:
>
>   I have been looking at the mathematics textbook you are using this year at Everyday Elementary School. I think that it is bad news! Why does it have all of that material on graphing and measurement in it? I thought that this was supposed to be a mathematics text! Why are

my children studying geometry instead of addition, subtraction, multiplication, and division? I thought that the move today was back to the basics. It appears to me that the planning and coordination for your program is dead—and has been for over fifteen years by the look of the text you are currently using. What plans are being made to resurrect it, so that you can meet the *real* needs of our children?

Sincerely,

Ima Always Watching

Ima Always Watchin

2. You have just been appointed to the textbook selection committee for your school district's adoption of a new K–8 mathematics series. What are the factors you would want to consider in the selection of the series? What topics should be included? What emphasis should be placed on each of the topics? What is the role of numerical calculation in relation to problem solving and geometrical learning? Sketch a brief outline of your plan for text selection. What features would be helpful to teachers dealing with special learning needs?

The development of an appropriate mathematics curriculum is based on many factors, the strongest of which is probably tradition. However, the ideas and influence of professional and pressure groups do help shape the direction of the mathematics programs for grades K–8.

The Back-to-Basics Movement and General Goals

It is a difficult task to state the general goals of the mathematics curriculum for grades K–8, specify learner objectives for these goals and list the methods by which these intended learnings might be maximized. The first step for a particular school district is to establish general goals. Once this step has been taken, defining specific learner objectives is much easier.

The statement of general goals for the mathematics curriculum is the central issue in the recent back-to-the-basics movement. This issue is not new, but it is exceedingly difficult to deal with. Part of the problem lies in the way general goals are stated and part lies in deciding the relative emphasis for each goal. Most people agree that students should be able to compute with whole numbers, fractions and decimals; to apply basic geometric ideas; to carry out everyday measurement situations; and to solve problems. Others add the additional objective of developing positive attitudes toward the learning of mathematics.

Perhaps the most publicized statement of goals for a basic mathematics program is that given by the National Council of Supervisors of Mathematics (NCSM) in 1977. Their list of ten basic skill areas appears in the accompanying box.

The Basic Skill Areas

Problem Solving. Learning to solve problems is the principal reason for studying mathematics. Problem solving is the process of applying previously acquired knowledge to new and unfamiliar situations. Solving word problems in texts is one form of problem solving, but students also should be faced with nontextbook problems. Problem-solving strategies involve posing questions, analyzing situations, translating results, illustrating results, drawing diagrams, and using trial and error. In solving problems, students need to be able to apply the rules of logic necessary to determine which facts are relevant. They should be unfearful of arriving at tentative conclusions and they must be willing to subject these conclusions to scrutiny.

Applying Mathematics to Everyday Situations. The use of mathematics is interrelated with all computation activities. Students should be encouraged to take everyday situations, translate them into mathematical expressions, solve the mathematics, and interpret the results in light of the initial situation.

Alertness to the Reasonableness of Results. Due to arithmetic errors or other mistakes, results of mathematical work are sometimes wrong. Students should learn to inspect all results and to check for reasonableness in terms of the original problem. With the increase in the use of calculating devices in society, this skill is essential.

Estimation and Approximation. Students should be able to carry out rapid approximate calculations by first rounding off numbers. They should acquire some simple techniques for estimating quantity, length, distance, weight, etc. It is also necessary to decide when a particular result is precise enough for the purpose at hand.

Appropriate Computational Skills. Students should gain facility with addition, subtraction, multiplication, and division with whole numbers and decimals. Today it must be recognized that long, complicated computations will usually be done with a calculator. Knowledge of single-digit number facts is essential and mental arithmetic is a valuable skill. Moreover, there are everyday situations which demand recognition of, and simple calculation with, common fractions.

Because consumers continually deal with many situations that involve percentage, the ability to recognize and use percents should be developed and maintained.

Geometry. Students should learn the geometric concepts they will need to function effectively in the three-dimensional world. They should have knowledge of concepts such as point, line, plane, parallel, and perpendicular. They should know basic properties of simple geometric figures, particularly those properties which relate to measurement and problem-solving skills. They also must be able to recognize similarities and differences among objects.

Measurement. As a minimum skill, students should be able to measure distance, weight, time, capacity, and temperature. Measurement of angles and calculations of simple areas and volumes are also essential.

Students should be able to perform measurement in both metric and customary systems using the appropriate tools.

Reading, Interpreting, and Constructing Tables, Charts, and Graphs. Students should know how to read and draw conclusions from simple tables, maps, charts and graphs. They should be able to condense numerical information into more manageable or meaningful terms by setting up simple tables, charts and graphs.

Using Mathematics to Predict. Students should learn how elementary notions of probability are used to determine the likelihood of future events. They should learn to identify situations where immediate past experience does not affect the likelihood of future events. They should become familiar with how mathematics is used to help make predictions such as election forecasts.

Computer Literacy. It is important for all citizens to understand what computers can and cannot do. Students should be aware of the many uses of computers in society, such as their use in teaching/learning, financial transactions, and information storage and retrieval. The "mystique" surrounding computers is disturbing and can put persons with no understanding of computers at a disadvantage. The increasing use of computers by government, industry, and business demands an awareness of computer uses and limitations.

These ten skill areas have been well-received by mathematics educators as a basic set of goals for the study of mathematics. While the ten areas are not totally disjoint, they do clearly point the direction for a minimal program in mathematics.

Your Time: Activities, Exercises, and Investigations

1. Write out the answers to the two questions posed in the box on pages 2–3.

2. Use two elementary-school mathematics series published since 1978 and skim them to see which of the ten skill areas mentioned above are covered in the texts. Which goals were emphasized? Which were only minimally covered or ignored?

3. Talk to at least three people: a parent of a school-aged child; a public school teacher; and a person over fifty years of age. Ask them what mathematics they think should be taught in grades K–8. What do they list as the goals? What comments do they have about the present curriculum? How informed do you think they are about the present school mathematics program?

4. Write out a personal reaction to the goal list given by the National Council of Supervisors of Mathematics. As you consider the topics they list, do you find any which you think should be stressed more than others? If so, which ones and why? What alterations, if any, should be made in the list as the needs of the special child are considered?

Stating Specific Learner Objectives

Once the general objectives for the mathematics program have been agreed upon, the next step is the consideration of specific learner objectives. These are statements that clearly indicate what a child is to be able to do at the end of a given period of instruction. Specific learner objectives are written for major strand topics for a given grade level.

Appendix A of this text gives a sample set of specific learner objectives for a K–8 program arranged under the strands of Number and Notation, Operations, Geometry and Measurement, and Problem Solving. There are many other such lists of objectives. These are the *minimal* objectives for such a program. Other available lists use headings such as Arithmetic, Number and Operation, Geometry, Measurement, Problem Solving/Applications, Probability and Statistics, Relations and Functions, and Logical Thinking.

An examination of these lists reveals that much more than mere computational ability is involved. Most lists cover essentially the same areas, though differences in emphasis from one topic to another are noted. An analysis of the specific learner objectives indicates the emphasis assigned.

Each learner objective usually contains three parts. The first is a *statement of the specific activity* that the learner is expected to perform. The second is the *conditions under which the student is expected to perform*. This may include using a calculator or not, using measuring instruments or not, and working with friends or not. The third portion of an objective is the *level of performance expected* from the child. An example of such an objective is the following: Given three minutes the student can correctly write the products to 95 percent of the one hundred basic multiplication facts arranged in random order.

Learner objectives written for each grade in a school system clearly define the expectations placed upon the students. In general it is not the job of the teacher to write such objectives. Rather it is the responsibility of the teacher to see that students can meet the expectations set out by the objectives. For teachers on school or district teams who are expected to write such objectives, there is a wealth of texts on the market that provide ample instruction and assistance.

Adapting the Curriculum for the Special-Needs Child

Until recent times little modification of the mathematics curriculum was made for children with special learning needs. Children were expected to adapt to the mathematics curriculum as it existed or take some sort of terminal sequence of courses developed for low-ability students.

Times have changed. Pressure brought by parents of children with special learning needs, the educational community, and the political programs of the 1960s and 1970s led to the formulation and passage of Public Law 94–142: The Education for All Handicapped Children Act of 1975. This law requires that school districts must provide an appropriate program of instruction for the children with special learning needs who reside within its boundaries. Further, this instruction must take place in an educational setting that places the least number of restrictions on the program of studies offered to the child.

By law, the school district must write and file an annual individualized educational program (IEP) for each handicapped learner. IEPs are required for each handicapped learner from age 3 to 21 who is enrolled in the school's programs. For district students not in the schools, the district must make provisions for appropriate instruction befitting the child's condition and needs. The IEPs constitute an individualized curriculum for each special-needs student. As such, they are required to contain, at a minimum, the following information:

1. A statement of the child's present levels of educational performance.

2. A list of annual goals for the child's progress.

3. A list of short-term instructional objectives.

4. Specific educational support services to be provided for the child, including starting dates and the duration of each service.

5. The percentage of time the child is to spend in regular classroom learning situations.

6. Objective criteria and evaluation procedures to determine if the short-term goals have been met on at least an annual level.

7. The physical education program in which the child will participate.

8. Any special media or materials needed for the child's special program.

9. A justification of the child's program and educational placement.

10. A list of individuals responsible for implementing the student's IEP.

As you can see, the construction of each student's IEP is intended to bridge the gap between the regular curriculum and the special-needs curriculum. It may also be the case that major efforts will be needed to create new materials and combine other materials that already exist in order to bring these areas together to satisfy the IEP. In any case, the teacher dealing with the special-needs child in mathematics will have to be familiar with both the regular and special-needs curricula.

Figure 1.1 indicates that the regular classroom is, generally, the least restrictive environment for the education of the special-needs child. Other more restrictive educational settings are pictured on the steps which lead up from the regular classroom. The IEP assessment must identify and justify the least restrictive program for each individual child. This frequently involves the *mainstreaming* of the child for mathematics instruction, that is, the assignment of the child to a regular mathematics classroom for full- or part-time mathematics instruction.

Such mainstreaming can never be done indiscriminately. For many children, especially those with severe handicaps, such a program would be totally inappropriate. Mainstreaming for mathematics, whatever the form, should be the product of informed decisions based on an understanding of the child's capabilities and self-confidence as well as the child's predicted ability to adapt to and succeed in a regular mathematics classroom. When such a placement is made, it should be with all of the support facilities and faculty required to make it a success.

Continual assessment of a student's progress and abilities is crucial to the IEP model developed for the education of special-needs students. This book is designed to prepare you to carry out an initial diagnosis of the child's mathematical abilities and then to provide mathematics instruction aimed at meeting these needs.

What Do *You* Say?

1. Why is it important for the special-needs teacher to understand both the characteristics of the regular classroom child and the characteristics of the special-needs child?

2. In gathering information about a child's handicap, what type of information is most helpful?

3. What types of people might serve on a child's IEP resource team in the area of mathematics instruction?

4. What types of special equipment might be needed for the successful completion of a child's IEP program of study in mathematics?

Figure 1.1

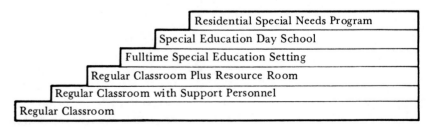

Residential Special Needs Program
Special Education Day School
Fulltime Special Education Setting
Regular Classroom Plus Resource Room
Regular Classroom with Support Personnel
Regular Classroom

THE TEACHER'S ROLE IN THE MATHEMATICS CLASSROOM

Within the classroom the teacher's job is to focus on each child's progress toward meeting the stated curriculum or IEP goals. For some students this will mean the mastery of basic daily living skills. For others it will mean mastery of topics that go beyond the regular K–8 curriculum in vocational applications. Teachers must be prepared to monitor each child's progress carefully, communicate the necessary mathematical understandings, and know the limits imposed by a child's special needs.

To meet these expectations effectively, the teacher must be skillful and knowledgeable. Several important characteristics identify the classroom teacher who can deal effectively with special children:

1. Knowing children

2. Knowing how to manage a classroom

3. Knowing the relevant mathematics

4. Understanding the child's special needs

5. Understanding the special-needs child

6. Understanding mathematical learning patterns

7. Knowing and understanding concrete teaching aids for mathematics

Having these skills and understandings will do much to ensure a successful teaching career. It will be well worth the time to give thoughtful consideration to these factors.

Knowing Children

One of the most important characteristics possessed by a classroom teacher is that of being a child-oriented person. A child-oriented person is one who can relate to children in a one-to-one, small-group, or whole-class situation. Child-oriented people often know what music groups are currently in vogue, what TV shows have captured the imagination, what is the latest school news. They also are sensitive to the individual student's interests. All of this is helpful in planning mathematics instruction. Such information can allow the teacher to use familiar analogies in teaching concepts and skills, a type of instruction effective with children of all abilities. Knowing a child's interests also helps the teacher understand the child's learning capabilities, and strengthens the instruction that can be offered in either a one-to-one or group situation. Understanding how children feel about things and how they react in a given situation distinguishes the child-oriented teacher from other teachers. Good child orientation is also important in keeping records on a child's growth and learning patterns.

Knowing How to Manage a Classroom

A second facet of a teacher's preparation is knowing how to employ classroom management techniques effectively. This includes planning instructional sequences, monitoring on-task behaviors, and dealing with discipline situations. Teachers with deficiencies in this area are constantly dealing with situations that have little to do with student learning. The lack of purpose and control pervades their classes.

The area of classroom management includes the ability to adapt classroom flow to:

1. Meet special needs without losing others in the class

2. Focus children's energies in fruitful directions

3. Provide boundaries for children's behavior

The development of these skills comes from observation, planning, and trial-and-error. There is no one set of management skills that will work for all teachers. However, a teacher who does not have a set of effective skills will have many problems. Children with special needs do not do well in a hectic, unfocused classroom.

Knowing the Relevant Mathematics

A third feature necessary for successfully implementing curricular objectives and managing classroom learning is knowing the necessary mathematics content. Many beginning teachers fear mathematics as much as they fear discipline problems. However, almost all teachers who have successfully completed a "Math for Teachers" course at a university are minimally prepared for the teaching of elementary school mathematics. Teachers of middle-school mathematics should have additional hours in geometry and pre-algebra mathematics. In all cases the teacher should have had a course in methods for teaching mathematics in K–8 classrooms.

After several years of teaching many teachers rate mathematics as one of their favorite subjects to teach. This may be due partly to the very structured nature of the subject and partly to the fact that progress in mathematics learning is easier to evaluate than that in other areas of the K–8 curriculum.

Understanding the Children's Special Needs

As teachers adapt the program to each child's needs, they need to know as much as possible about the various handicaps their children have. In addition, the teacher needs to know enough about other common handicaps to identify problems that may arise during their contact with children. When problems do occur, the teacher will then be prepared to refer a child for special diagnosis or help.

Special knowledge of the major handicapping conditions also

prepares the teacher to do a more effective job of planning instructional sequences in mathematics. Several questions can guide a teacher in studying children's handicaps: What are especially effective modes of instruction? What learning abilities are suppressed by each handicap? What learning/teaching styles are more effective with specific learning needs?

Understanding the Individual Special-Needs Child

The teacher must know the limits and special characteristics of each child's handicaps. The child's educational history will provide information concerning effective instructional approaches, gaps in content background, and levels of achievement and aptitude. The family history will provide valuable information about the child's self-image, the degree to which the child may have been protected or challenged, and the attitude of the family toward mathematics. Finally, the teacher must observe the quality and quantity of the interactions between the child and his or her peer group. This type of information reveals the child's approach to life, attitudes about school work, and feelings about mathematics.

Understanding Mathematical Learning Patterns

Along with knowledge of handicaps, the teacher needs to understand the patterns of mathematics instruction that lead to effective learning. Knowledge of developmental patterns in counting, number recognition, and measurement skills is essential, and it helps the teacher to plan and sequence lessons that promote mental models as well as concrete understanding of the concepts. Knowledge of learning processes is strengthened by a basic knowledge of the works of Piaget, Gagné, Dienes and others who have written on the mastery of mathematical concepts and skills.

Knowing and Understanding Concrete Aids for Teaching Mathematics

A successful classroom teacher of mathematics is one who can effectively use concrete teaching aids. Can the teacher choose an appropriate sequence of models for teaching the concept of numeration? Can the teacher select or adapt teaching aids appropriate to the child's special learning needs? In addition to knowing that the aids exist, the teacher needs to know how to use them. Effective use must take into account the students' level of mathematical maturity and development.

These special skills will do much for the success of the teacher. The following chapters cover each of the mentioned areas of teacher competence in mathematics. However, the text cannot totally prepare you for the tasks ahead. You must constantly observe, integrate, and test ideas concerning the teaching of mathematics.

Your Time: Activities, Exercises, and Investigations

1. Observe a special mathematics class and a regular mathematics class at a local school. Write a report on both classes, noting the following points:
   (a) Name of the school and grade levels
   (b) Description of students (regular class, special class—special needs observed, grade or ability levels, number of students in each class)
   (c) Topics taught
   (d) Materials used (if any) to demonstrate, reinforce, or drill the mathematical skill or concept.
   (e) General goals of the lessons
   (f) Any special efforts you noticed the teachers making to deal with the special needs of individual learners
   (g) Anything that impressed you or troubled you about either class

2. Observe a special child for a day's time. Note the following facts about the child's special program and the child's reaction to it.
   (a) What school subjects did the child study?
   (b) What percent of the school day was devoted to each subject?
   (c) What percent of each period was the child actively involved in the study of the subject at hand? What percent of the time was spent daydreaming and handling nonsubject activities?
   (d) What special accommodations have been made to meet the child's special needs in each subject area?
   (e) To what extent is the child mainstreamed? What nonregular class support does the child have?

3. Read "On Improving One's Ability to Help Children Learn Mathematics" by Phares O'Daffer in the November 1972 issue of *The Arithmetic Teacher*. Discuss how the five-step lesson plan helps meet some of the needs and characteristics of the special-needs mathematics teaching situation. Show also how this plan helps meet the criteria for successful teaching listed in the chapter.

4. Examine a recent copy of the *Stanford Achievement Test* or the *Metropolitan Achievement Test*. What accommodations do these tests make for the special-needs child? What areas from the NCSM list of basic skills are emphasized on these tests? What areas of the basic skills are ignored?

5. Interview an elementary school student in a regular class and a child in a special needs classroom. Ask them why they study mathematics, what they like about it, what they don't like about it, and what general comments they would like to make about it.

6. When you recall your own elementary-school experiences with mathematics, what things proved helpful to your learning? What things hindered your learning? Did you study anything other than computational algorithms and facts? Who was your best math teacher? Why do you classify this teacher that way? Did you ever carry out an experiment in mathematics class?

7. How closely do the grade-level objectives listed in Appendix A match the content of contemporary elementary mathematics series?

Reference

National Council of Supervisors of Mathematics. *Position Paper on Basic Mathematical Skills.* Minneapolis, Minnesota: National Council of Supervisors of Mathematics, 1977.

2 A GLOBAL LOOK AT INDIVIDUAL DIFFERENCES

Every teacher is faced with a class of smiling faces that mask a multitude of individual differences. It is these differences that make the teaching profession a challenging and rewarding one.

This chapter examines the range of individual differences commonly found in elementary and middle school mathematics classrooms and emphasizes the manner in which these differences can be identified and described. It presents the educational problems of sensory or physically handicapped students, learning disabled or "slow" learners, mentally retarded pupils, and behaviorally disordered children and considers the impact of these handicaps on the teaching-learning process. Each of the handicaps is discussed separately, with special suggestions for classroom identification, sources of professional help, and possible paths of dealing with the individual special needs child. Illustrations of specific cases blend the diagnosing of educational difficulties, the carrying out of associated teaching, and the evaluating of the resulting learning outcomes.

A LOOK AT CHILDREN'S ABILITIES

Evaluating a child's abilities is often accomplished through the use of a standardized achievement test. A child's performance on this test is used to show where the child is in comparison to a norm group of students of the same age who took the same test. The results of the various subtests (spelling, social studies, reading comprehension, mathematics computation, mathematics concepts, mathematics application, . . .) indicate that the child is at various percentile levels in the same age group of students. If a child scores at the 50th percentile or above, the child is ranked as performing at or above the average expectation for a child of that age. If the child scores below the 50th percentile, the child is ranked as having performed below average on that test or subtest.

Evaluation of a child's abilities on a basis of a standardized achievement test has certain dangers. First, the content of the group achievement test is seldom checked to see how closely it matches the curriculum of a school district in each of the test's scored areas. Second, these are paper and pencil tests with specified time limits. Therefore, they provide minimal accommodations for a child's ability to apply a large amount of the learning which takes place in school. Third, such tests also discriminate against the student who is a poor reader, a slow worker, or suffers from any one of a number of physical or learning disabilities.

As a result the teachers must gradually develop, consistently apply, and constantly reassess their methods of student evaluation. Methods for pinpointing individual differences and their educational implications are the heart of effective teaching of mathematics to the special needs student. Such methods must account for the students' verbal and nonverbal abilities, motor skills, attitudes, and overall approach to life and learning. In addition, an effective teacher considers information sources such as other teachers and parents. The analysis of this information gives the well-prepared teacher a base for the diagnosis of educational difficulties and guidelines for the remediation of the identified problems.

To be able to carry out such diagnoses, the teacher should know and understand the following.

The modes of learning and behavior. Every teacher should understand the expected behaviors and styles of learning exhibited by students of a given age, along with the possible variations due to developmental differences. This equips the teacher to notice a significant deviation from the expected. A knowledge of expected behaviors also frees the teacher to observe details of individual learning styles. Some children learn best in situations where information is presented visually, others react positively to listening, while still others react to doing and touching. Seeing, doing, and touching are sensory modes through which children like to express their feelings and findings. Some children express themselves through drawings and dramatizations. The sources of communication vary greatly among classroom populations. An effective teacher notices student's individual differences whenever possible. Variations from the norm need to be noticed and recorded. If it is a positive deviation, the skill can be used to strengthen other areas of learning. If it is a negative deviation, plans for strengthening the area can be made.

Teachers need to note differences as they interact with various concepts and areas of study within mathematics itself. Does the child react positively toward geometric (spatial) concepts while rejecting verbal application problems? Effective teaching, resulting from such baseline information, involves the children in activities using their strengths while gradually modifying instruction to address their weaker areas. In the above case, an effective teacher might have the child visualize the verbal problems in a spatial-geometric mode and then work toward the solutions.

Each child's special needs. Children's prior experiences contribute greatly to their present needs and abilities. Since all of these experiences have an impact on the child's approach to mathematics, it is helpful if teachers understand the child's home environment as well as its possible impact on attitudes about math.

In addition to focusing on a child's special needs, teachers should know children's likes and dislikes. What types of experiences have they had in the community? What types of travel experiences have they had? Do they have brothers and sisters? Who are their friends? What are their hobbies? This information can provide a background for helping the child model mathematical concepts and skills in terms of familiar surroundings.

Knowing a child involves observing and relating. How children feel about themselves and the world about them will influence how they feel about mathematics and their ability to succeed in it. What types of expectations have adults placed upon the child? Are they reasonable or not? How has the child reacted to these imposed goals?

The available resource personnel. Knowledge about specific resources available to each special needs child is vital to the teacher. Are there special materials available for the child's use? Are there special training procedures for the specific need? Is there a special facility nearby for children with a particular problem? What trained consultants, resource people, or community resources are there in a given area? Cooperative communication between the regular classroom teacher, the teacher's aide, the resource teacher, and the parents of the special needs child all work for successful learning experiences.

Active intervention. One role of the teacher in facilitating mathematics learning is to use the classroom environment to involve the child in mathematical activities. Often the special need is limited to a particular physical handicap. In such instances the teacher's active intervention is necessary. Careful planning will enable the child to participate more fully. For example, special efforts may be needed to include large-print directions so a visually impaired child can participate in learning center activities. In other cases special adaptations may be needed in the social structure of the classroom to include children with behavioral disorders.

Often teachers of special needs children must become programmers who pace the introduction of necessary activities, tasks or stimuli. This timing allows children to master one step before being overwhelmed with new tasks. Imitation, instruction, coaching, or the direct structuring of an activity to accommodate special learners may be required to elicit their participation in certain mathematical activities.

Sometimes intervention techniques need to be applied to the peers of a special child. A teacher might discourage other children from doing what the child must master himself or herself. The object here is to strengthen the special child's ability and self-confidence in each new task.

Interventions require careful and thoughtful planning based on ongoing observations and reviews of each child's development. There is no one way to set up a classroom for special needs children. There is no packaged set of activities that will effectively teach mathematical concepts to all children with maximal efficiency. Intervention requires a competent teacher who is sensitive to the child's needs and who can balance flexibility with structure.

Special needs. It is important that teachers dealing with special needs children know as much as possible about the particular problems or handicaps of these children. In general the teacher should know about the range of behaviors and the variation to be expected from each handicapping condition. In addition, the teacher should know of any special symptoms of a handicap. A teacher does not need to be a speech therapist or an audiologist to know what happens when children have hearing losses and how these factors affect their language and communication skills. To meet the special needs of a child with impaired vision, a teacher should have at least a general idea of how such a child might deal with sensory stimuli and how distractions in the environment affect the way a child might react to various learning activities.

General information, too, is needed about children with a special handicap. Will the child need more time to master certain things, or need more repetition and practice than others to acquire the same level of skill? Will the child ever be able to master a given set of concepts and skills?

Knowledge about a child's handicap is also important for others with whom he or she will come into contact in a schoolroom setting. For instance, will the child always be confined to a wheelchair? Or, with therapy and training, is there a possibility that the child will walk eventually? The answers to these and related questions will help the teacher and other children more effectively relate to the special child.

TESTING CHILDREN'S ABILITIES

One of the first points in establishing appropriate baseline behavior and achievement levels is the recognition of a child's intact abilities in relation to those of other children of the same age and background. Several methods can be used to ascertain these abilities and their range.

In the area of cognitive abilities a child's level of functioning is often estimated by some form of IQ test, achievement test or diagnostic test. The IQ tests deal with a child's native information processing abilities. Achievement tests deal more with what the child has acquired through classroom learning. Diagnostic tests are aimed at placing the child along the continuum of the mathematics curriculum of concepts and skills. Another way of examining differences would be to say that IQ tests are more process-oriented while achievement and diagnostic tests are more product-oriented. The following paragraphs examine each of these test forms.

What Do *You* Say?

1. What are the advantages and disadvantages of IQ tests in diagnosing a child's problems?

2. What types of other tests are available for use in diagnosis of classroom learning problems in mathematics?

3. (a) What is the *KeyMath* test?
   (b) What kind of information does it give the math teacher about a student's abilities?

4. What is the difference between a norm-referenced test and a criterion-referenced test? Which is more like a checklist?

5. How would you measure a child's attitude toward mathematics?

IQ Tests and Their Uses

The measure of one's IQ score gives a rough indication of the potential for dealing with subject matter content. An IQ score of 100 indicates that an individual has an average score on the test for person of his or her age. A score above 100 usually indicates an above-average performance for a given age, while a score below 100 indicates a below-average performance.

In practice, IQ scores are determined by consulting a table of scores for performance on a given test. Traditionally, the IQ score was determined by applying this formula:

$$IQ = (MA/CA) \times 100$$

In the formula, *IQ* stands for intelligence quotient, *MA* for mental age, and *CA* for chronological age.

IQ scores have sometimes been given much more importance than they deserve in the determination of a person's abilities to function in an academic setting. There are several recorded cases of individuals with very high IQs who could not process information well. On the other hand, there are many individuals with low IQs who have quite adequate information processing skills. Thus, one must be careful when describing a child's potential for success in learning when using IQ scores.

A fundamental use of IQ scores is the rough categorization of individuals into levels of special educational needs. The score guidelines for these categories follow:

IQ	Category of Need
0–24	Profound retardation
25–39	Severe retardation
40–54	Moderate retardation
55–69	Mild retardation
70–84	Borderline retardation
85–116	Normal range of functioning

These ranges of IQ scores are sometimes used as legal definitions for levels of need and corresponding required services.

IQ scores can be used in two different but related ways to guide teachers as they diagnose students' difficulties in mathematics. First, the IQ score gives a test measurement which can be translated into an expected grade level of performance. This expected level can be used to check agreement with other scores from achievement and diagnostic tests. The IQ score can also be used as a global measure along with other scores and information to form a composite picture of a child's overall ability.

For example, consider a child who has an IQ of 87 and a chronological age of 11. This child might be expected to perform well at the third-grade level. This level was determined by substituting the IQ and the CA given into the IQ formula and solving for the corresponding mental age. The result was 9.5 years of age for a mental placement. Using average ages by year in school, the middle third-grade level was identified. In doing this, one must be aware that the process outlined for determining a grade placement level makes the assumption that one's IQ score does not change with age. Research reviews show that IQs do change, in many cases due to emotional and sociological changes in a child's environment.

This information, along with other scores, informs a teacher of some consistency of performance or of wide-ranging fluctuations. In the case of a consistent low level of performance, the teacher can move to develop a program of studies at the level indicated. If the levels of performance are widely scattered, the teacher should consult a person trained in the interpretation of tests and learning needs to make a further diagnosis of the child's special needs. Such wide ranges of performance levels often indicate a learning disability which may require further testing and observation.

Some of the commonly used IQ tests are the *California Short Form Test of Mental Maturity*, the *Kuhlman-Anderson Intelli-*

gence Test, and the *Otis-Lennon Mental Ability Tests*. All three of these tests are designed to be administered to entire classes at a given time, so that many people can be tested in a short period of time. In addition to these forms, there are more lengthy and reliable individually administered IQ tests. Foremost among these are the *Stanford-Binet* and the *Wechsler Intelligence Scale for Children*.

The latter two intelligence tests require a trained examiner and a longer period of administration time. They give a more reliable estimate of a person's ability to process information through their various subtests. The *Stanford-Binet* at the twelve-year-old level has subtests including: vocabulary, verbal absurdities, picture absurdities, digit repetition, abstract words, and sentence completion. The absurdity tests involve verbal and visual problem-solving to indicate the incorrect information. The abstract word subtest involves vocabulary words at a higher level than found on the vocabulary scale. These tests require a child to exhibit past learning, verbal and visual interpretation skills, short- and long-term memory, and other processing skills.

The *Wechsler Intelligence Scale for Children* (WISC) is a collection of verbal and performance subscales. The verbal scales consist of general information, general comprehension, arithmetic reasoning, similarities, digit repetition, and vocabulary. The performance scales consist of mazes, digit-symbol substitution codes, picture completion, block designs, picture arrangements, and object assembly. The first group of subscales focuses on the verbal components of intelligence, calling on memory of past learned information, numerical reasoning skills, information storage and retrieval skills and the ability to see similarities in verbally presented information. The second set of subscales focuses on the subject's ability to deal with visually presented information in various processing modes.

Achievement Tests and Their Uses

The second general class of tests used to gather information on a pupil's level of performance is standardized achievement tests. These tests consist of a collection of subtests on such subjects as reading, writing, spelling, mathematics, science, and social studies.

Some of the achievement tests have subtests for work-study and listening skills. Each of these subtests is further divided into subscales. In the case of mathematics, the composite mathematics score is usually a weighted average of scores on concepts, computation, and application subscales. The scores for the test as a whole and for each of the subtests and subscales are given in terms of number right, the related percentiles for an age equivalent national norming group, and the related stanine for an age equivalent national norming group. In addition, a grade placement score is usually given for the work on the test and for the work on each of the subtests and subscales. This information can be used, along with IQ scores, in the development of a profile of student strengths and weaknesses.

There are many different achievement tests commercially available today. Among those most commonly used are the *Comprehensive Tests of Basic Skills*, the *Iowa Tests of Basic Skills*, the *Metropolitan Achievement Tests*, the *SRA Achievement Series*, and the *Stanford Achievement Tests*. Further information about each of these tests can be found in educational and psychological testing books and in test files at university libraries.

Modern curricular movements have caused some revisions and improvements in the design of achievement tests. Many now have specific instructional objectives related to individual test items. Tests that only give the child's placement in terms of performance level compared to some nationally selected age equivalent sample are called *norm-referenced tests*. Tests that report a student's scores in terms of how the child did in relation to a set of educational objectives are known as *criterion-related tests*. Some tests give both kinds of information. The norm-referenced tests give a picture of how the child is doing compared to other children of the same age who have completed the same examination; they do not necessarily show the causes for the level of performance or what particular items contributed to it. Criterion-related tests provide the teacher with a specific list of skills and concepts the child has either mastered or not mastered. This provides a baseline for further learning or for remediation.

Tests that provide both types of feedback are the most helpful in making an educational diagnosis. Most of the norm-referenced tests have a profile chart of the child's performance across the various subtests and subscales that gives the teacher a feel for the child's overall strengths and weaknesses. An example of one such chart is shown in Figure 2.1.

Figure 2.1

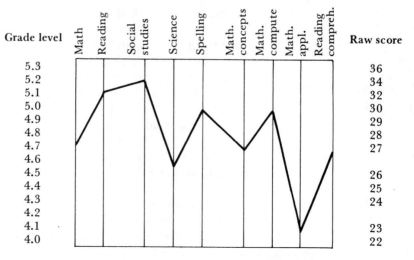

Depending on this student's grade placement level at present, this profile could represent the work of a gifted or a slow student. Suppose the student is currently a seventh-grader in the third month of school (a current grade placement of 7.3). This profile then clearly represents the work of someone with some special educational needs. The fact that the scores are all at about the 4.2 to 5.2 grade placement range indicates that the student is uniformly weak. The reading and social studies scores indicate that the student may derive information well from verbal and written forms. The lower science and mathematical applications scores indicate that problem settings may cause difficulty. The relative highs in spelling and mathematical computation may indicate good memory and algorithmic skills when these processes are compared to other forms of processing. The drop in reading comprehension may be tied to the low mathematics application score due to the written presentation of the problems on the test form. These findings may then be compared with other input about the student's strengths and weaknesses. The individual items on the test can also be studied to determine what skills and facts that child has mastered or not mastered.

Achievement tests, like IQ tests, also have their limitations. Prominent among these is the possible discrepancy between the test and the curriculum from which the child has studied. In mathematics, this match may depend on the relative emphasis given certain areas of the curriculum in the classroom and the coverage of that content on the test. One area that always seems to be at odds with the tests is the geometry content of the math program. Other areas of discrepancy are the symbols used or the presentation form of problems on the test. If there is not a good fit between the curriculum and the test, the results may not reflect the child's real ability.

Achievement tests may not report accurately when the students taking the test are slow workers or poor readers; because these tests are timed, these students are discriminated against. The tests are highly verbal, and this discriminates against visually handicapped and hearing-impaired children. While the resulting scores show how a child performs compared to a nationally selected norming group, it does not really show how the child compares in mastery of the content. Hence achievement tests, like IQ tests, must be treated with caution in making firm educational diagnoses. They provide valuable profile information as a basis for action, but they alone do not hold the key to the source of the problems or the needed remediation.

Diagnostic Tests

In addition to IQ and achievement tests, there are several other types of tests that may aid the teacher and special education specialist in determining the bounds of a child's abilities. These tests include diagnostic tests, screening tests, and checklists. A diagnostic test differs from an achievement test in that it is not interested in a survey of what the child knows, but rather takes a detailed look at what the child does and does not know in a specific area. A diagnostic test is given to derive information for the remediation of a child's learning problems. In using a diagnostic test, a teacher is not interested in whether the student is in the top 15 percent of an age equivalent group or not. The teacher is interested, for example, in which of the subtraction problem types the student has mastered and which still needs work.

There are several arithmetic tests on the market. Some are produced by publishing companies and designed to accompany their mathematics textbook series. Others are published by commercial test publishers for use across all different kinds of mathematics curricula. Two of the most commonly used diagnostic tests for elementary mathematics are the mathematics portion of the *Peabody Individual Achievement Test* and the *KeyMath Diagnostic Arithmetic Test*. Both of these tests are designed to be administered individually. The *Stanford Diagnostic Mathematics Tests* are designed to be used either with a group or an individual.

Diagnostic tests are developed by selecting from a particular group of mathematical skills those problems that represent the various levels of each skill. Problems are then arranged according to level of difficulty, starting with the easy problems and working up to the hard problems. The various skill strands are then arranged on the test so that individual strands may be diagnosed separately or several at a time. The individual diagnostic tests are designed so that the examiner can carefully monitor a child's progress. Questioning in a subtest is usually stopped before a frustration level is reached. In addition, individual diagnostic tests reduce the role of reading skills in the diagnosis of mathematical skill.

The *Peabody* and *Stanford* tests give both grade level and age equivalent norms. In addition, they relate a child's performance to a set of objectives designed to show mastery of various levels of skill in the mathematics curriculum. In examining the results of either of these tests, the teacher must still make many decisions concerning the areas of probable student deficiencies.

The *KeyMath Diagnostic Arithmetic Test* gives more specific information about the child's level of performance in the various areas of the mathematics curriculum. The subtests in the *Key Math* test are grouped under three main categories called Content, Operations, and Applications. These correspond roughly to the achievement test areas of Concepts, Computation, and Applications.

These three main groups of subtests are then subdivided into a total of fourteen different scales. These scales and the categories they fall under are as follows:

1. Content
    a. Numeration
    b. Fractions
    c. Geometry and Symbols

2. Operations
    d. Addition
    e. Subtraction
    f. Multiplication
    g. Division
    h. Mental Computation
    i. Numerical Reasoning

3. Applications
    j. Word Problems
    k. Missing Elements
    l. Money
    m. Measurement (Metric or English)
    n. Time

The *KeyMath* is perhaps the best commercially available diagnostic test for a classroom teacher's use. It indicates the level at which the child can be reasonably expected to perform, a diagnostic profile showing the child's relative performance in each of the fourteen areas, a description of the skill each item tests, and a sample of the child's computational work with each of the four basic operations.

Even with this, the *KeyMath* test still has some difficulties. The size of a step between the individual items is sometimes too big, as can be seen in the Numeration scale. The Geometry and Symbols scale uses some outdated symbols and then mixes the student's knowledge of these with the student's performance on geometric and spatial items. Finally, the test gives no measure of a child's ability to handle the basic facts.

The use of the *KeyMath* test in the classroom is most effective between grades three and seven. It can be administered in a short time (usually forty-five minutes or less) and requires little formal training in administration or interpretation. The questions are easily asked and visual aids are available in the test kit for the child's use. When a child misses three successive items on a particular scale, questions on that scale are finished and the examiner moves to the next scale. This technique helps reduce frustration and attitudinal problems that might result from repeated failure on a set of items.

The examiner may incorporate additional diagnostic techniques with the administration of the *KeyMath*. For example, since the test has no time limit, the teacher is free to ask additional questions in a given area and ask more probing questions about the responses to particular items. One might ask a child to explain how he or she arrived at an answer just given orally. Such questioning, of course, must be kept to a minimum, otherwise the session will become too long and the child will lose interest in the testing situation. However, the responses to these interjected questions can provide as much or more information concerning the current status of the child as the rest of the *KeyMath* test.

Screening Tests

Screening tests are a special form of diagnostic tests which provide teachers and schools with early warnings of special needs or handicaps children may have. The results from such tests allow the proper placement of children in special preschool or compensatory programs to strengthen the areas of weakness.

Special diagnostic screening instruments also aid in identifying children who may have problems in speech, gross or fine motor skills, self-help skills, social skills, or cognitive processing abilities. This latter area contains factors related to aptitude and readiness for mathematics. Some examples of these tests and related information are given in Appendix B.

For older preschool children, screening often includes the administration of the *Kraner Preschool Mathematics Inventory*, an individualized test; the *Arithmetic Readiness Inventory*, which can be individually or group administered; or the *Level K Mathematics Test of Basic Experiences*, a group test. The general purpose of such tests is to assess a child's mastery of fundamental concepts and skills that relate directly to subsequent success in mathematics. Subtest items frequently assess a child's understandings and skills in the following areas: the ability to classify, compare, and order objects on a number of variables; the ability to conserve number, length, and mass; the ability to handle early numerical and geometric vocabulary; and the ability to deal with simple size comparison and measurement experiences. Such diagnoses serve as a basis for prescriptive teaching within the child's future educational program. They also point the way for specific activities and experiences to strengthen a child's mathematical development.

A more advanced screening test is the *Illinois Test of Psycholinguistic Abilities (ITPA)*. This test, which requires a specially trained psychological examiner, is very useful for pinpointing specific information-processing deficiencies. The test and its subscales focus on the following cognitive processing abilities:

1. Reception skills
   a. auditory reception
   b. visual reception

2. Association skills
   a. auditory association
   b. visual association

3. Expression skills
   a. verbal expression
   b. manual expression

4. Sequential memory skills
   a. auditory sequential memory
   b. visual sequential memory

5. Closure skills
   a. visual closure
   b. auditory closure
   c. grammatical closure

These eleven areas are most often cited in research dealing with the special needs student—especially the learning-disabled student. The *ITPA* test and related materials provide the teacher with some very specific suggestions for classroom processes and procedures. Teachers who work entirely with special needs children should become familiar with the *ITPA*.

Checklists

A final form of evaluation used to pinpoint a child's level of performance is a checklist. Several school districts use a checklist of objectives (derived from their curricular objectives) to record an individual's progress through the curriculum. For example, consider the set of minimal objectives contained in Appendix A. A checklist form could be constructed to include each of these performances. At the point a child shows mastery of an objective, the teacher places a checkmark on the form to the right of the objective. Periodic tests are then used to see if the performance level has been maintained. An example of such a checklist is given in Figure 2.2. This checklist is taken from a portion of a school district's second-grade objectives.

Attitude Tests and Their Use

In some cases children's troubles with learning mathematics are attitudinal. Several attitude inventories have been developed for use in such situations. These forms give the teacher some information concerning the child's willingness to approach mathematics.

At the lowest level of attitude evaluation, the teacher can use a checklist that might include items on the percent of on-task time in mathematics class as compared to other classes, the number of student disruptions in mathematics class as compared to other classes, and the number of times the child approaches the mathematics corner or uses mathematics games as compared to corresponding activities from other subject matter areas. This anecdotal information often gives a fairly accurate picture of a child's attitude toward mathematics.

Figure 2.2

Arithmetic Achievement Record

_____
Student

Grade 2

_____
Teacher

KEY:
S--Satisfactory
N--Needs improvement
--Blank indicates not covered

Module I – Numbers, Numerals, and Two-Digit Place Value

S N

A. Reading and writing of numerals
B. Place value of two-digit numerals
C. The number line

Module II – Counting, Order, and Inequalities

A. Counting skills, spanning a decade
B. Greater than >; less than<

Module III – Money

A. Penny, nickel, dime and quarter
B. Making change: 25 cents or less
C. Value of a collection of coins less than one dollar

Module IV – Telling Time

A. Roman numerals
B. Time to the hour and half hour
C. Written notations for indicating time
D. Vocabulary for telling time
E. Time in intervals of five minutes

Module V – Addition and Subtraction to 10 -- Power Skill

A. Concepts of addition and subtraction
B. Symbols for addition and subtraction
C. Addition and subtraction on the number line

Module VI – Missing Addends and Differences

A. Concept of missing addend
B. Subtracting using missing addends
C. Story problems

Module VII – Basic Principles

S N

A. Commutative principle of addition
B. Use of parenthesis
C. Associative principle of addition

Module VIII – Sums to 18 -- Power Skill

A. Sums to 18 (power methods)
B. Regrouping to make 10
C. Solving story problems

Module IX – Differences to 18-Power Skill

A. Subtraction combinations 11 to 18
B. Subtraction in making comparisons

Module X – Measurement

A. Length, perimeter, and area
B. Units: inch and centimeter
C. Volume with liquid measure

Module XI – Two-Digit Addition and Subtraction-- Without Regrouping

A. Addition of two-digit numbers
B. Subtraction of two-digit numbers

Module XII – Three-Digit Numbers

A. Numbers from 100 to 999
B. Concept of place value to hundreds

Module XIII – Three-Digit Numbers

A. Names for the numbers 100 to 999
B. Place value through 999

Module XIV – 3-Digit Addition and Subtraction With Regrouping

A. 3-digit addition and subtraction

Module XV – Geometry

S N

A. Point, line, line segment
B. Square, circle, triangle, and rectangle

Module XVI – Addition and Subtraction Skills to 18

A. Sums of 18 or less
B. Differences using missing addends

Module XVII – Multiplication

A. Concept of multiplication
B. Multiplication--number line
C. Multiplication--repeated addition
D. Symbol for multiplication

Module XVIII – Sums to 18 -- Speed Skills

A. Speed skills for sums to 18

Module XIX – Sums and Differences to 18

A. Sums and Differences to 18
B. Subtracting using missing addend
C. Solving story problems

Module XX – Fractions

A. Concept of fraction
B. 1/2, 1/3, 1/4, 2/3, 3/4, 4/4

Module XXI – Addition and Subtraction With Re-Grouping-Power Skills

A. Adding and subtracting 2-digit numbers
B. Adding and subtracting with regrouping--number line

More formal instruments are available in which the child marks, writes, or otherwise indicates feelings concerning a particular statement or problem area of mathematics. Some examples of these forms are:

Type 1: Children circle a "Smiley Sam" to tell how they feel about doing a problem like that shown in Figure 2.3.

Type 2: Children tell how they feel about math by marking the extent to which they *agree* with a given statement (Figure 2.4). Positive and negative statements are included so that a teacher can check the consistency of children's responses. The statements can

Figure 2.3

Figure 2.4

Statement	Mark response that describes you			
	All the time	Most of the time	Some-times	Never
1. When I work carefully I do well in math.	○	○	○	○
2. A good grade in math is just "luck" for me.	○	○	○	○
3. I get lots of math work wrong even when I try.	○	○	○	○
4. When I get math problems right, it's because they are easy to do.	○	○	○	○
5. I like math better than any other school subject.	○	○	○	○

be pre-recorded on a tape cassette to be played back later by the child.

Type 3: Children indicate the extent to which they agree *or* disagree with a given math related statement. Again both positive and negative statements are included. The child makes one response for each statement (Figure 2.5).

Type 4: Children mark which of two words best describes how they feel about a topic from mathematics. When an exact decision is hard to make between the two adjectives, a mark is made in one of the circles between the words to indicate the amount of feeling present. The word pairs are mixed with some positive terms in the left-hand column and some negative terms in the left-hand column (Figure 2.6).

The use of one of these types of attitude forms, or the use of an anecdotal form, will supply valuable information that you might otherwise miss. Often the attitude level is a determining factor in the selection of a method of remediation or of an approach to a particular topic.

Figure 2.5

Statement	Mark one response				
	Strongly agree	Agree	Neutral	Disagree	Strongly disagree
1. Math is fun.	O	O	O	O	O
2. Nobody should do math unless they have to.	O	O	O	O	O
3. Doing math is boring.	O	O	O	O	O
4. There are too many chances to make an error in math.	O	O	O	O	O
5. Math is needed in everyday life.	O	O	O	O	O

Figure 2.6

Mathematics Word Problems

Interesting	O	O	O	O	O	O	O	Boring
Happy	O	O	O	O	O	O	O	Sad
Bad	O	O	O	O	O	O	O	Good
Easy	O	O	O	O	O	O	O	Hard
Useless	O	O	O	O	O	O	O	Needed
Long	O	O	O	O	O	O	O	Short

**Case Study:
Krista**

Mathematical Diagnosis—A Case Study

The cumulative information from all sources combined with individual knowledge of classroom performance provides the framework for diagnosing a child's difficulties. The various parts fit together to suggest a plan for teacher action. The total picture shows whether the problem is cognitive, physical, attitudinal, or some combination of the three. It also breaks down the causes within each of the three main areas.

In the cognitive domain, the information will show whether the problem results from low mental ability, lack of organizational skills, slow development, inability to relate concepts to real-world objects, or some other reason. Attitudinal problems may include low motivation, poor attitude toward mathematics, inability to function in social settings such as the classroom, or a lack of educational encouragement at home. Physical problems may include poor health or permanent disability.

Let us now turn to consider the case study of Krista, a fifth grader who was tested in May (5.9). Krista has a very timid personality. She will answer direct questions, but does not initiate any other contact with the examiner. This conduct is very similar to that which she shows her teacher.

Krista seems to get along fairly well with her peers, both in the classroom and outside on the playground. The school she attends is a rural elementary school about ten miles from a city of about 60,000 people. She is in a class of twenty-four students. She lives on a dairy farm and is somewhat interested in being a farmer when she grows up.

When observed in class and measured on an attitude form, Krista's approach to mathematics seems to be about the same as her approach to other subjects. She works diligently for about the first five minutes after homework is assigned, but then puts down her work and daydreams for the rest of the period. The teacher does little to try to bring her back to the study at hand; it appears that the teacher has given up. An analysis of Krista's cumulative folder shows that at the last testing period she had an IQ of 85. She is 11, having had her birthday in February. Her fifth-grade achievement test scores, taken from the *Stanford Achievement Test* given in April, were as follows: math computation, 2.7 grade level; math concepts, 4.3 grade level; and math application, 3.2 grade level.

The figures on pages 32–35 contain the results of Krista's performance on the *KeyMath Diagnostic Arithmetic Test*. Figure 2.7 presents the overall profile for Krista. Figure 2.8 shows the results of Krista's work on the computation items on the test. Look at all of this information and make your diagnosis of Krista's problem. What would you do?

An analysis of Krista's scores gives the following information:

1. Her IQ score indicates that with her current chronological age (11.3) Krista should be doing work somewhere about the level of a student with a mental age of 9.6. This would place her level of expectation at the mid-fourth grade level.

2. Her expected grade placement level is consistent with the results of the *Stanford Achievement Test* grade placement level in concepts (4.3), but far ahead of the results in computation (2.7) and applications (3.2).

3. Her overall grade placement level from the *KeyMath* test puts her at the 3.1 to 3.2 grade level. This rating is fairly consistent with her scores in computation and applications.

4. Her work on the individual scales of the *KeyMath* indicate that she is performing below her own average in Numeration, Subtraction, Multiplication, Division, Numerical Reasoning, and Time. An analysis of her work on the Numeration and Subtraction scales indicate that she has problems in simple number progressions and basic numeration concepts. She made many basic subtraction fact errors. Facts knowledge may be an area for further probing.

As a result of these observations, Krista was given the Number Facts scale of the *Stanford Diagnostic Mathematics Test*. Results confirmed the doubts about her mastery of basic facts. In addition she did fairly well (34/40), but still showed a need for review. In subtraction (12/40), multiplication (14/40), and division (2/40) the results showed weak command of the basic facts.

Based on these findings, the examiner developed a program of study for Krista. This program of study will be discussed in the next chapter. What would you prescribe? Where would you start? It will help you to have a little history of Krista that was not shown by the test scores.

Krista had significant problems in the first grade, and her teacher suggested that she be retained. Her parents refused, and Krista was passed on to the second grade. Here the picture repeated itself. Krista was tested by the school psychologist, who recommended that Krista be placed in a special program for students with low mental abilities but potential for normal success. Again her parents refused. Since that time Krista has been passed on by the school system, each year falling more out of step in the mastery of academic work and in her enthusiasm for it. While this may sound like an isolated case, it is not. Children with special needs in mathematics and other subjects are often just passed along without the root cause of the malady ever getting any attention.

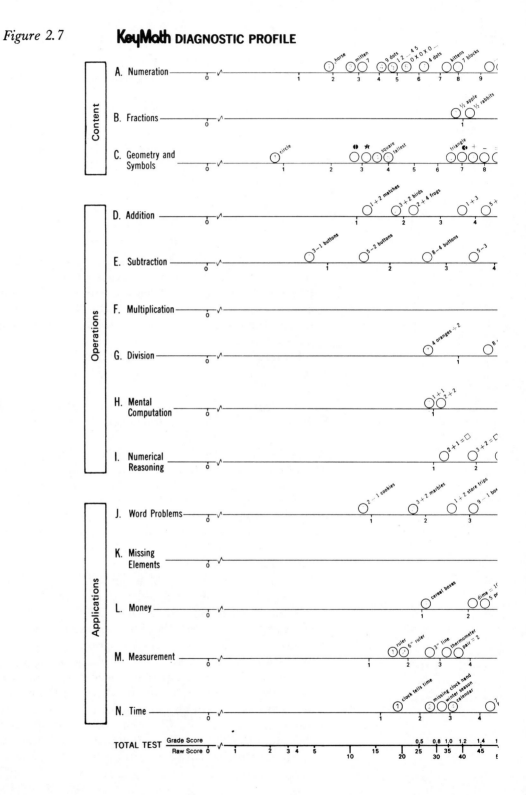

Figure 2.7

KeyMath DIAGNOSTIC PROFILE

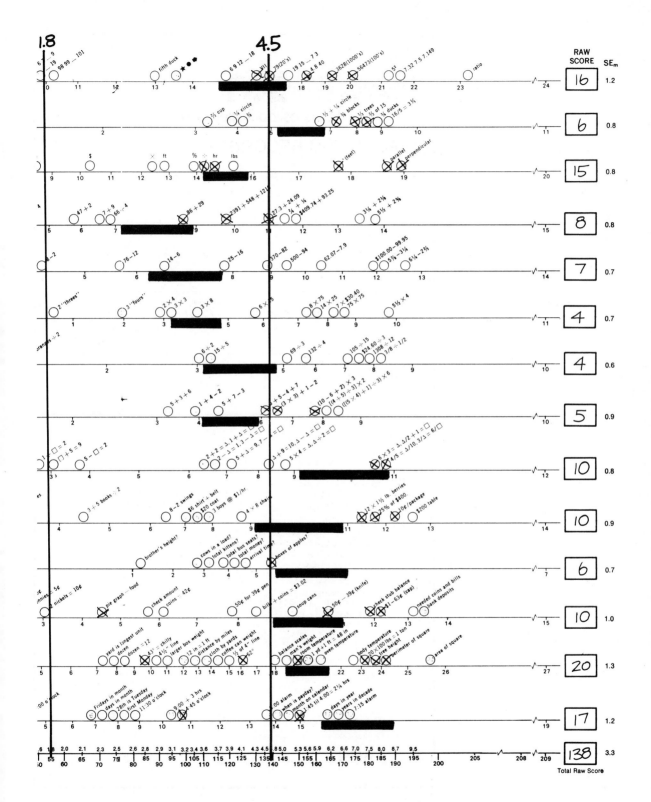

Figure 2.7 (cont'd)

KeyMath
diagnostic arithmetic test

DIAGNOSTIC RECORD

Austin J. Connolly, Ed.D; William Nachtman, Ed.D.; E. Milo Pritchett, Ed.D.

NAME _Krista J. Smith_   SEX: M _ F ✓ GRADE _5_
SCHOOL _Longfellow_
EXAMINER _Dossey_   DATE _5/10/81_
MATHEMATICS TEACHER _K. Jones_

OTHER TEST DATA

Test	Date	Results
Stanford Achievement	4/3/81	Computation — 2.7
		Concepts — 4.3
		application — 3.2

Name or description of pupil's mathematics program
Unit Curriculum

TEST BEHAVIOR

Did the subject appear to be performing at his or her best level?
shy

Briefly describe subject's test behavior (rapport, speed of responses, work habits, etc.).
works quickly at first, then slows down and becomes quiet — seems a little nervous

TEST PERFORMANCE

Summary statement regarding subject's strengths and weaknesses in arithmetic:

INSTRUCTIONAL RECOMMENDATIONS

Suggested Retest Date:

PUBLISHED BY **AGS** AMERICAN GUIDANCE SERVICE, INC., PUBLISHERS' BUILDING, CIRCLE PINES, MINNESOTA 55014

Figure 2.8

D.   Addition

4)
$$\begin{array}{r} 1 \\ +\,3 \\ \hline 4 \end{array}$$

5)
$$\begin{array}{r} 5 \\ +\,4 \\ \hline 9 \end{array}$$

6)
$$\begin{array}{r} 47 \\ +\,2 \\ \hline 49 \end{array}$$

7)
$$\begin{array}{r} 7 \\ +\,9 \\ \hline 16 \end{array}$$

8)
$$\begin{array}{r} 66 \\ +\,4 \\ \hline 70 \end{array}$$

9)
$$\begin{array}{r} 86 \\ +\,29 \\ \hline 107 \end{array}$$

10)
$$\begin{array}{r} 2{,}391 \\ 548 \\ +\,1{,}210 \\ \hline 3/49 \end{array}$$

11)
$$\begin{array}{r} 27.3 \\ +\,24.09 \\ \hline \end{array}$$

12)
$$\begin{array}{r} \tfrac{2}{4} \\ +\,\tfrac{1}{4} \\ \hline \end{array}$$

13)
$$\begin{array}{r} \$409.74 \\ +\,93.25 \\ \hline \end{array}$$

14)
$$\begin{array}{r} 3\tfrac{1}{4} \\ +\,2\tfrac{1}{8} \\ \hline \end{array}$$

15)
$$\begin{array}{r} 5\tfrac{1}{2} \\ +\,2\tfrac{3}{8} \\ \hline \end{array}$$

E.   Subtraction

4)
$$\begin{array}{r} 5 \\ -\,3 \\ \hline 2 \end{array}$$

5)
$$\begin{array}{r} 8 \\ -\,2 \\ \hline 6 \end{array}$$

6)
$$\begin{array}{r} 76 \\ -\,12 \\ \hline 64 \end{array}$$

7)
$$\begin{array}{r} 14 \\ -\,6 \\ \hline 8 \end{array}$$

8)
$$\begin{array}{r} 25 \\ -\,16 \\ \hline 29 \end{array}$$

9)
$$\begin{array}{r} 370 \\ -\,82 \\ \hline 3/2 \end{array}$$

10)
$$\begin{array}{r} 500 \\ -\,94 \\ \hline \end{array}$$

11)
$$\begin{array}{r} 62.07 \\ -\,7.9 \\ \hline \end{array}$$

12)
$$\begin{array}{r} \$100.00 \\ -\,99.95 \\ \hline \end{array}$$

13)
$$\begin{array}{r} 5\tfrac{7}{8} \\ -\,3\tfrac{1}{4} \\ \hline \end{array}$$

14)
$$\begin{array}{r} 6\tfrac{1}{4} \\ -\,2\tfrac{3}{4} \\ \hline \end{array}$$

F.   Multiplication

3)
$$\begin{array}{r} 4 \\ \times\,2 \\ \hline 8 \end{array}$$

4)
$$\begin{array}{r} 3 \\ \times\,3 \\ \hline 9 \end{array}$$

5)
$$\begin{array}{r} 8 \\ \times\,3 \\ \hline \end{array}$$

6)
$$\begin{array}{r} 15 \\ \times\,6 \\ \hline \end{array}$$

7)
$$\begin{array}{r} 75 \\ \times\,8 \\ \hline \end{array}$$

8)
$$\begin{array}{r} 25 \\ \times\,14 \\ \hline \end{array}$$

9)
$$\begin{array}{r} \$30.40 \\ \times\,7 \\ \hline \end{array}$$

10)
$$\begin{array}{r} 75 \\ \times\,75 \\ \hline \end{array}$$

11) $5\tfrac{1}{2} \times 4 =$

G.   Division

3) $2\overline{)\,6}$ quotient 3

4) $5\overline{)\,15}$ quotient 3

5) $3\overline{)\,69}$

6) $4\overline{)\,132}$

7) $15\overline{)\,105}$

8) $3\overline{)\,\$24.60}$

9) $12\overline{)\,1{,}308}$

10) $\tfrac{3}{8} \div \tfrac{1}{2} =$

Your Time: Activities, Exercises, and Investigations

1. Prepare answers to the "What Do You Say?" questions on page 18 of this chapter.

2. Select one preschool inventory (from those listed in the chapter or those listed in Appendix B) and then do the following exercises.
   (a) Write a critical analysis of the inventory, indicating:
      (i) Appropriate age or functional level of children for which it can be used
      (ii) Specific developmental lags or disabilities which it assesses
      (iii) Usefulness of the instrument in assessing readiness or aptitude for mathematics. (Some instruments are more general in nature and do not focus on mathematics readiness.)
   (b) Follow the manual directions and administer the inventory to a child. If a child of the appropriate age is not available, use one of your classmates (who should role play a child of the appropriate age level).
   (c) Write a short report of the screening, giving specifics of the diagnosis and any suggestions for followup diagnosis.
   (d) If you tested a real child, discuss your analysis with a classmate and with an experienced preschool teacher. Discuss it with the child's teacher if the teacher is available. Submit a report of your discussion.

3. Do a comparative analysis of two preschool inventories. Choose from those listed in the text or the Appendix, or those listed by your teacher. Focus on comparative strengths and weaknesses with respect to: age restrictions, limits in identifying specific developmental lags, disabilities, concept or skill proficiencies (and deficiencies) in mathematics; practicality and ease of administration and scoring; and usefulness of results.

4. Using Chapters 5 (all) and 6 (first part only, on Number) of the *National Council of Teachers of Mathematics 37th Yearbook, Mathematics Learning in Early Childhood* as a guide, develop a preschool mathematics diagnostic inventory to assess a child's aptitude or readiness for primary mathematics. Write an introduction to your inventory which indicates:
   (a) The functional age limit for which your test is designed
   (b) The specific mathematical objectives for each item
   (c) Minimal acceptable behavior expectations for each item or group of items
   (d) (optional) Followup questions for in-depth probes on critical items

5. After your teacher has examined your diagnostic inventory, use it with a child of the appropriate age. Write up the results of the interview and the changes you would make in your test as a result of the trial.

6. Obtain a copy of one of the standardized norm-referenced achievement tests from your teacher or from the library. Select one of the subtests for mathematics: concepts, computation, or applications. Then write a critical analysis of the subtest in terms of its match with a particular textbook series over the age range for which the test is designed. Comment on the items selected, their treatment (symbolism, format, size, spacing, . . .), and how the anticipated scores based on the textbook analysis compare with the test norms.

7. Do a critical analysis of a diagnostic or achievement test written to accompany a mathematics textbook series. If possible, choose a unit, module or chapter test rather than one which covers an entire book. Focus on:
   (a) Whether important concepts and skills are assessed
   (b) How adequately they are assessed
   Also indicate whether a child's ability to apply those concepts is assessed.

8. Select a topic for a short one- or two-day mathematics lesson. Design a set of teaching-learning activities related to that topic which focuses on the levels of processing and learning associated with the *ITPA* model. In your writeup of the activities, sequence them in the order they would be presented to a child during instruction.

9. Construct an attitude inventory similar to those given in the text. Check with your teacher, then administer it to a child of the age level for which the inventory was designed. Write an analysis of the results of your field test.

10. Plan to observe a mathematics class while instruction is taking place. Write up your observation including the following information:
    (a) Name of school and level of class being taught
    (b) General overall class ability level
    (c) Description of the students in the class (social, economic, overall behavior, attitude, etc.)
    (d) Select a particular student to observe for a period of ten minutes (part of the time during teacher talk and part of the time during supervised study). What percent of each time period was the student actively engaged in following the teacher or in working on the assigned work?
    (e) What did the child's attitude toward mathematics seem to be?
    (f) Were there any disruptive incidents in the class? How were they handled if they occurred?
    (g) How was the transition time into mathematics, from teacher talk to supervised study, and from supervised study to the next class handled? Was there wasted time, or were the transitions smooth and orderly?

SPECIAL NEEDS CHILDREN

In this section we examine the various conditions—and their effects—that cause children to have special needs in mathematics. In particular, we consider the slow learner and the following handicapping conditions: sensory or physical impairment (hearing, language, visual, or neuro-muscular), learning disabilities, behavior disorders, and mental retardation. These overviews are presented with the characteristic problems and basic modifications that may be necessary to deal effectively with special needs in the mathematics classroom.

Slow Learners

In many cases students experiencing difficulty in mathematics are referred to as slow learners, but this name is not sufficient to really describe what their problems are. For example the words *slow learner* may refer to a student's ability to acquire information, to the fact that the student is minimally educationally retarded, or to one of a thousand other classifications. When we refer to the slow learner in mathematics we are talking about students who are at least one year behind on their placement scores.

The source of these students' problems may be cognitive or affective. The cognitive problems include those that result from low mental ability, lack of organizational skills, slow developmental patterns, inability to relate concepts to the real world, and a host of other reasons. In none of these cases does the extent or nature of the deficit allow the student to be classified educably mentally handicapped. The affective (or attitudinal) causes include low motivation, poor attitude toward mathematics, inability to function in class or in small group activities, poor health, or lack of educational encouragement at home.

Sensory or Physically Handicapped Learners

Mathematics teachers have had physically handicapped learners in their mathematics classes for some time; their presence did not first come with Public Law 94-142. However, in many classrooms there has been little or no adaptation of the instruction to these students' special needs. In this section we consider the handicaps of hearing, speech, visual, and physical impairment as they relate to the learning of mathematics. With each area, we discuss the relationship of the disability to mathematics and then suggest ways of dealing with it. These suggestions will be general in nature; in Chapter 3 specific adaptations to mathematics content will be considered.

Hearing Impaired Students

While only a small percentage of the age 5–21 population are deaf (0.0075 percent) or hard of hearing (0.5 percent), the problems of the hearing-impaired child are readily understood. Many such students cannot understand speech well enough to imitate it or use it in any intelligible way. Generally these individuals have great difficulty in expressing ideas and in understanding language. Not all use hearing aids or learn sign language. Because of associated communication problems, the average hearing-impaired child is below grade level in mathematics achievement. Language deficiencies often cause the arithmetic reasoning scores to fall below the related computation and concept scores on an achievement test.

Most hearing problems are detected by parents, by preschool screening efforts, by teachers, or by the hearing tests given in the primary grades. Teachers can be cued to undetected or recently acquired hearing impairments through the observation of consistent behaviors such as the following. The student:

1. has trouble paying attention;

2. has trouble following directions and constantly asks for them to be repeated;

3. has to watch lips carefully when being talked to or when someone is reading;

4. seems to do better in academic situations where eyes or hands play a greater role;

5. has an attention-level drop as the lesson period progresses;

6. has frequent earaches;

7. talks more loudly than others in the classroom; or

8. has more trouble pronouncing consonants than vowels.

Once a hearing-impaired child has been identified and resource people have confirmed it, there are several avenues of action open to the mathematics teacher. Some of these suggestions are as follows:

1. When talking to a hearing-impaired child, maintain eye contact. Do this especially when giving directions or an important explanation.

2. Use an overhead projector when possible, as the hearing-impaired child will be able to see your face more easily than when you are facing the chalkboard.

3. Speak in a natural tone of voice. Do not exaggerate tone, pitch, or lip movements.

4. Teach from a position where your face is well-lighted. Do not stand with your back to a glaring light or a window, as this makes it difficult for the student to see your face and lip movements.

5. If there is a lot of interaction in your class, seat the hearing-impaired child near the front where it is easy to turn and look into the face of another student who may be talking. Nod or indicate the direction of the next student to speak so that the hearing-impaired child knows where to look.

6. Write key words and phrases on the board whenever possible. Use diagrams and models to introduce new concepts and mathematical ideas.

7. If a child wears an amplification device, keep the level of noise down. This enables the child to sort out the important sounds.

8. When necessary, especially in upper grades, arrange for a note-taker. This allows the hearing-impaired student to focus on the teacher's explanations and not on having to shift from the teacher's face to paper and back again.

Speech or Language Impaired Children

To be successful in mathematics and other language-using situations, one must be proficient in the language of the subject matter. Acquiring new mathematical concepts, principles, and skills requires effective communication and language-processing abilities. Excluding the hearing-impaired, approximately 3.5 percent of children and students in the 5–21 age range have some form of speech or language impairment. Communication disorders represent the highest proportion of handicapping conditions in the school-aged population. Foremost among these are articulation problems; students having such problems are often unable to pronounce words correctly. If these students can produce words it is often with much effort; hence they have a low rate of communication. Such problems can result from hearing impairments, malformations of the speech mechanisms, or some form of central nervous system disorder.

Conditions which may signal such problems are indicated by the following behavior. The student:

1. speaks in incomplete sentences, expressing incomplete thoughts, using choppy syntax;

2. uses telegraphic speech, i.e., uses incomplete sentences with limited vocabulary and nonverbal gestures to complete a communication;

3. lacks fluency in expressing ideas—repeats or prolongs certain syllables or shows an unwillingness to speak at all;

4. stutters;

5. has disorders of pitch, intensity, or quality of voice; or

6. evidences improper breathing patterns or physical disorders of the speech mechanisms.

When one or more of these or associated problems leads to the diagnosis of a speech or language problem, the mathematics teacher can adjust the classroom and instructional patterns to help the child. Foremost among these adjustments are the following:

1. Try to tune into the child's speech pattern and work with the child in improving it.

2. Provide the child with corrective and supportive feedback as he or she works to master the language of mathematics.

3. Supply the child's speech correctionist with mathematical words and symbols for use in the therapy sessions when appropriate.

4. Encourage the student to verbalize mathematical conceptions.

5. Avoid interrupting language-impaired students. Allow them to finish their thoughts verbally before responding. They need to learn to finish their own sentences.

6. Try to understand children's problems and support attempts made to overcome them. Remember that many speech disorders have a psychological basis and supportive efforts can help the child overcome these difficulties.

Visually Impaired Students

Visually impaired students in the 5–21 age range account for about 0.1 percent of the population. Some visually impaired students are partially sighted while others have no sight. The partially sighted students often also have very restricted fields of vision and defective color vision. Many other classroom students wear glasses, but they are not classified as visually impaired because their vision is good enough that they can function in the classroom quite effectively. However, they may still have problems that a concerned teacher can resolve in order to make learning a more pleasurable experience.

There are several signs that a student may be undergoing a change in vision or have an undetected visual impairment. Notice if students:

1. Hold their textbooks close to the face or bury their heads in books when reading;

2. Hold their heads forward or tilted to the side (over 15°) when reading;

3. Move their heads and shoulders during a visual activity rather than just moving their eyes;

4. Rub their eyes often when engaged in reading or other visual activities;

5. Complain often of headaches;

6. Seem to learn better when things are explained verbally than when they are expected to garner the information from reading; or

7. Use facial contortions or squint when reading.

Unless a more serious condition is present, most of these situations can be corrected with the use of prescription corrective lenses. For partially sighted and sightless students, the following classroom arrangement and instructional techniques may make mathematics class a much more meaningful experience:

1. Encourage the children to use whatever vision abilities they have.

2. Place the students' working space in a well-lighted area of the classroom. (Some partially sighted youngsters, especially albinos, are very light-sensitive and need to be placed in a shaded portion of the classroom.)

3. If possible, provide the children with a slant-top desk and a book holder that keeps the book at a comfortable angle for reading. (Such holders are available from the American Printing House for the Blind.)

4. Consult with resource teachers for the visually handicapped to find appropriate writing materials and paper that ease the visual strain associated with communicating thoughts in writing. Often a cream-colored paper and a black felttip pen make effective writing tools for the vision impaired student.

5. Check to see if large-print versions of your mathematics text are available through your state office of education or from the American Printing House for the Blind. These versions of the text, or a cassette tape of the text, will often ease the reading strain on the student.

6. Give careful and complete explanations when you write something on the blackboard. Use a thick yellow or orange chalk whenever possible. Stress thinking and accuracy over speed.

7. Special drawings and diagrams can be made on pressure sensitive materials. The students can then feel the shapes and relationships through a guided discussion of the drawing. These special materials are also available from the American Printing House for the Blind.

When dealing with children having severe visual impairment, additional modifications are sometimes necessary.

1. Work with the children in establishing mobility patterns in the classroom. In particular, establish obstacle-free paths for the children's travel and determine special places for the storage of mathematics and other content-area materials.

2. Involve the children actively in the study of mathematics. Use a label tape printer and a brailler to adapt mathematical games and activities for their use. For example, one could use dice with raised dots or having different textures on the various faces of the cube.

3. Prompt the children to use whatever residual vision they have through the use of shading or other graphic techniques.

4. Structure learning activities and provide the children with aids such as a cookie sheet with raised edges or a commercial working

tray to define a systematic and bounded working area for activities.

More modifications of regular classroom procedures are needed to accommodate the visually impaired student than for many other handicaps. A creative teacher willing to devote time and thought to the matter can almost always modify a classroom environment to accommodate these special needs.

Physically Handicapped Children

An additional 0.5 percent of the students in the 5–21 age group have some special form of orthopedic or neuromuscular handicapping condition. Children with severe and profound physical problems normally require special mathematics programs and adaptive mathematical learning materials; rarely will such children be found in the regular classroom. Most physically handicapped students are quite capable of functioning in a regular classroom at an average or above-average rate of achievement. In a sense, they are most like regular students in that their potential for information processing and mathematical learning has not been impaired.

Depending on the disability, special provisions (e.g., a standing table, an electric typewriter, grab rails, or a desk that will accommodate a wheel chair) may be used to remove learning and environmental problems in a classroom having a physically impaired student. In addition to their academic studies, many of these students will also have a period of physical therapy scheduled in their programs. The therapist can help the classroom teacher develop any special learning materials (e.g., magnetic trays, counting devices) that might be necessary to help the child progress.

Your Time: Activities, Exercises, and Investigations

1. Why is it important that teachers of special-needs children understand the general conditions that signal special needs? What adaptive measures might be taken for children having sensory or physical handicaps?

2. Plan to visit a mathematics class having physical or sensory handicapped students. Report on your visit noting:
   (a) The age level of a student
   (b) A description of the handicap
   (c) The relationship of the student to the rest of the class
   (d) Special accommodations made for the student
   (e) The topic taught
   (f) Other changes you might suggest for dealing with the child in the mathematics class

3. Consider the following two case studies. For each of the students, develop an individual educational plan which:
   (a) Identifies the grade level which best matches the student's achievement level in mathematics. Use the curriculum outline in Appendix A as a guide.

    (b) Outlines a year's individualized mathematics program listing:
        (i)   Priorities you would recommend (*major* mathematical concepts and skills from those which normally are taught at this level)
        (ii) Any mathematical topics you'd definitely omit (from among those normally taught at this level)
        (iii) Any additional mathematical topics (not normally taught at this level) which you'd suggest receive special review or emphasis
        (iv) Any special adaptations in teaching techniques, activities, or materials you feel will be necessary because of a child's handicap.
    (c) Discuss your design with one other classmate, revise it, and turn it in at the specified time.

Student I: Jerry is a five-year-old sightless boy. His parents have worked carefully with him since birth to help him cope with and compensate for his lack of sight. Bright and cooperative, Jerry has good travel mobility in familiar places. Jerry's language skills are average for his age, and he can correctly make object comparisons (tall, taller; long, longer; heavy, heavier) for objects he has the ability to feel and judge. He can count by rote to 10, but has only demonstrated a real *understanding* of numbers through four (zero excluded). He has never learned to use a ruler. His experience with money is extremely minimal. He cannot distinguish coins from noncoins, though he has some concept of the value of money.

Student II: Alice is a fourteen-year-old hearing-impaired girl. She plans to work as a bookkeeper or clerk in her aunt's craft store when she graduates from high school. Alice seems to be quite bright, and she enjoys working with numbers. Her addition and subtraction skills are quite good. Her multiplication and division skills are not quite as strong, as she can only use one-digit multipliers and divisors. She doesn't always remember her multiplication and division facts. Money-counting skills are good, but she still has some difficulty in making change. Numeration skills for two- and three-digit numerals are generally weak, so she has difficulty comparing and ordering numbers. Her estimation skills are also weak.

Learning Disabled and Behaviorally Disordered Children

Two remaining classes of children with special needs who are commonly enrolled in regular mathematics classes are the learning and/or behaviorally disabled child. These classes of children seem much like the regular child in the class in both their appearance and much of their behavior. Students with learning disorders have as much, and in many cases more, natural ability to learn than any of their peers. The major difference is the fashion in which they learn. Behaviorally disordered children process information as well as anyone else in the class. Their limited ability to interact socially and educationally is the factor that interrupts their learning.

Teachers often have difficulty in differentiating between slow learners and learning-disabled children. This is due to the lack of information about common problems faced by learning-disabled students and how they relate to the teaching of mathematics. The recognition of possible problems extending beyond the rate at which a student can acquire information must rest with the classroom teacher. From that point professionals can be consulted to determine the extent and exact nature of the learning problem. Similarly, classroom teachers may have difficulty differentiating between students who cause classroom disruptions and those who have behavioral disorders. Like the diagnosis of learning disabilities, the identification of behavioral disorders is a job for a professional, but the classroom teacher may be the first step toward identification.

Learning Disabilities and the Mathematics Classroom

About 3 percent of the 5–21 school population has some form of learning disability severe enough to require special educational considerations. A learning disability causes difficulty in learning but does not stem primarily from mental retardation, emotional disturbances, physical handicaps, sensory or motor impairment, or cultural or educational deprivations. In many cases learning disabilities do coincide with other handicaps. The basis of such learning disabilities is not always easily identifiable. Known causes include some form of trauma to the central nervous system or other neurological disturbances during childbirth or early childhood.

Learning-disabled students having difficulty in mathematics are not considered slow learners, due to their IQs and achievement patterns. Learning-disabled children usually have average to high IQs and an accompanying pattern of potential for high achievement. In many cases their achievement profiles show work quite advanced beyond their classroom peers. The one factor that sets them aside from the rest of the class is their inconsistent learning patterns. Some educators have even characterized these patterns as being "consistently inconsistent."

The "strong in some areas, weak in other areas" pattern of performance is due to a wide variety of problems. In the following pages we discuss several of these learning disabilities. Their sources and cures are often beyond our grasp, adding to the frustration felt in dealing with children afflicted with them. The problems that surface in the mathematics classroom and are classified as learning disabilities include the following:

Perceptual Impairments. Perception is the process by which the intellect recognizes and derives input from what it receives from the senses. If sensory organs are intact and the information is still not being conveyed, a dysfunction of the central nervous system is assumed. Visual perception problems are quite common young children with learning disabilities. Inversions, rotations, and distortions of symbols, signs, and words occur. Some children, for example, confuse the + and − signs; others cannot distinguish between the =, + and − signs. Others confuse the 6 with 9 and 3 (or

5) with 8. Fractions are often distorted in the child's perception. Some confuse 17 with 71, and so on. Some cannot distinguish a square from four unrelated line segments. Some children are unable to read numerals written in sequence correctly.

Figure-ground problems are quite common. The figure is the desired focus of attention and the ground is the other visual stimuli in the field of vision. Children with figure-ground problems are unable to separate irrelevant stimuli from the important details. Because of this, they cannot scan for information and become perplexed when looking for material on a page. They may be unable to solve familiar problems when they are presented on a crowded page or worksheet. They may also find it difficult to deal with a large number of concrete learning materials.

Another common visual perception problem is the correct alignment of place value columns found in the addition and subtraction algorithms and the more complicated alignments found in the multiplication and division processes.

Auditory perception problems seem to be less prevalent in mathematics. This may be due to the fact that math is highly visual. But for those students having auditory problems, verbal instructions in mathematics class seem like a hopeless jumble of instructions and disconnected symbols. Intervention by a trained specialist is needed to overcome these problems.

Memory Disturbances. The mathematics performance of other learning-disabled students can be severely impaired by not remembering things which they have seen or heard. Visual memory deficits, for example, hamper the child who cannot remember the appearance of numerals. Such children may be able to copy numerals out of books and repeat number names, but they may not be able to reproduce them from memory. They may not be able to recall the shape of a square or triangle in order to reproduce it on paper. This ability to revisualize, to recall mentally the picture of objects previously viewed, is an important aspect of success in geometry and many other branches of mathematics. Visual memory, from the point of recall, influences the responses to questions such as these: "Were there two or four beads in the pile?" or "Did the triangle have a right angle?"

Auditory memory problems make it difficult for the affected child to follow oral directions or explanations given by the teacher. Children with this difficulty may not remember what was said during instruction or what they were asked to do. They may not be able to work story problems that are presented orally, since they cannot hold and assimilate the facts necessary for a correct solution. Oral drill on basic facts will not help this child.

General Orientation Defects. Basic ideas such as time, space, quantity, size, amount, order, distance, and length are nonverbal concepts, and relationships which are important to the learning of mathematics. On standardized tests of intelligence and achievement, children with learning disabilities tend to score low in these areas. Reasoning processes for many early quantitative thinking tasks have a visual-spatial frame of reference. Some disabled children have short attention spans, poor perception, poor motor coordination, and lack of experience in appropriate manipulative activities which provide the background for dealing with these concepts.

Children with spatial problems may have little feel for or understanding of relational concepts such as *up* and *down*, *over* and *under*, and a host of others. These children have trouble finding distances on a number line, have prolonged inversion-reversal tendencies, and have difficulties with left-right orientations. These orientation defects are particularly evident in the area of place value and numeration skills. Directionality and orientation skills are of the utmost importance here. These defects also surface in situations involving time and sequence of events.

Inability to Integrate Processing Skills and Information. Many learning-disabled children cannot coordinate various learning capabilities and related processing skills. For example, there are times in a mathematics class when one must employ visual-auditory, visual-motor, or visual-spatial integrations. Some disabled children lack the ability to exhibit these integrations upon demand. They cannot put visual, motor, and spatial skills together to work problems such as writing numerals in a correct order and of the same size. Computational algorithms often require numbers to be placed carefully, in correct sequence and proper positions, for the correct answer to be developed. Some students have difficulty copying from a written page; they lose their place when returning to the text. The integration of such skills is vital to personal achievement.

Abstracting, Generalizing, and Conceptualizing Difficulties. Quantitative thinking and abstract reasoning are important activities for success in mathematics. The learning-disabled student in mathematics is often unable to infer, draw inclusions, or form generalizations concerning number relations and concepts apart from instruction which is slowly, carefully, and sequentially developed from concrete to more abstract ideas. Such children must progress through the same steps as other children, but at a slower rate and with more emphasis on the steps than other children. It is not enough to present examples to such a child. The child must examine these examples in greater detail and then be asked to work on producing and verbalizing other examples at each step along the way.

Hyperactivity and Perseveration Problems. Two types of attention disorders are especially frequent among learning-disabled children. Many are hyperactive and easily distracted. These children characteristically have short attention spans, are in continuous motion, and are generally not goal-directed. In addition, they tend to have difficulty integrating motor acts, language, and perception. Many also have very short memory spans and act on impulse. These children are often distracted, rather than motivated, by the colorful pages of the elementary-school mathematics text.

A separate problem, characterized by impulsive behavior, is *perseveration*. Students afflicted with perseveration get into patterns of behavior and persistently repeat the pattern on every activity they face over a short period of time. For example, if the first problem in a set is an addition problem, they treat every other problem in that set as an addition problem.

Summary. This section has looked at the major types of learning disabilities and the major problems resulting from them. The learning disability problems most often noticed in mathematics include:

1. Place value-numeration confusions

2. Positioning of digits in algorithms

3. Transpositions and/or rotations of the digits

4. Auditory and visual memory problems in oral drills and problem-solving situations

5. Visual-spatial geometric tasks

Global Suggestions. Some children, because of the severity of their problems, will never completely overcome their disabilities. Hence the teacher must help them compensate for their handicaps and learn to work within their abilities. In doing this, the classroom teacher, with the help of a resource person, can look for strengths to teach toward. Special help can come from the psychological examiner and results from the specialized tests such as the *Illinois Test of Psycholinguistic Abilities (ITPA)* or the *Wechsler Intelligence Scale for Children (WISC)*.

There is much that the classroom teacher can do to ease the learning problems of these students.

1. Start with the children at the level at which they can comfortably function. Study their previous work and find where they are and what they can do. Document their strengths and weaknesses and note their particular interests. Develop anecdotal records to keep these facts and others in mind as you work, and continually evaluate these students' work. Pinpoint as closely as possible the bounds of what they can and cannot do. Note their predominate learning styles, reading difficulties, recollection abilities, spatial skills, and expression abilities.

2. Capitalize on what a child does best. Praise the good you see in the students' work. Avoid problem areas until the child has developed some self-confidence. Use the child's preferred mode

of learning during instruction and then work on the deficiency during practice and extension exercises. The value of such activities cannot be overestimated in helping the students grow in their areas of deficit. Some teachers have found the use of dramatization and body movement to be very effective in reaching many learning-disabled students in the mathematics classroom.

3. Involve the children in the establishment of short-term goals. Make them aware of their progress and help them to develop methods for their own self-appraisal. Comparing learning-disabled children to themselves rather than to others is the best standard of progress in mathematics learning. Give children a sense of success by helping them see the progress they make from day to day. In addition to doing this by setting attainable goals, help the children develop methods by which they can monitor their own behavior in mathematics. This will minimize the dependence on others for feedback and monitoring. It will aid in curtailing undesirable patterns and create a more positive attitude toward mathematics.

4. Forestall problems in mathematics. If the children lack certain skills and are not prepared for an area of study, develop a carefully controlled and sequenced introduction to the area that minimizes the students' problems as they encounter it for the first time. Make sure that these activities keep the children cued in on the principal focus and provide the followup activities necessary for permanent learning. Allow students time to "talk" the mathematics so they also develop the necessary communication skills.

   Some children need concrete learning aids to help formalize abstract relationships, while others can develop the same level of learning with only an oral explanation. A careful analysis of the children's abilities will point the effective way. With children having figure-ground problems, avoid cluttered worksheets and provide 2-cm graph paper to help organize computational work. For students who are easily distracted, establish work carrels in a corner of the room to give them some privacy. For students who are hyperactive, establish routine, keep unstructured time to a minimum, and reward good work. These maxims and constant attention to each child's progress will maximize the students' learning while minimizing problems.

Behaviorally Disordered Children
Approximately 1 percent of the total 5-21 school population is affected with severe behavior disorders. Problems such as excessive fears, aggression, shyness, reclusiveness, lying, sadness and crying, self-stimulation, and destructiveness can clearly interfere with successful learning of mathematics. Behavior disorders vary from child

to child and no two problems are alike. Every child at one time or another has behavior problems. The ideas that follow are designed for dealing with *severe* emotional or behavior problems in the classroom, but many are effective in dealing with minor disturbances as well.

The basic rules for helping children with behavior disturbances are simple and adhere to common sense. They consist of developing simple, clear, and reasonable rules and procedures for classroom behavior; reinforcing the appropriate positive behaviors; and giving immediate followup for infractions of rules and procedures. Every mathematics classroom has procedures that must be followed in order to maintain a smoothly functioning environment beneficial to all members of the class.

Besides the three basic rules mentioned above, there are other things a mathematics teacher can do to stabilize classroom behavior. They are:

1. Follow an organized daily schedule. This does not mean that the teacher cannot take advantage of the chance happening, but that everybody knows the schedule, it is usually adhered to, and the child can feel confident that the classroom functions predictably.

2. Arrange the physical environment to create a calming effect on the students. The room does not have to be sterile or non-stimulating, but it should be arranged to focus attention on the centers for activities, for small groups, and for private work. Keep distracting influences to a minimum.

3. Plan for children to spend most of their time on learning activities focused on objectives. Minimize transition times between classes and at the beginning and end of the day.

Following these guidelines will produce a classroom where learning takes place and disturbances due to behavioral disorders are minimal. Additional techniques are available through the use of behavior modification. Teachers who exercise control over their classrooms are successful and their students make real progress. The secret to such successful teaching ventures lies in advance planning by the teacher. It does not happen spontaneously. The teacher who makes these efforts will be rewarded by noticeable student intellectual growth in the area of mathematics.

Mentally Retarded Learners

The mental impairment of retarded children is shown in their intellectual development, mental capacity, adaptive behavior, and academic achievement. In each of these areas the mentally retarded child's progress and growth are markedly delayed. Such mental impairment may be mild for students in the IQ range from 55 to 69, moderate for those having IQs from 40 to 54, severe for those in the 25 to 39 IQ span, or profound for those in the 1 to 24 IQ span. For educational purposes, mentally handicapped individuals have traditionally been referred to as educable, trainable, or profoundly handicapped. These names have changed from time to time and from state or local district to another.

Education of the severe and profoundly handicapped is a new and growing field. No mathematics education research has been done in this area.

Trainable individuals (IQs from 35 to 50) account for 0.3 percent of the 5–21 year range. The students in trainable mentally handicapped programs deal with mathematics which normally tops out at a first- or second-grade level. Some individuals, partly due to their capabilities and partly due to the program they are enrolled in, never reach this peak. Most trainables require some form of supervision throughout their lives. Many do well in sheltered workshops and semi-residential homes where some sense of responsibility can be fostered.

Educables (students with IQs of 50 to 75) represent about 2 percent of the 5–21 age grouping. Mathematically educable retarded individuals tend to achieve between third- and fifth-grade level. Though it does not often happen, some students go beyond this expectation. Because of developmental lags and other factors these students might be expected to cover about 1/2 to 3/4 of the material that regular students would cover in a given year. For this reason educable students who are mainstreamed for mathematics are often older than their peers. Thus the advisability of mainstreaming an individual must take into account a consideration of the student's size as compared to peers, his or her social maturity, attitude, ability, and confidence in ability to achieve.

Regardless of the placing for mathematics or the educational classification used, the following guidelines for instruction may prove helpful.

1. Keep the step size for instruction small. Carefully develop one concept or skill at a time and teach it to some level of mastery before moving on to the next concept or skill in the sequence.

2. Provide the student with a concrete and physical basis for the learning of new mathematical concepts and skills by effectively using materials to illustrate and reinforce mathematical ideas.

3. Encourage children to verbalize their understanding of concepts.

4. Provide for overlearning through motivational activities and games *in addition to* worksheets. (Mathematics is more than a paper and pencil activity.)

5. Foster children's ability to think, to deal effectively with the quantitative data of their environment by providing many

instances for oral and physical problem situations and the applications of the related skills in everyday situations.

6. Provide frequent, motivational review of previously learned materials.

7. Be selective in the mathematical content taught. Consider the relative importance of a concept or skill for helping the child deal, in daily life, with the mathematics of common situations. Minimal expectations should include mastery of the topics included in the box.

Math Basics for Daily Living (For children with mild mental or with severe sensory or physical impairments)

Number meanings (at least through three-digit numbers);

Number sequence and its practical applications (e.g., finding house and page numbers; understanding *more*, *less*, and *enough* with respect to money);

The basic operations of addition and subtraction (and, in some cases, multiplication and division) in order to know what to do in simple situations requiring computation; what button to push when using a calculator;

Basic addition and subtraction facts—to mastery! (The value of learning basic multiplication and division facts is a function of the child's ability, interest, and career goals);

Addition and subtraction computations involving two- and three-digit numbers;

Measurement and basic applications in cooking, sewing, carpentry;

Time, Money, and practical applications of both;

Basic fractional understandings (e.g., half a dollar, half an hour, one-third of a cup, a quarter or fourth of an apple);

Problem solving (at least through those posed orally or through pictures and those representing common, real-life situations);

Mental estimation and its applications (checking the reasonableness of computational or calculator answers; deciding whether one has enough money for the purchases one wishes to make in a store);

Percentage, to include at least the understanding of its use in finance charges (18% interest means 18 dollars on $100 borrowed per interest period); in interpreting newspaper ads (20% discount means you will pay less); in recognizing what the interest amount on a loan or on a savings account means; and in understanding the nature of payroll deductions.

While some educables will remain in the job force or sheltered workshops, others will function at least semi-independently in adult life. The mastery of essentials such as those listed in the box make a great difference in the job options open to educably handicapped individuals in life beyond school.

DISCUSSION   This chapter introduced ways of diagnosing specific mathematical deficiencies found in resource as well as regular classroom situations. In addition, it introduced various areas of special needs and some modes of teacher adaptations to these needs. The reader might wish to elect additional coursework in special education, or in educational or psychological testing.

The global introduction to the diagnosis of special needs and possible teacher responses provides a base for the remaining chapters of this book. In them you will see the emphasis shift to specific methods for teaching the special child, for classroom organization and control, and for teaching the concepts, skills, and principles that make up the elementary mathematics curriculum.

The most important thing for you to remember from this chapter is that no one piece of information or one method of diagnosis will ever solve the problem of the special needs child. It is only through the careful study of the student's profile, based on several pieces of information, that you can begin to get a picture of a student's strengths and weaknesses and of possible plans for remediation. Attention to the student's special needs will lead to success in teaching.

**Your Time: Activities,
Exercises, and
Investigations**

1. Plan to observe a classroom which has either learning-disabled or mentally retarded students in it. Then write up a report of your visit detailing the following information:
   (a) Name of the school and teacher involved
   (b) Description of the students observed (area of handicap, age, orientation of the student to work, ability level, . . .)
   (c) Description of the class as a whole (number of students, ability levels, regular or special class, . . .)
   (d) Topic being studied
   (e) Materials used (text, worksheets, manipulatives, games, activities, . . .)
   (f) General nature of the lesson for the day (pre-book motivational, development of a new concept or skill from activities, development of new concept or skill from textbook, application, practice, or review based on book, followup of material from non-book sources, . . .)
   (g) Any adaptive techniques or materials you noted for children with special needs
   (h) Special comments about the student on which you focused your observation
   (i) What impressed you the most
   (j) What activity could be developed for the topic being discussed which might serve the special student's needs in addition to the ways they were served in the lesson you observed.

2. Case Study IEPs. Here are detailed descriptions of three special students. For each student develop in outline form the material necessary for an IEP, giving the following information:
   (a) The grade level that *best* matches the student's achievement level in mathematics. Use the material in Appendix A as a guide to the grade placement of topics.
   (b) A summary of the individualized mathematics program you would prescribe for a year by listing: Priorities you would recommend (*major* mathematical concepts and skills from those normally taught at this level) o Any mathematical topics you'd definitely omit (from those normally taught at this level) o Any additional mathematical topics (not normally taught at this level) which you'd suggest receive special review or emphasis o Any special adaptations in teaching techniques, activities or materials you feel will be necessary because of the child's handicap.
   (c) Discuss your writeup with one other classmate and, if possible, with an experienced teacher, before turning it in.

Student I. Billy is a seven-year-old learning-disabled student. Due to severe visual perception problems that were not diagnosed until he was six, Billy is below his peers in arithmetic achievement. He still has consistent reversal problems which affect his ability to sequence two-digit numbers. While he has a firm grasp of number meanings and of the concepts of addition and subtraction, his addition and subtraction skills are extremely poor. He can correctly answer fact problems (sums to 10) only by counting on his fingers. He can usually read time on the hour and half hour (though reversal of clock hands is a common tendency). Otherwise he cannot read clock times (e.g., minutes after the hour), nor can he measure accurately with a ruler (even to the nearest centimeter). He likes to work with money, however, and can count amounts to 50 cents, and even make change for small amounts up to 25 cents. His work on his brother's paper route has paid off in this area.

Student II. Danny is a ten-year-old hyperactive and highly distractable child. He repeated second grade, and was in a regular classroom until third grade (9 years old) when he was placed in a special resource room. He was on drugs for a year (8½ to 9½), but has been off them for a half year. He is responding well to the structure and security of the resource room and its teacher. Arithmetic is his best subject, but even here he is still having some problems. He does not know all of his basic addition and subtraction facts, has difficulty borrowing in subtraction and in sequencing two- and three-digit numbers. His strong area is money, though he still has some difficulty making change. He's just beginning to show an understanding of the concepts of multiplication and division. He has never worked formally with fractions.

Student III. Ted is a fourteen-year-old educably retarded student. He has been deprived of many of the educational experiences most children have had due to his growing up in a remote rural area and having only attended school for about five years on a very haphazard basis. His addition and subtraction skills are good. If he forgets a fact, he knows how to help himself. He has learned to use a hand calculator to compensate for his computational difficulties. (Addition is strongest. Borrowing still causes some difficulties, especially if zeros are involved.) He never really mastered multiplication or division concepts or skills. Money causes no problems up to $100. Both counting and change-making skills are very good. He can read time from both a digital and "face" clock (time after the hour only). Number sequencing and estimation skills are weak. Measurement concepts and skills are developed to approximately early fourth-grade level. He has just a basic feel for fractions, no concept of decimals apart from money, and no understanding of percent. Next year he will enroll in the school's occupational-vocational program, and train for custodial and light repair work with a local firm.

3 A BALANCED INSTRUCTIONAL PROGRAM

CONTENT

The single most important characteristic of an effective instructional program is balance. The appropriate combination of concepts, skills, and applications in teaching is essential to a successful educational program. The omission of, or even slight emphasis upon, any one of the three will guarantee an inferior program of mathematics instruction. A program with heavy stress on conceptual learning and light stress on skills and applications will leave a student unable to apply his knowledge of concepts in computation or problem solving. A program with light emphasis on concepts and heavier emphasis on skills and applications will handicap children who are not able to work out conceptual understandings on their own. Similar statements can be made concerning the relative emphasis given to the other two areas and the various combinations of all three. In the following sections, we examine the three areas and the methods by which they are taught in the elementary school classroom.

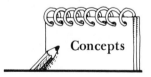

In simple terms, concept development consists of doing two things:

1. *Forming a classification.* (Knowing what is included and what is not included in some group of things or ideas.)

2. *Associating a label with the class.* (The label *may be* a formal term such as "rectangle," but all that is *really necessary* is that the label provides a way of referring to the group.)

For example, when looking at the four-sided polygons in Figure 3.1, students may notice that certain of these polygons have some things in common and can be grouped together (the ones circled). A mathematics teacher would note that the indicated figures are polygons with four sides, two pairs of parallel sides, four congruent angles, and symmetry about both diagonals (which happen to bisect each other). Children may recognize that the figures are all alike in a way that they understand but cannot verbalize. The child has developed the concept when he or she can tell you whether any new object belongs to the class and can refer to that class using a label. The label the child attaches to the identified class of objects might be *rectangle*, *quadrilateral with square corners*, or merely *these shapes*.

As a second example of concept learning, consider the items shown in Figure 3.2. A child may first notice that A, C, E, and F all have something in common (even though they also have some differences). After deciding that A, C, E, and F "go together," the child can determine whether G, shown in Figure 3.3, also belongs to the class. In the course of instruction, attempts should be made to give the child methods for testing whether G belongs or not. In most cases the procedure would be some sort of matching or attribute checking.

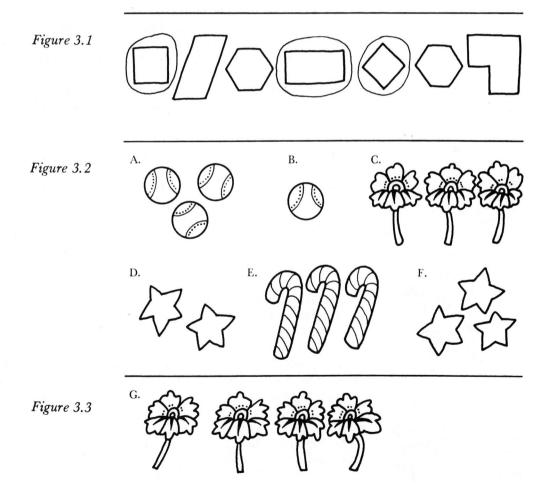

Figure 3.1

Figure 3.2

Figure 3.3

Although the label used by the children may at first be an informal one like *this many*, here we would want the child to label the concept *three*. When a concept has a commonly accepted name, as in this case, it is a good practice to have the children agree on the same label that the rest of the world uses. Note that the formal name for a concept need not be used immediately. Indeed, it is more effective to let the children use their own vocabulary when they originally begin talking about a concept and to supply the formal name (the label used by others) only after the concept is firmly established. Of course, the formal label may be so common that it is natural for the child to begin using it almost immediately.

The teaching of class concepts such as those shown above requires several skills on the part of teachers. These skills center around the various ways in which such content can be presented to children. To teach a concept to a class of children, or to an individual child, the teacher can either talk about the characteristics of the objects in the group or give examples of objects in the class. At the elementary school level, the teacher usually presents children with several examples of the members of the class and with several examples of objects that do not belong to the class. This presentation of examples and nonexamples is usually followed by a period during which students decide about additional cases presented by the teacher. At this point the teacher, or a student, may attempt to provide a definition of the concept using the children's own vocabulary.

Once the children have gained a certain feel for the concept, they can be expected to sort examples and nonexamples of elements of the concept's set and state why the examples are members and the nonexamples are not members. This activity naturally extends into discussions of the similarities and differences of the various examples and nonexamples and distinctions between the concept in general and other concepts. After some experience with the concept, a formal definition may be presented to the children to conclude study of the concept.

Higher Order Concepts—Rules

Sometimes the term *concept* may be applied to other types of groupings. These classifications deal with examples that show relationships between simple concepts such as those we have considered to this point. Most of these concepts result from a study of patterns. *These higher-order concepts, sometimes called rules, describe a pattern on the basis of similar results or spatial configurations.*

For example, when you combine a group of tens and ones with another group of tens and ones, the result will be a new group of tens and ones. When you add the same number to both sides of an equation, the result will still be an equation.

When you multiply 47 x 10, 28 x 7, 24 x 10, 5 x 6, 83 x 2,

15 x 10, 61 x 10, 71 x 6, and 7 x 10, you notice that

$\underline{47}$ x 10 = $\underline{470}$,
$\underline{24}$ x 10 = $\underline{240}$,
$\underline{15}$ x 10 = $\underline{150}$,
$\underline{61}$ x 10 = $\underline{610}$, and
$\underline{7}$ x 10 = $\underline{70}$

all have a common characteristic that the other results do not. In this set of sentences, the common property is the possessing of 10 as a factor. Furthermore, the digits in the product to the left of the one's place are identical to the digits of the other factor.

The analysis of these instances shows the child that when you combine the concept of multiplication with the higher-order concept of multiplication of 10, the yet-higher-order concept results: *When you multiply a one- or two-digit number by 10, the product is the one- or two-digit number followed by a 0 in the one's place.*

The development of these higher-order concepts, sometimes called rules or principles, is a task that increases with importance as one moves upward through the mathematics curriculum. Activities involving them, plus their related applications, are all firmly based on the student's thorough understanding of the simple concepts contained within them.

The teaching of higher-order concepts requires special skills. The first of these is the ability to focus students' attention on the instances of the higher-order concept and motivate them to search for the pattern. In addition, the teacher must provide the necessary structuring of the pieces of information provided by the students in bringing the statement of the pattern to its final form. Once this has been accomplished, the transfer of the rule to many different instances falls on the teacher's shoulders. The teacher needs to examine the use of the rule in many cases, especially cases where the students often make common errors. Finally, the teacher should assist students in differentiating one higher-order concept from another when they are closely related.

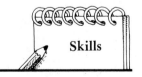
Skills

While a concept can be thought of as something you understand, a skill can be thought of as something you do. You can do things very well or not so well. Things can be done easily or with great difficulty. Things can be done quickly or slowly. It follows, then, that *a skill tends to develop by degrees.* Rather than think of a skill as something that you have or do not have, it is a bit more useful to *think of a skill as something that you can improve.* Of two children who both have a skill, one may be very fast with it while the other may be very slow. One may be very accurate, the other error-prone.

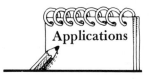

A child's skill in a certain area can be improved in many ways. *Sometimes a stronger conceptual base will bring about greater skill.* A child who knows that he can add in any order may think through the accompanying example more rapidly and accurately than a child who always adds the numbers in the given order.

Sometimes, the memorization of additional facts can improve a skill (see the example of 28 + 15).

Sometimes repetitive practice to habituate a procedure can improve a skill. Habituation of a procedure may save mental steps which may be both time-consuming and error-inducing.

The teaching of skills requires several things of the teacher. The teacher must carefully develop the skill based on the students' conceptual knowledge. This usually involves relating skills to physical models and earlier number patterns, facts, and skills—those already mastered by the students. Also, the teacher usually spends some time demonstrating the correct application of a skill in problems and applications. In other cases the teacher should interpret what the skill is related to, step by step, and how the performance of those steps in a given order will give the desired correct answer.

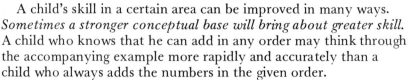

An application is using a skill or concept to accomplish some task or solve some problem. Almost never is a concept or skill applied all by itself. Applications, as a rule, involve the selection and use of some combination of concepts or skills, very often in a new or different setting.

Perhaps *one of the more important emphases in teaching applications is the identification of analogous situations—the recognition that this situation is similar to another situation in certain ways and that those similarities allow one to apply the same concepts and skills in both situations.*

Applications come in many forms. For example, in doing long division, one uses (applies) multiplication and subtraction skills. When finding the distance from the floor to the ceiling in the middle of a room by measuring up the wall, one applies, among other things, the concept of parallelism (i.e., the floor and ceiling are parallel, making the distance between them the same wherever it is

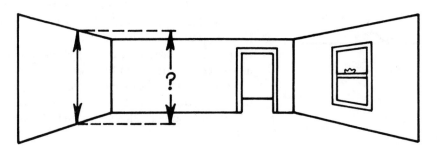

measured). In some applications, the problem solver must apply many widely varied skills and concepts. For example, when planning a vacation trip, one applies a combination of skills and concepts from measurement, computation, reading, social studies, and other areas.

Your Time: Activities, Exercises, and Investigations

1. Consult the sample curriculum given in Appendix A. From the objective list:
   (a) Identify two concepts from each of the grade levels, 1–6.
   (b) Identify two skills from each grade level, 1–6.
   (c) Identify four applications that are taught in elementary school.

2. Choose any concept that is taught in the first three years of school.
   (a) List the concept and give the grade level at which it is taught.
   (b) Find a skill that uses (requires as a prerequisite) the concept that you gave in part (a). Give the grade level at which the skill is taught.
   (c) Find an application that uses the skill that you listed in part (b). Indicate the grade level at which the application is taught.

3. Choose any application that is taught in elementary school and list all the concepts and skills that are required by the application.

4. Select one category of special needs children. Select a mathematics topic from a chapter in an elementary school text. In view of functional skills, list two skills that you believe are of high priority and one skill of low priority. Select or create an application problem.

ACTIVITIES

A second way to achieve balanced instruction is to strive for balance in the kinds of activities that students are asked to do. By activities, we mean all the things that children are asked to do in the school setting. These activities vary from games to pages of exercises, from building a model to listening to an explanation, from participating in a group discussion to working quietly and independently.

Certainly, there should be a balance in this sense, for a child who only plays games will develop little of the discipline that is needed for other kinds of activities. The child who hardly ever works independently may become overly dependent on other people for simple tasks. The child who seldom works with his or her hands will not easily develop a feeling for the spatial relationships that are so important in many facets of daily life.

All of the activities in which students participate can be placed in at least one of the following classes:

1. Developmental
2. Practice

3. Application
4. Diagnostic/Evaluative
5. Transitional

Appropriate balance and sequencing of activity types 1, 2, and 3 will generally result in effective instruction.

Developmental Activities

Developmental activities in mathematics can be separated into two subgroups: exploratory activities and consolidating activities. *Exploratory activities are those activities which allow the child to experience his world.*

Exploratory activities can be thought of as the opportunity to gain background experiences and information that will be needed to learn a new concept or skill. Exploratory activities may be very structured or directed. There may be very specific things that you want each child to have experienced before attempting to develop a new idea.

On the other hand, exploratory activities may be unstructured and undirected, such as a period of "free play" with a new set of instructional/learning materials like number rods, geoboards, or a balance scale. At times a child will have already had many experiences in the course of his everyday life so that no additional exploratory activities, structured or unstructured, are needed.

When sufficient exploration has taken place, it is time for the teacher to begin pulling loose ends together, pointing out important relationships and asking leading questions to see that relevant patterns are noticed and appreciated by the child. The teacher moves from a rather passive kind of instruction during the exploratory phase of the development to a more active kind of instruction. The activities that take place during this phase of development are consolidating activities.

Consolidating developmental activities are those activities which are used to help a child visualize a new idea, to verbalize a useful rule, to see relationships between ideas already acquired, to agree on terminology, to standardize a system of procedures, to relate what's seen to what's written.

One type of developmental sequence that is worthy of special consideration is that which can be called a laboratory approach. Other terms that may be used for what is essentially the same approach are scientific approach, inquiry approach, and inductive approach. There are six basic steps that students are led through in a laboratory approach. They are:

1. Experience/observe
2. Gather data
3. Organize data/identify patterns/see relationships

4. Hypothesize
5. Test hypothesis
6. Communicate findings

Note that the above sequence may be followed formally, using data sheets and carefully verbalized hypotheses. However, the laboratory approach can be just as effective if used informally, with the gathered data merely what's remembered, the hypothesis stated in "kid talk," and the hypothesis testing simply "Let's try it again to see if it turns out the same way!" A laboratory development may be independent or teacher-directed. A laboratory development may be used by an individual or a group. A laboratory development may be used in a special room with special equipment (a laboratory) or virtually anywhere.

Figure 3.4 indicates the relationship of the five-step laboratory sequence to the more general two-part developmental sequence.

Sample Developmental Activities

The following sample activities illustrate the two types of developmental activities just discussed. Examples 1 and 2 are exploratory activities; Examples 2 and 4 are consolidating activities.

Example 1: Volume
Give the student an assortment of differently shaped boxes and a package of cubes. Have the student figure out how many cubes will fit into each box.

Example 2: Volume
Choose a box. Have the student place one complete layer of cubes in the box and tell how many. Have the student place a second layer of cubes in the box and tell how many. Help the child to see that every layer has the same number of cubes.

Example 3: Addition
Have the child, using colored rods, make as many as possible trains of the same length.

Figure 3.4

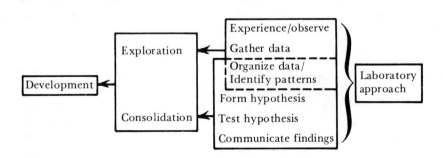

Example 4: Addition

Using colored number rods, make trains the same length, assigning number names to the trains. Help the child to see that numbers can be named in many ways.

$$2 + 2 + 2 + 1 + 1 = 1 + 1 + 1 + 4 + 1 = 3 + 5 = \ldots$$

Practice Activities

Once a new mathematical idea has been developed, it is necessary to provide sufficient practice with that idea for the student to become efficient and comfortable with it. That practice doesn't just happen. It must be planned and scheduled by the teacher and then carried through. It is not enough, however, that students have a lot of practice. That practice must be with the right materials or activities and with the right emphasis at the right time.

As with developmental activities, *we can separate practice activities into two subgroups: procedural practice activities and habituation practice activities.*

The purpose of procedural practice is to standardize procedures, gain confidence in procedures, and gain accuracy in the results produced by procedures. Procedural practice is generally characterized by a lack of pressure. Plenty of help is given, since we want the procedures being practiced to be correct. Plenty of time is allowed, since we want the student to be sure of results. Even though correct procedures will produce correct answers, the emphasis is on getting the *procedures* right rather than on getting the answers right.

When sufficient time and effort have been spent on procedural practice so that, given enough time, the student can be confident of producing correct results virtually 100% of the time, the emphasis of the practice begins to shift from the procedure to the result. *This second type of practice activity, habituation practice, is used to make procedures quick and efficient.* Now the student can work independently. The student may have limited time and be required to work more quickly.

Habituation practice helps the student reach that point where processes become almost automatic. Little thought needs to be given to how or why a particular step in an algorithm is taken when it is. If sufficient procedural practice has been given, the student will be confident that each step is correct. Habituation practice changes the emphasis from figuring out facts to remembering them.

Sample Practice Activities

The two activities that follow are examples of practice activities. Example 1 is a procedural practice activity that might be used after the notion of regrouping has been taught using bundled sticks. The students practice the procedure until they understand it and can do it correctly. Example 2 is a typical practice activity that might be used to help the student habituate the written algorithm.

Example 1: Addition

Have the child do several addition examples by representing the two numbers with bundled sticks, combining the modeled numbers and naming the resulting number after making any needed trades. Record the steps as they are performed.

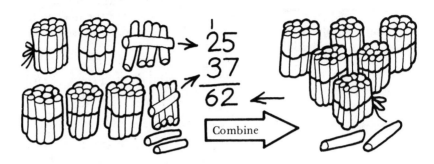

Example 2: Addition

Have the child complete the six addition examples shown in the accompanying illustration.

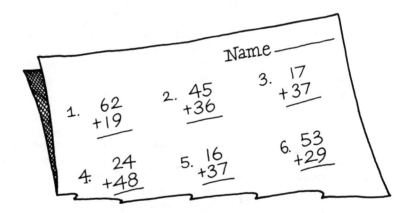

Application Activities

It is common for a student to have a good understanding of a mathematical concept or skill and still forget it. Also, it is usual for a student who has practiced thoroughly with a concept or skill to forget it. Indeed, one human characteristic common to most people—young and old, student and teacher alike—is the tendency to forget concepts, to lose skills not maintained through regular use.

If mathematics is important enough to remember, if it is more than an intellectual exercise, we need to provide regular opportunities to apply it. *Application activities help the child to see that the mathematics learned has daily practical use.* It is not enough to tell the child that "You're going to use this someday." We must provide activities that let the child use mathematics as it is used in the real world.

Sample Application Activities

The examples that follow are fairly typical application activities. Note that in each case the child is placed in a "real" situation where it is necessary to gather the needed information and then use it, in a way left to the student, to solve the problem.

Example 1: Addition
Have the child select a school wardrobe from a mail order catalogue and find the total cost.

Example 2: Area
Have the child estimate the cost of carpeting his bedroom if the cost of the carpet is $10 per yard.

Example 3: Time
Have the child keep a record of how long it takes to do homework for each subject. Using the information, have the child plan a daily schedule, allowing appropriate time to complete school work.

Effective balance of instruction implies effective balance of instructional activities—developmental, practice, and application. Developmental activities help the child to *know* mathematics. Practice activities help the child to *do* mathematics. Application activities help the child to *use* mathematics.

Diagnostic/Evaluative and Transitional Activities

The remaining two activity types, diagnostic/evaluative activities and transitional activities, will be discussed more fully later. At this point it is sufficient to note that without learning, there is no real teaching. Diagnostic/evaluative activities are essential in that they help the teacher to know what is needed and what is not, what is helpful and what is not, what is learned and what is not, what needs to be taught and what does not.

Transitional activities are what we do in the classroom to get from one thing to another. How do we get children grouped for a game? How do we get children to put away their library books? How do we get children to straighten their desks? How do we accomplish the transition from one instructional activity to the next? Transitional activities are almost incidental, but they are extremely important to the teacher. Many teachers who have given little thought to transitional activities find that their day is so hectic that little of real substance seems to get done. Chapter 5 includes ways to accomplish these transitions more effectively.

Your Time: Activities, Exercises, and Investigations

1. Choose a teacher's guide from a commercial textbook series. Select five suggested instructional activities and classify them as developmental, practice, or application activities.

2. Write a lesson plan that includes at least one developmental activity and at least one practice activity.

3. Write a lesson plan that includes an application activity.

4. Choose a topic and a grade level at which that topic is taught. Using ideas that are appropriate at that level for that topic, describe each of the following:
   (a) A developmental activity
   (b) A practice activity
   (c) An application activity

5. Examine a teacher's guide from a textbook series and choose a developmental activity suggested by the authors. Then tell how you might adapt the activity for:
   (a) A deaf child
   (b) A blind child
   (c) A sighted child who cannot read.

Reactive Teaching

Planning is important. There is no substitute for a thoroughly planned lesson. A good teacher will *over*plan, trying to consider every contingency that might occur, trying to anticipate student questions, trying to have available all the aids, models, and materials that might be needed. However, the effective teacher realizes that it is necessary to let student reactions adjust a lesson.

Expect the child to react to the teacher, to school, to instruction. It is equally important that the teacher, the school, and the instruction react to the child. Nothing is more dehumanizing than a situation where children must react, must do as directed, and yet never have an impact on their world. From earliest infancy children strive to cause change in their environments. It is almost as if they are saying, "See the motion! Hear the noise! Feel the vibration! *I* did that!"

What begins as the enjoyment of causing sound by banging a hand on a high chair and progresses with the desire to extend one's reach by crawling and walking to make things happen "over there," continues into the elementary classroom. Children want to make things happen—not just have things happen to them. If the teacher does not react to children, it is almost as if the teacher is saying, "You are not very important. You are not real. You don't exist."

Reactive teaching is a process which allows both the teacher and the student to react. It is a dynamic process of adjustment which allows a lesson to become more appropriate than what was originally planned. Figure 3.5 shows reactive teaching as a process of constant adjustment. Note from the figure that, although the student has a constant impact on the direction that instruction may go, there are constraints that keep the process from being taken over by a clever child, that keep the process from losing direction.

The most important constraint is that, even though the teacher reacts to the child, it is the teacher who adjusts instruction. It is the teacher who adjusts the learning environment. It is the teacher who is in charge. Even in reactive teaching the teacher should remain in control. A second constraint that is nearly as important is that the teacher should make adjustments in light of the established curriculum goals. As shown in Figure 3.6, when a necessary adjustment in

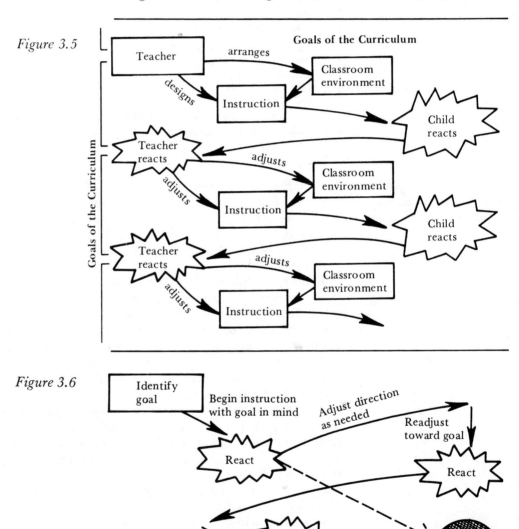

Figure 3.5

Figure 3.6

instruction may seem to move one away from the established curriculum goals, there should be a corresponding adjustment back toward the goals. The teacher needs to maintain sight of the goal in reacting to a student. At a given point, progress may not appear to be toward the goal, but *overall* progress is in that direction.

Diagnostic/Prescriptive Teaching

Diagnostic/prescriptive teaching is a formalized plan of reactive teaching. Student reactions are gathered through a system of diagnostic procedures. The teacher then reacts to that student input and designs instruction (prescribes treatment) to deal with the specific student needs discovered during the diagnosis.

Since much of the language and many of the procedures used by diagnostic/prescriptive teaching proponents has been adopted from the medical profession, we will examine the analogy between medical and educational applications of that process. First, when is a diagnosis called for? Two situations call for diagnosis. A patient may be experiencing alarming symptoms such as pain, dizziness, weakness, inability or difficulty in performing some function. Similarly, a student may display alarming symptoms, usually in the form of difficulty or inability to perform with mathematics. Such symptoms are often easily recognizable by the teacher.

Unfortunately, many symptoms experienced by the student are not obvious to the teacher and are often overlooked: discomfort when attempting to do mathematics, dislike of school in general, a need for peer approval which seems greater than the need for teacher approval. Such symptoms are important. Indeed, they are at least as important as the more obvious symptoms, but are, alas, often overlooked.

The second situation that calls for a diagnosis is the regular checkup. In the early stages of many diseases there are few noticeable symptoms. Often the early symptoms are not too alarming and are consequently ignored. As a result, in order to better discover diseases in their early stages, when they are still treatable with a degree of success, the medical community has encouraged regular medical checkups. Since many diseases are more prevalent at certain stages in a person's lifetime, the regular medical checkup is usually designed to discover those diseases most likely to occur during those years. Of course, some things are basic and are included in examinations regardless of the age of the individual.

In education, too, the regular checkup has become an established part of most school programs. Most school districts have adopted a program of standardized testing. Most individual teachers use some kind of regular checkup in their own classrooms to see if the students are learning well. The wise teachers use those same regular checkups to see if they are *teaching* well.

Whether a problem has been discovered because of alarming

symptoms or as the result of a regular checkup, the process has only begun. The next step in the process is the analysis. *Analysis is the discovery, compilation, and classification of symptoms.* The doctor talks to the patient, questions the patient, observes the patient, and conducts whatever tests seem appropriate to identify relevant symptoms. The doctor notes all of the obvious symptoms and discovers many symptoms which are less obvious. The observation of one symptom may indicate the need for further testing to see if other, related symptoms are evident.

The teacher also talks with the student (his patient), questions, observes, and conducts tests that seem appropriate. There are three levels at which analysis of learning can be performed: the *abstract* level, the *concrete* level, and the *verbal* level. At the abstract level we observe whether the student can demonstrate skill or understanding by doing written examples. At the concrete level, the question becomes "Can the student demonstrate skill or understanding by manipulating physical materials?" At the verbal level, we note whether the student is able to demonstrate skill or understanding in conversation with the teacher and other students (Figure 3.7).

We should take care that we do not limit ourselves to the use of pencil and paper tests and forget that much useful information can be gathered by observing and listening, information that may very well be more accurate than that gathered using conventional tests. All three levels are important in analysis. Not only should we be interested in what the student writes. Of equal interest is what the student does and what the student says. Some combination of the three approaches is normally indicated for the most thorough analysis.

The next step in the procedure is diagnosis. *The analysis indicates, in detail and in a systematic way, what's wrong. The diagnosis is an attempt to explain why.* Few doctors claim that their diagnoses are exact. They recognize that *a diagnosis is merely the identification of patterns of symptoms followed by an educated guess as to the cause.*

Of course, with experience and appropriate analysis those educated guesses can become very accurate. The more thorough the

Figure 3.7

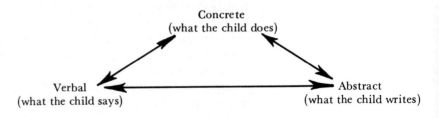

Three Levels of Problem Analysis

analysis and the more relevant the physician's experience, the lower the error rate in diagnosis. Similarly, with relevant experience and thorough analyses, teachers can become quite accurate in making their diagnoses.

We should remember that our goal is not diagnostic perfection. Our goal is to correct the difficulty being experienced by the student. Diagnosis should be thought of as a tool to improve the direction of our instruction, not as an end in itself. We must remember that a greatly detailed diagnosis is only of value if it contributes to the correction of the problem.

As soon as diagnosis is complete enough for some treatment to be indicated, that treatment is prescribed. Even though treatment begins, the diagnosis may continue. Often a physician will begin treatment of certain symptoms immediately while continuing to analyze or diagnose the problem. It is not unusual for a doctor to prescribe treatment based on a tentative or incomplete diagnosis. The rule, however, is that the prescription should do no harm to the patient even if it proves to be inappropriate.

Similarly, in education, diagnosis and remediation are seldom separated completely. Most teachers will naturally make a tentative diagnosis and begin some instruction immediately, making a finer diagnosis as they observe how the student reacts to their instruction. There are times, however, when the teacher should refrain from teaching (showing, telling, asking leading questions) until a fairly complete diagnosis has been made.

Some children are very good at following cues given by others and may appear to know, when in fact they are merely following subtle unconscious leads given by the teacher and giving apparently meaningful responses without real understanding. With such children, the teacher may need to refrain from doing any teaching in order to get a more accurate diagnosis.

Once the analysis and diagnosis have been made it is necessary to choose a treatment. The treatment itself may be designed to correct the symptoms. At other times it may be more appropriate to treat the apparent cause of the symptoms. And sometimes, for many less serious problems, the best treatment is no treatment at all. Many difficulties will correct themselves without any overt instruction. It may be most appropriate simply to offer support and encouragement while keeping an eye on the problem.

In both medicine and education, treatment is sometimes successful—the problem is corrected, the symptoms disappear. Sometimes the treatment is only partially successful—things may be better than they were but not as good as we might wish. And sometimes the treatment is unsuccessful.

In either of the latter circumstances, a number of things might be done. The situation may call for further analysis. A revised diagnosis may be in order. If it is determined that the analysis was complete enough and that the diagnosis is not in need of revision, then a different treatment might be prescribed. Finally, there is one other course of action that may be most appropriate. It may be decided that further analysis, diagnosis, or treatment are unlikely to be effective and the patient (or the student) should learn how to live with a less-than-ideal situation.

The medical analogy of the analysis, diagnosis, prescription, and treatment approach to remediation is summarized in Figure 3.8.

Figure 3.8

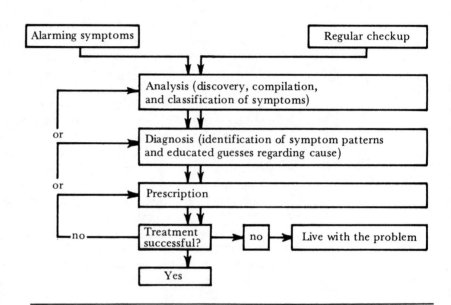

Case Study:
Jonny

Jonny R, a third-grade student, is a relatively cooperative child with a positive home influence. During a parent-teacher conference early in the school year the discussion revealed that Jonny has often received poor marks on addition papers. The teacher decided that the problem was consistent and serious enough to warrant extra time and attention.

The Initial Analysis:

As a first step in the analysis of the problem, the teacher went to his files. In the files were copies of daily papers and tests that the children had completed so far during the school year. On examination of Jonny's papers, the teacher compiled a list of the addition problems that Jonny had done incorrectly.

83 +16 98	43 +55 97	$8 + 11 = 91$ $5 + 11 = 61$ $3 + 35 = 65$	$3 + 44 = 74$ $4 + 15 = 55$ $63 + 5 = 67$	$26 + 53 = 78$

Tentative Diagnosis:
Jonny's teacher decided that Jonny had problems in two areas—basic facts (3 + 6, 3 + 5) and horizontal addition. Before beginning instruction, however, he double-checked his conclusions with further analysis.

Further Analysis:
The teacher went back to the file to see if the observed errors were consistent. The teacher noted that, although Jonny had given 6 + 3 = 8 on two occasions, he had given the fact correctly on five occasions. Although he had given 3 + 5 = 7 twice, he had given that fact correctly once.

A more complete look at Jonny's attempts at horizontal addition showed a rather mixed and complicated picture. Jonny consistently missed problems like these.

8 + 11 = 91   4 + 15 = 55   3 + 44 = 74

Jonny consistently answered problems like these correctly.

51 + 7 = 58   23 + 2 = 25   63 + 5 = 67 (correct except
                                                        for the fact
                                                        error, 3 + 5 = 7)

He also consistently got this type of problem correct.

20 + 6 = 26   5 + 40 = 45   7 + 10 = 17   8 + 20 = 28

Next, Jonny's teacher gave him the following problems and observed his work.

30 + 6 =   5 + 70 =   24 + 5 =   8 + 11 =

Jonny gave correct answers on all but the last problem. When he wrote the incorrect answer 91, his teacher stopped him and said, "Tell me how you get that." Jonny's response was, "You just add the 8 and the 1 because they are next to each other."

Revised Diagnosis:
Jonny's teacher recalled that Jonny had done well on problems about tens and ones and decided that Jonny answered problems like 20 + 6 and 7 + 40 correctly by simply thinking about tens and ones.

He decided that Jonny's problem with horizontal addition was caused by his following an incorrect system of rote rules that were not really meaningful but which gave enough correct answers to be reinforced. He decided that Jonny apparently did not "see" what was going on when two-digit numbers were being added.

Prescription:
The teacher felt that the problem with basic facts was not a serious one but one that should be watched carefully.

He decided to reintroduce Jonny to addition involving two-digit numbers using the following sequence.

1. *Model numbers using 10-rods and 1-cubes.*
   When shown a number with the model, Jonny should be able to say the number and write the number. When shown a written number Jonny should be able to show it with the model and write it. After hearing a number spoken, he should be able to write it and show it with the model.

2. *Model addition by combining numbers represented by the model.*
   When shown a written addition problem, Jonny should be able to both read the problem and show the problem by representing the numbers with the model and combining them. When shown the models of two numbers being combined, he should be able to write the addition problem and say it.

3. Emphasize that we add like units (ones plus ones and tens plus tens).

4. Practice using item 3 in both horizontal and vertical addition examples. When difficulties or questions occur, use physical models to reinforce correct procedures. Continue until skill is mastered.

Treatment and Results:
   Jonny's fact errors decreased in regularity without any special instruction or practice on facts. (Such errors will often be corrected when the student is using a physical model, since those errors are seen to be inconsistent with what really happens.)
   Jonny now has no difficulty in understanding how to do addition examples like those that previously gave him trouble. He still will make occasional errors of a random or careless nature. However, his error rate is now low enough to cause little real concern.
   When Jonny makes an error he is able to go to the physical model, look at what happens when the two "numbers" are combined and correct his own mistake.
   Note. Because Jonny's teacher had the foresight to accumulate a file of papers from each student, the process of analysis and diagnosis required no special diagnostic instruments, and took little of his own and virtually none of Jonny's time.

Case Study:
Krista

Let's review the case study of Krista that was introduced in Chapter 2. You will recall that by considering her age and IQ we were able to determine an expected level of achievement.

Age (11.3) ⎫
IQ   (85) ⎭ ⟶ About mid-fourth grade

Recall, also, that her math concepts score on the *Stanford Achievement Test* was near her expected level of achievement (4.3), while her computation score was well below what we would expect her to achieve (2.7).

Ability to apply mathematics would seem, to a large extent, to depend upon level of ability in computation as well as in concepts. If we were to average Krista's concepts and computation scores, the result would be quite close to her application score (3.2), so that score should not be too surprising. Furthermore, it would appear that if we wanted to improve her ability in applications, instruction in computation would be one obvious place to start.

Instruction in computation has three major phases (see Chapter 8). They are: (1) the development of the concept of the operation (i.e., what is addition, what is subtraction); (2) teaching of basic facts; and (3) the development of the algorithm. Krista seems to have some difficulty in (2) and (3) and may possibly have difficulty with (1).

Since what we do in computational procedures depends almost entirely on the application of the concepts of numeration, Krista's deficiency in the area of numeration as indicated by the *KeyMath Test* is probably a major contributing cause of her difficulties in computation. Several of the specific problems missed by Krista, particularly the type of errors made, seem to indicate that she has not developed a system of mental imagery. She cannot "see" (imagine) what the numbers look like and cannot, therefore, "see" what happens to those numbers during the computation process. So one of Krista's obvious needs is to develop mental imagery for numbers.

The following list outlines a program of study for Krista to correct her deficiencies in addition and subtraction. Notice that some of the steps overlap. The arrows show how the steps overlap. Many of the instructional procedures referred to here are illustrated in greater detail in Chapters 7 and 8.

Step 1: Use a base-ten model (base-ten blocks or money: $1, $10, $100) to develop mental imagery for two- and three-digit numbers. Krista should be able to read, write, and show such numbers with the model (see Chapter 7).

Step 2: Perform a *specific* fact analysis to determine precisely which addition and subtraction facts Krista knows and doesn't know.

Step 3: Develop a thinking strategy for figuring out facts that are not memorized. Krista should use the facts that she knows to figure out the ones she doesn't know (see Chapter 8).

Step 4: Make even trades to find many names for numbers using the base-ten models and a place-value chart.

	hundreds	tens	ones
	4	3	12
same number	4	4	2
	3	14	2

Step 5: Use the thinking strategy (step 3) to work on memorization of remaining facts in clusters (see Chapter 8).

Step 6: Use base-ten models to demonstrate addition and subtraction of two- and three-digit numbers without regrouping.

Step 7: Review step 4.

Step 8: Use base-ten models to demonstrate the regrouping process for addition and subtraction (ones to tens and tens to ones).

Step 9: Review step 4.

Step 10: Use base-ten models to demonstrate the regrouping process wherever needed for addition and subtraction.

Dangers of Diagnostic/Prescriptive Teaching

As with every other teaching technique, diagnostic/prescriptive teaching should be treated as one model for instruction, a technique that can be effective in many instructional situations. The quality of diagnostic/prescriptive teaching can quickly deteriorate, however, if the process becomes so programmed that the teacher spends more time in keeping records and managing the "system" than in teaching.

Remember that no test is truly diagnostic. At best, a test can be an effective tool for analysis. That intangible element called *professional judgment* is important and should not be shoved aside in favor of impersonal charts. When this happens, instruction almost invariably becomes inferior.

The process of diagnosis can be very interesting and intellectually challenging to a teacher. However, diagnosis of itself is not really important. Instruction and learning are important, and only as diagnosis contributes to these does it become important.

Your Time: Activities, Exercises, and Investigations

1. You suspect that a second-grade child knows the easy addition facts (sums of ten or less) but does not know the more difficult basic addition facts. Design a short test that can be used to see if you have identified the problem.

2. A child thinks that 1/3 is more than 1/2, and 3/8 is more than 3/4. What instruction would you provide?

3. Design a *short* (10–15 minute) test to analyze a fifth-grade student's strengths and weaknesses in whole-number multiplication. (Consult a commercial textbook series to determine normal fifth-grade expectations.)

4. Examine a "diagnostic" test that is part of a commercial textbook program. Does the test provide adequate and appropriate information to assist in diagnosing student learning problems? What additions or changes (if any) are needed?

SUMMARY

Good instruction is well-balanced instruction. The things that we teach must include *concepts*, *skills*, and *applications*. To teach these things effectively teachers need to use a combination of *developmental activities*, *practice activities*, and *application activities*.

Reactive teaching is a dynamic process of adjustment. The adjustable teacher is the most effective teacher. It is as important for teachers to react to their students as for students to react to their teachers. As the teacher reacts, however, it is important that adjustments be made in light of curriculum goals.

Diagnostic/prescriptive teaching is a process that is, in many ways, analogous to medical diagnostic/prescriptive procedures. The process can be described in four steps using medical terminology. *Analysis* is the process by which one discovers, compiles, and classifies symptoms. *Diagnosis* is the identification of symptom patterns and educated guess work regarding cause. *Prescription* is the plan of instruction. *Treatment* is the carrying out of the prescription.

Diagnostic/prescriptive teaching can be effective, but it must be remembered that analysis is always incomplete. Diagnosis should always be considered tentative. Treatment may be completely successful, partially successful, or may not work at all. A problem may appear in one area while the cause may be in a different, seemingly unrelated area. In medicine, many symptoms are psychosomatic, having no apparent physical cause. In education, problems may appear that seem *not* to be caused by a deficiency in concepts or skills. Psychosomatic problems (those arising out of affective deficiencies) should not be overlooked in education.

For the classroom teacher, the most appropriate guideline for diagnosis is just "keep it simple." Never give a sophisticated test where a simple question will give you the needed information. Never use elaborate diagnostic plans when simple observation will tell you what you need to know. Never use thirty minutes in diagnosis when five minutes will do. If you find yourself spending too much time managing the system, then the system is probably managing you, and you need an easier system. Remember that your goal is for the student to learn. Use only those tools that help you reach that goal.

4 PSYCHOLOGIAL ASPECTS OF MATHEMATICS LEARNING

I hear, and I forget,
I see, and I remember,
I do, and I understand.

—Ancient Chinese Proverb

This Chinese proverb epitomizes the processes by which humans internalize and understand the ideas and events encountered in their daily lives. The proverb's simplistic lines make the problem of learning seem easy. However, it is not always so. This chapter examines various theories that have been developed as models for mathematics learning and teaching. The relationship between doing and understanding is examined in detail. The chapter concludes by examining several findings from educational research that relate to increased student learning in the classroom.

These theories are often challenged on the basis that they are only suppositions. Critics ask for *evidence*, and question the manner in which they were developed. They also question what relationship the theories have to the classroom teaching of mathematics. These considerations provide a framework for our investigation of the theories of Piaget, Bruner, Gagné, and others.

What Do *You* Say?

1. What are the intellectual stages a student passes through in developing from early childhood to adulthood? What impact do these stages have on the learning of mathematics?

2. What role do concrete aids used in the mathematics classroom play in the development of mathematical knowledge?

3. How should concrete models be introduced and sequenced in the classroom in order to maximize student learning?

4. What is a task analysis and how is one implemented? What relation

---

does such an analysis have to student learning and the measurement of student abilities?

5. What factors of a teacher's classroom activities might be related to student achievement in mathematics?

THEORIES OF MATHEMATICS LEARNING

Several individuals have developed theories to explain the manner in which humans acquire, develop, and store information, especially information dealing with the content of mathematics. As we examine the major points of the works of Piaget, Bruner, Gagné, Guilford, and Ausubel keep the following questions in mind.

What does each model for learning have to offer in:

- Planning for instruction?

- Planning for diagnosis?

- Individualizing for classroom instruction?

- Meeting special needs of children in the classroom?

The answers to each of these questions are still being formed by mathematics educators, experimental and applied educational psychologists, and classroom teachers. There are no firm answers to these questions. Each of the theories has something to offer. Some fit better in one phase of learning or teaching and others fit better in others. It is the task of the effective teacher to integrate the implications of these theories into a viable program for communicating mathematics to children with special needs.

Jean Piaget's Theory

The Swiss philosopher Jean Piaget (1896–1980) has probably contributed most to discovering how mathematics knowledge develops within, is transformed by, and is used by human beings. Piaget became interested in the development of the human intellect through his work with Binet and the development of IQ testing. The processes Piaget observed in the thoughts of children working with reasoning items on the tests fascinated him. This fascination led to the development of the theories we are about to consider.

Upon the birth of his first child, Piaget kept detailed notes, from which later emerged the early form of his theories on the development of human intelligence. His early works were later expanded by researchers around the world.

The central theme of Piaget's analysis of intellectual development is a sequence of four stages of mental development. Various factors affect a child's growth through these stages, the most important of which is the growth of logical reasoning abilities. These information-processing abilities develop at different rates in different individuals, but the evidence is that the order in which they develop does not differ. In particular, the special needs child experiences these same stages of mental growth, but over a greater length of time than the normal learner.

The four major stages of development detailed by Piaget are the following:

- Sensorimotor state (birth to 1½ years)

- Pre-operational thought (1½ to 7 years)

- Concrete operations (7 to 12 years)

- Formal operations (12 years through adulthood)

The age cited for each of these levels of development is an estimate of the age by which 75 percent of the appropriate population has achieved the criteria marking one stage and has entered the next stratum of intellectual development. The ages are cited as guidelines to assist in the comprehension of Piaget's works. The actual passage from one level to another is marked by command of intellectual skills, not by chronological age.

The growth of these patterns is due to the child's internalization of stimuli from his environment and experience into a schema (or model of the situation and appropriate reactions) which is constantly revised over a period of time. The process of schema development is illustrated in Figure 4.1.

Figure 4.1

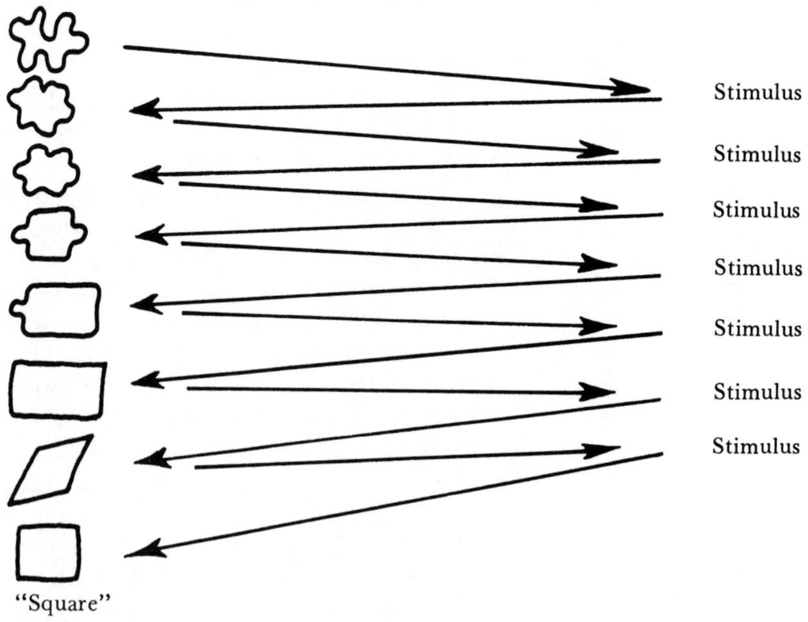

"Square"

The processes by which these schemata mature are known as *assimilation* and *accommodation*. The process of assimilation involves ingesting information and perceptual material into the present form of the appropriate schema. In many cases this new material fits the existing schema.

In other cases the incoming information contains some elements that cannot be explained by the existing schema. The process that governs the adjustment of the schema to this new information is called accommodation. Accommodation refers to the modification of the schema to include the new material. Through this process the schema adapts and grows so that it can meet, deal with, and appropriately react to new situations.

As a child grows and develops, the myriad of schemata that constitutes the child's evolving intellect pass through a series of assimilation-accommodation patterns or cycles. With each pass through one of these cycles, the schema grows and presents a more accurate picture of the child's view of reality. The correlative relationship between these processes is known as *equilibration*. It is the intellect's method of dealing with the stimulus-reaction-testing-modification sequence.

The formalization of the various manners in which an individual deals with the environment and its stimuli leads to the notion of mental strategies which Piaget refers to as *operations*. It is the development, maturation, and expansion of these operations that leads to the four stages of mental growth mentioned earlier.

The first stage of intellectual growth, according to Piaget, is the stage of *sensorimotor development*. This stage can be likened to the development of intelligence prior to language. It is one in which reactions to the senses are developed and command of motor skills progresses from the gross level to finely coordinated actions. This growth is characterized by a child's gradual change from uncoordinated responses to stimuli to predictable, acquired patterns of response. The sensorimotor child is characterized by:

- The development of eye-hand coordination

- Application of prior successful actions to attack new problem situations

- Recognition that an object out of sight is not necessarily out of existence

- Growth in seeing spatial relationships between self and the objects in the immediate environment

- Seeing oneself as the source or cause of all things happening in the environment.

These actions are seen in the general motor development of the child and the ways in which the child applies these skills in new situations. A child adapts to the immediate space of the crib by first noting the objects that bring pleasant results—bottles, toys, and musical devices. The child notes the position of the objects relative to its own position, is able to perform the appropriate motor skills to reach the objects, and then to perform the necessary motor skills to activate, or use, the object of attention.

The coordination of these actions to achieve a goal is the apex of the sensorimotor stage of actions. This level shows the growth of elementary recognition of particular objects and a rudimentary understanding of the relationships between actions and effects.

The second stage of intellectual development is the state of *pre-operational thought*. The age range for this stage covers the span from 1½ to 7 years. The major feature of intellectual development is the growth of language and other forms of representing objects and actions.

Mathematically, the important features of this age are the formation of simple concepts and the gradual enlargement of these concepts to the total class of objects to which the concept refers. For example, the child may originally attach the term *ball* to a particular favorite ball, but not allow the term to refer to other balls in the child's environment. As the child develops in this state, the concept of ball grows and more and more objects are included under the umbrella to which the term *ball* applies. In addition, a child begins to abstract the properties of *ball* from the class of objects and form a simple definition of *ball* from the various appropriate attributes.

The child also begins to develop some rudimentary ideas about number in this stage. At the end of this period of development, the child can usually rote count to ten, but still does not have a concept of number in terms of a class of sets all of which have the property of being placed into one-to-one correspondence with each other. A child at this stage would agree that sets A and B in Figure 4.2 have the same number of elements. However, when the perceptual state of set B is changed, as in Figure 4.3, the student would then respond that set B has more elements. The same type of quantity comparison responses also hold when the child is confronted with situations involving length, volume, mass, and capacity.

A child in the pre-operational stage is unable to reverse actions and to center on the critical elements in a particular situation. Further, the child in this stage is still bound heavily by perceptual cues. The intellectual activities of a child in this period are marked by slow, and restricted, thoughts built about concrete objects. While the child has yet to develop far-reaching intellectual patterns, it is actively growing in the use of language and the ability to represent real-world objects in thought and play. In addition, the child has moved from a self-centered view of the environment to an approach based on social interaction.

The third stage of Piaget's development levels is the stage of *concrete operations*. Most children in the 7- to 12-year range are in this stage. This phase of development moves away from perceptual cues toward a marked increase in the importance of the roles of abstraction, generalization, and testing as a means of determining the true state of objects and their relationship to one another. This growth is due to the child's ability to decentralize thoughts and to allow alternate views. A child has more ability to apply and coordinate simpler intellectual functions, such as classification and seriation. Tied to this is the growth of the ability to reverse actions, both in physical and mental matters, and to study the effects of transformations. Hermine Sinclair characterizes this stage of development and contrasts it with the following stage of formal operations by saying

> *Concrete*, in the Piagetian sense, means that the child can think in a logically coherent manner about objects that do exist and have real properties and about actions that are possible; he can perform the mental operations involved both when asked purely verbal questions and when manipulating objects. The latter situation is far preferable to the former, mainly for the reasons of clarity, but the actual presence of objects is no intrinsic condition. Nor is the reverse—that is to say, the absence of objects—a condition for formal operations; these may indeed involve the solving of problems dealing only with propositions, but they may, and usually do, apply to quite concrete situations. (Sinclair, 1971)

Mathematically, the child makes great strides in the period of concrete operations. In particular, methods of classification and seriation develop (see Chapter 6), the concept of number forms and grows (see Chapter 7), and basic patterns of inductive thought develop and grow.

The fourth level of intellectual development is called the stage of *formal operations*. The mental acts that define this stage are those

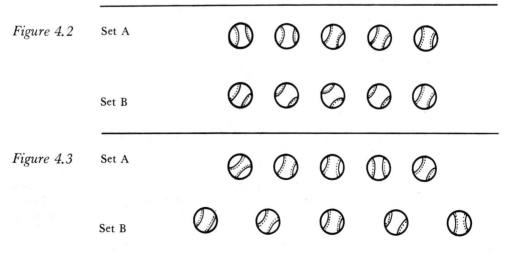

Figure 4.2   Set A

Set B

Figure 4.3   Set A

Set B

found in people aged 12 and older. Recent research indicates that some college-age students have yet to move into the stage of formal operations, at least with respect to mathematical reasoning. Actions in this period of development are characterized by the ability to focus on the main concepts and principles in a given situation and then manipulate these factors when they are divorced from the physical setting. This frees the individual in the formal-operations stage to deal with complex verbal problems, hypothetical problems, combinatorial situations, and conjectures about future happenings. It also allows for the development of formal logical arguments and the tests of hypotheses and statements related to the hypotheses. Formal operations allow the intellect to deal with ideas that do not have concrete representations or cannot be observed directly.

These stages provide the framework for Piaget's analysis of the development of the intellect. This model sees intelligence as consisting of a set of internalized actions. These actions have been developed through the process of equilibration as a result of continual interactions with stimuli from the environment, peers, and other sources. These actions are modified by experience, by maturation, and by interaction with others in school and other settings.

Implications of Piaget's Model for Mathematics Instruction

While Piaget's search for patterns in the intellectual development of humans was not based on attempts to improve teaching methods or to develop a theory of learning, it has had a massive impact in each area. In the area of learning theories, Piaget's model gives a picture of intelligence as a set of internalized actions which constantly undergo changes through continued repetitions of the assimilation-accommodation cycle. Further, intelligence is not simply a complex of simple stimulus-response bonds, but rather a highly developed system of plastic schemata which allow both response to stimuli when appropriate and the modification of schemata in cases where the stimuli and other forms of input require new reactions.

Another contribution of Piaget's work is the development of a mathematically oriented view of intelligence focused on mental acts closely tied to the handling of concrete objects and abstract ideas. This sequential development of intellectual functioning provides a basis for research both into learning and into the development of diagnostic instruments.

From the vantage point of education, Piaget's work suggests that different methods are necessary as one deals with students of differing ages from early childhood through collegiate and adult levels. This is due to the stages of growth as one moves through the four stages of development. The needs and capabilities of learners at all levels differ appreciably and require different materials and methods for effective learning.

The increased role of concrete materials in the learning of mathematics is also an outgrowth of Piaget's theories. These concrete learning experiences allow for the development of schemata through experimentation and concrete actions. The materials and their role in the growth of the intellect are not limited to students in the pre-operational and concrete operational stages, but extend into the formal operation stage.

A third implication of Piaget's work is the vital role of social interaction in the growth of the ability to deal with knowledge. This process allows for the filtering, restructuring, resymbolization, and restating of information in forms that are more readily assimilated and accommodated by developing schemata.

These implications have left their mark on mathematics education in the form of mathematics laboratories and activity learning aimed at the development and nurturing of concepts, principles, and skills. Additional references to these implications, and sources focusing on Piaget's work, are listed at the end of the chapter.

Jerome Bruner's Theories

A second individual who has had an impact on the development of mathematical teaching theories is Jerome Bruner (1915-   ). His works *The Process of Education*, *Toward a Theory of Instruction*, and *A Study of Thinking* have provided guidance to educational psychologists and educators as they have searched for effective methods of instruction.

Bruner's impact comes from several directions. Possibly his greatest contribution is his model of ways of knowing. According to Bruner, knowledge can be represented in the three forms: *enactive*, *iconic*, and *symbolic*. These three forms of knowledge serve as models for the storage and retrieval of useful information.

Enactive knowledge has its representation in action, but not at the verbal level. It is knowledge of how to throw a ball, how to touch the keys on the calculator, and how to bundle sticks in tens. Enactive knowledge is knowledge at a concrete-action level. It can be demonstrated in action or by the direct manipulation of objects. Enactive knowledge is not displayed in pictures, symbols, or words.

The second level or representation of knowledge is the iconic level. Knowledge at this level has a visual or perceptual organization. It is communicated by pictures and forms. An example of iconic knowledge is the mental manipulation of images of concrete objects, such as takes place in modeling and solving simple verbal problem situations. Another such example would be the information about the number of units present in the particular set of arithmetic blocks pictured in Figure 4.4. One might view this information as being semi-concrete in nature, as it has ties with real-world objects, but it does not have the concrete representation of enactive knowledge.

The third stage of knowledge is the symbolic level. This type of knowledge represents information about concrete and semi-concrete situations in a purely symbolic fashion. The use of symbols allows for easy transmittal of the knowledge and opens the door to higher level thought and problem-solving actions. For example, the units in Figure 4.4 could be encoded in symbolic form as 2134. Symbolic knowledge is knowledge at its highest and most formalized. At this level we manipulate the symbols representing knowledge. One example of such manipulation is the use of symbols, in the form of numerals, in the calculation of an answer to a particular arithmetic operation.

Bruner, like Piaget, sees knowledge developing in an evolutionary sequence. Knowledge first enters in the form of some enactive (non-verbal) form. This information is digested and recorded. Later the information is recalled in a mental image or seen in a visual representation at the iconic level. Finally the information reaches a symbolic level through the use of spoken words or written symbols. This sequence, or set of stages, of the development of knowledge corresponds with the levels of internalization and intellectual growth of Piaget's model. First we have sensorimotor, or concrete, knowledge. Second we have knowledge which is perceptually bound, the state of pre-operational or concrete operation knowledge. Finally we have knowledge that transcends concrete models and their mental images. This is knowledge at the formal or symbolic level.

Educational Implications of Bruner's Model

Bruner proposes that each learner must go through the three levels of knowledge in "knowing" a concept. Thus, one must start a manipulation of objects or ideas in a rough state, develop mental imagery, and then move to the symbolization of the concept. This mental-concept development period might be represented by the triangle-like model shown in Figure 4.5.

The idea of developing concepts in terms of their three stages of representation relates to the design of curricular materials. Consider

Figure 4.4

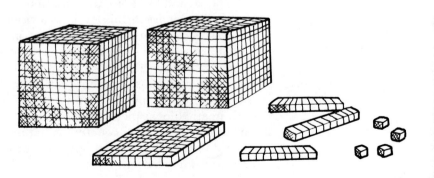

a model that first introduces concepts in a concrete manipulative stage and involves students in activities modeling these concepts. Next the curriculum introduces the learner to classroom situations that stress the internalization of the concept through mental imagery or the abstraction of the main properties through its presentation in picture formats. Finally the concept reaches the stage where it is referred to only through symbolic or verbal forms. This process leads to a curriculum model known as the *spiral curriculum.*

In this model each object is considered at the three levels Bruner mentions: the enactive, the iconic, and the symbolic. The period of transition between the stages varies with the concepts and skills under consideration. The variation in these periods reflects the stages of development each of the concepts goes through. It might be modeled as shown in Figure 4.6.

Further, with each level of learning, the curriculum must provide opportunities for all three levels to be considered. Each new step in the learning of a concept has new factors which in turn must have their own little cycles of growth. This leads to the development of curricular guides which might take the form illustrated in Figure

Figure 4.5

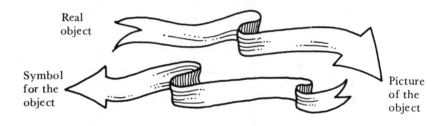

Figure 4.6

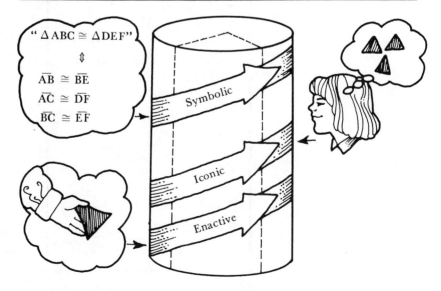

4.7. Here, each objective at each level has a space provided for learning situations at each of the three levels of representation. It is not always possible to think of related activities for each objective at each level, but the more the teacher can provide, the more effective the students will be in acquiring new concepts and skills and in developing ones already acquired.

Robert Gagné's Work

A third individual who has had a large influence on the mathematics curriculum and theories of mathematics learning is Robert Gagné (1916–   ). Gagné's views are quite different from those of Piaget and Bruner, in that Gagné's work focuses on the steps which must be mastered in order for an individual to acquire a given skill, concept, or principle. Gagné's analyses, and the resulting sequencing of steps to achieve a given academic or other goal, are often referred to as *instructional or learning hierarchies*.

The foundation of an effective instructional hierarchy is a carefully sequenced and tested task analysis. A task analysis is a careful listing of the various components of a skill or concept which are involved in the mastery of the more complete skill or concept. For example, what are the specific skills children need to find the sum in a five-column whole-number addition problem, given that they can handle problems like 24 + 57? The process for attempting to

Figure 4.7

Objective	Enactive mode	Iconic mode	Symbolic mode
a. *[illegible]*	1. *[illegible]* 2. *[illegible]*		1. *[illegible]*
b. *[illegible]*	1. *[illegible]* 2. *[illegible]*	1. *[illegible]* 2. *[illegible]*	
c. *[illegible]*		1. *[illegible]*	1. *[illegible]* 2. *[illegible]* 3. *[illegible]*
d. *[illegible]*	1. *[illegible]* 2. *[illegible]*	1. *[illegible]* 2. *[illegible]*	1. *[illegible]* 2. *[illegible]* 3. *[illegible]*

derive this solution can be visualized as shown in Figure 4.8. The peak of the pyramid represents the terminal goal, or the ability to add two five-digit numbers. Each of the supporting levels below the peak represents a stage of learning that is required to have success at the terminal level. These skills are called *enabling skills or behaviors*. The base level of the pyramid consists of those behaviors the student is assumed to have already mastered and available for immediate recall. These skills or knowledges are known as *intact skills or knowledge*. Each of these intact skills is at the apex of some earlier learning hierarchy aimed at the development of each of these skills.

An example of a completed instructional hierarchy for the development of the skill of finding the sum for a pair of multi-digit whole numbers if shown in Figure 4.9. This hierarchy starts with the intact skills of basic addition facts through 9 + 9, knowledge of base-ten place value and regrouping, and a concept of the operation of addition as it refers to whole numbers.

The processes involved in completing such an instructional hierarchy are as follows:

1. *Establish the terminal goal.* This initial stage involves carefully stating what the student is expected to be able to do at the conclusion of a sequence of learning or instructional activities. This

Figure 4.8

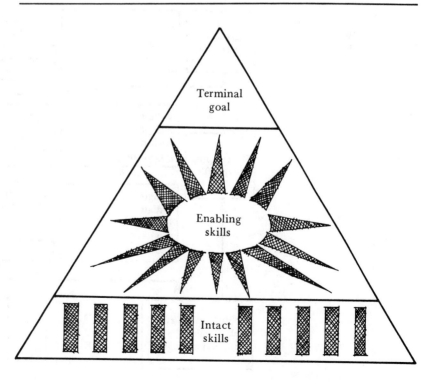

Figure 4.9

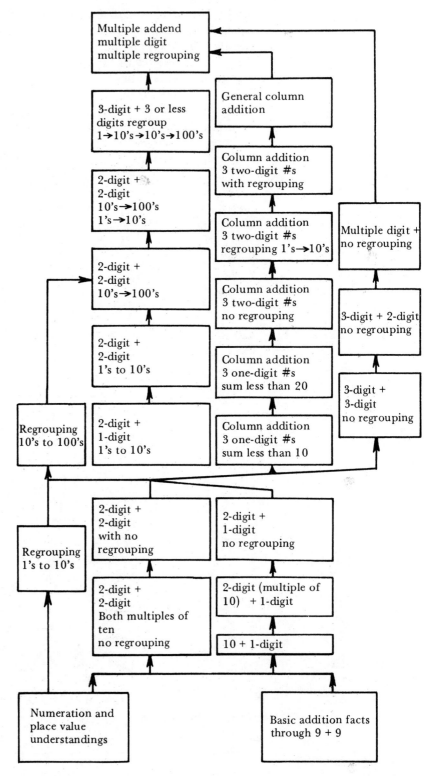

statement of the terminal skill does not have to be a polished work of instructional intent at the outset, but it does have to convey the desired learning goal in unambiguous terms, as well as the conditions under which the desired behavior is to be performed.

2. *Outline the enabling skills and related intact skills.* In this second phase an instructional blueprint for the development of the terminal skill is outlined. To reach this goal, the planner must consider the final skill, decide what enabling skills are needed, and arrange these skills in their order of development. In addition, the planner must decide what assumptions will be made about the learner's intact skills. This stage of the development of an instructional hierarchy is usually facilitated by actually working through several examples of the terminal skills with someone at the age level of the intended learners. In addition, keep in mind the sequence and the problems you would anticipate.

   In this process, carefully note all of the skills that comprise the achievement of the terminal goal and their relationships to the intact skills and to each other. Be sure to consider the performance level of these enabling skills and how recently they have been covered. This careful listing of intact, enabling, and terminal level behaviors for the development of a particular skill, concept, or principle, as shown in Figure 4.9, serves as the heart of the analysis of an instructional hierarchy.

3. *Sequencing the intact, enabling, and terminal behaviors.* The third stage is the careful sequencing of the behaviors identified in step 2. The analysis given in step 2 should provide a lot of information about the order and relationship existing between the intact, enabling, and terminal understandings of the skills, concepts, and principles involved. The proper sequencing of these behaviors is the most difficult task in instructional designing, for the ordering which may seem to make the most sense at the adult logical level may not be the same as that which makes the most sense for a young child or a special needs learner. In many cases, other sequences are just as effective as those based on logic and an analysis of the content of the subject matter from a sterile viewpoint. These other sequences take into account other aspects of the nature of the learner, the child's background, and special needs.

   Once this order of skills has been set for a specific instructional goal, it is shown in a flow chart (Figure 4.9). The order moves from the intact skills at the base through the enabling skills up to the terminal skill at the apex. Arrows connecting a box at a lower level with a box at a higher level indicate that the content of the box at the lower level is a direct prerequisite for the skill on the upper tier.

The relationships indicated by these arrows are the connections that must be developed in teaching the terminal goal. In addition, they are the crux of planning a diagnostic assessment of a student's capabilities for moving from intact skills to the terminal goal. Gagné's research indicated that when a student experiences difficulty in mastering the terminal objective, it is at one of these boxes that the difficulty resides—not in the student's motivation or mental level.

Gagné's work also includes the specification of a number of levels of mental skills ranging from stimulus-response learning to complex problem-solving skills. The study of these levels is beyond the scope of this chapter, but you can read more about the development of the instructional hierarchies in Gagné's book *The Conditions of Learning*.

Educational Implications of Gagné's Work

Gagné's task analysis and instructional hierarchy procedures and their related research findings have many implications for dealing with special-needs children in mathematics. First, the outline of skills and concepts that must be mastered in order to reach a terminal objective provides a teacher with a diagnostic teaching outline for evaluating individual student capabilities. This outline provides a basis for both test and interview assessments of a student's level of progress and knowledge in mathematics.

The task analysis also provides an outline for analyzing a student's classroom and test work in mathematics in order to pinpoint the child's level of functioning for placement and for instruction. Such analyses also provide a basis for the development of, or selection of, instructional materials.

Gagné's emphasis on the analysis of instructional steps which will lead to the mastery of the terminal skill does not include the methods by which the material should be presented to the learner. One is only interested in the fact that the learner's behavior is modified to achieve the final objective. Gagné does not focus on the developmental patterns or the precise instructional delivery system. This emphasis on the changing of the learner's behavior leads to the branding of Gagné as a *behavioralist*.

Gagné views readiness as the student's having mastered the prerequisite skills. This can be contrasted with Piaget's readiness concept of developed schemata and Bruner's model of completed levels of representation.

Other Models for Mathematics Learning: Guilford and Ausubel

Two other psychologists who have had some impact on mathematics education are J. P. Guilford (1897–   ) and David Ausubel (1918–   ). Guilford's contribution is in the form of an additional model for the structure of the intellect, while Ausubel's provides a framework for the delivery of what he terms "meaningful verbal learning" in the classroom.

Guilford's Model

After studying student and adult responses to a battery of test items, Guilford classified the response patterns into sets of performance or content items, which led him to the formulation of the cube model for the human intellect (Figure 4.10). Guilford labels this model as the *SOI model*, the initials standing for *structure of the intellect*.

The three named edges of the cube—*contents*, *products*, and *operations*—refer to the three main forms in which the intellect structures knowledge. *Content* deals with the forms of information capable of being discriminated by the individual. *Products* gives the organizational form of the information. *Operations* refers to the processes the intellect uses in working with the information. Thus, the terms *content*, *product*, and *operation* might be viewed as the specific terms denoting the raw materials, the piece size, and manufacturing processes of the mind's operation.

Mastery of the 120 subcategories contained within the model, by the forming of the combinations of contents, products, and operations and determining their ramificiations for education, takes far more study than can be provided here. Extensive work

Figure 4.10

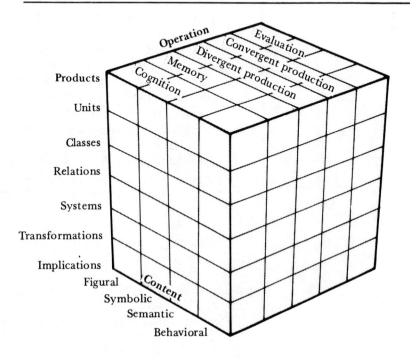

has been done with this model, its diagnostic possibilities, and its educational significance. For a total view of the model and its use in diagnostic work, consult Mary Meeker's text *The Structure of Intellect: Its Interpretation and Uses*. In it, the model is related to performances on the Binet and WISC, as well as to teaching activities for the specific intellect patterns noted.

Ausubel's Model

David Ausubel's contribution is a theoretical model that may promote what he terms "meaningful verbal learning." The concept of "meaningful" learning in mathematics goes back to the 1930s in a verbal form, but back to Socrates in theory. It is the purpose of the mathematics teacher to provide instruction in mathematics at a level readily understood by the learner. This instruction should relate the new material to prior learnings and at the same time fit into the overall development of the concepts and skills at hand. To meet this goal, Ausubel proposes a method for enhancing learning through the standard teacher-delivered verbal explanation, or lecture, on new material.

Ausubel does this by using what he terms *advance organizers*. An advance organizer is a lead statement in a sequence of instruction. This statement provides a framework for the handling of what follows. As such, it is a more general statement than the specifics that follow in the main body of instruction. It provides the links to join the previous knowledge base to the new material while establishing a framework for further learning.

These advance organizers can be thought of as the steel beams in a new skyscraper—they show the direction work is heading. They are built upon the material that has gone before and at the same time provide a comprehensive forward look. Advance organizers provide the framework for the following instruction.

Ausubel argues that verbal instruction is a very effective, if not the most effective, means of providing instruction. Research results support this fact. Well-designed studies have shown careful verbal instruction based on a structured outline to be very effective. It is the most used form of classroom instruction. Ausubel argues that it can be made more effective through advance organizers that illuminate the path ahead in terms of the material already mastered.

Input from Research on Teaching

An analysis of recent research in teaching may help in developing effective patterns of instruction in mathematics. The following findings were established in a multi-year Beginning Teacher Evaluation Study conducted in elementary-school classrooms by the Far West Research Laboratories.

The results that may hold some promise in dealing with special needs children are:

1. An increase in the amount of time allocated to students' direct engagement with content will be reflected in increased achievement. The crucial point to this finding is the focus on direct engagement, not just increased time. The study showed that teachers who kept their students actively on task for longer periods of time had students who showed greater mathematics achievement.

2. Teachers should see that students spend the majority of their time in activities that lead to success. Successful learning experiences, and the increase in successful new ones, lead to improved achievement rates in young learners. The increase in ratio of successful to nonsuccessful learning experiences of all types is a factor a teacher can control through careful material selection and cautious grouping within the classroom.

3. The teachers in the study who had better diagnostic skills also had higher student achievement rates in their classes. One could surmise that teachers with better diagnostic skills spend less time off task and are able to supply the correct instruction more quickly.

4. The percentage of time spent giving students direct feedback increased both time on task and student achievement. Hence small-group instruction may be a little more effective, as it gives the teacher a better opportunity to provide larger numbers of students with direct feedback faster. It also provides students with the opportunity to get quick feedback from their peers as well as increase social interaction.

5. The use of aids, parent volunteers, and cross-age and peer tutors is also related to increased time on task and increased student achievement. All of these factors provide quicker feedback, more social interaction, and more opportunities to integrate the material.

6. All of the higher-achieving classes had some form of positive reward system (e.g., token systems, "bionic handshakes," or teacher praise). The use of such methods for noting and providing feedback on student habits of work and achievement seems to have a positive effect.

These are but a few of the many maxims that might be drawn from the roles of feedback, motivation, and control in the classroom. It is clear that teachers who are capable of keeping their students directly on task and who minimize the wasted time between activities, subjects, and classes are more effective in increasing student learning. More specific comments will be made about methods of classroom control and management in Chapter 5.

SUMMARY

In this chapter we have examined the basic theories of Piaget, Bruner, and Gagné as they apply to mathematics learning. Alone they do not provide us with sufficient knowledge for effective instruction. However, they do show how manipulation of content and instruction may lead to increased student achievement and success.

Piaget shows how mathematical content might be ingested and modified with maturation. In addition, he has given us a model that hints that different approaches and materials are appropriate for students of varying ages and developmental levels. He also provides us with a basis for the careful development of concepts and principles from the concrete to the formal level.

This last theme is repeated in the work of Bruner, as he discusses the representation of knowledge and the sequence of stages for the growth of understanding from the enactive to the iconic to the symbolic levels of thought. These ideas, when combined with those of Piaget, strengthen the rationale for involvement with concrete materials and the careful sequencing of instruction from the concrete through the semi-concrete to the abstract level. Both Piaget and Bruner are concerned with the development of cognitive processes and methods of representing knowledge. Gagné, on the other hand, focuses on the development of instructional hierarchies through task analysis and the related pinpointing of learning problems. His small-step flow charts for the learning process have impact in both curriculum planning and in diagnostic work with special-needs children. His input in diagnostic methods, student assessment, and overall study of the learning process is aimed at modifying the learner's behavior in the desired manner.

The blending of these theories with others, Guilford and Ausubel among them, combined with the findings of researchers in teaching, provide a basis from which the individual classroom teachers can begin their lifelong study of the learning process. This study of each child's patterns of acquiring information, the resulting reactive teaching, and the assessment of progress is the same for all students in the class, for they are all individual learners. When viewed from this standpoint, all children have their special needs. The various models for mathematics learning provide the form for the analysis of each child's strengths and weaknesses in learning, as well as in mathematics itself.

Your Time: Activities, Exercises, and Investigations

1. Prepare answers to the *What Do You Say?* questions on pages 78–79 and be prepared to defend your answers in class.

2. In reflecting on Piaget's levels of intellectual development, what evidence have you seen in children's actions that would support or refute his theses?

3. Carry out an individual interview with a four-year-old child concerning the conservation-of-number experiment discussed in the chapter. Then repeat the interview with a six-year-old and an eight-year-old. What differences did you note in their responses and in their actions?

4. Select a first-grade-level mathematics textbook. What evidence do you see in it for the following:
   (a) A conscious attempt to note the fact that the child may be a non-conserver of number
   (b) Any note of the use of enactive, iconic, and symbolic representations of subject matter material
   (c) A sequential development of material within a chapter according to a possible task analysis

5. Observe a class of school children in a mathematics situation. Note their use of enactive, iconic, and symbolic forms of knowledge. For the enactive, note the children as they use their bodies (tongues, fingers, etc.) in representing mathematical ideas.

6. Using the task analysis presented in the chapter for the addition of whole numbers, diagnose the following student's work and suggest where the child is in meeting the enabling skills on the way to the terminal task.

   Fred:

17	52	53	62	87	3	1 5
+23	+31	+ 4	+29	+11	+24	+ 3
30	83	57	81	98	27	9

7. Construct a task analysis for making change from a given amount less than 25 cents to a quarter. Assume that the student can count to 25 and recognizes the numerals to 100 with their place values.

8. Observe an intermediate-level mathematics classroom during a regular period. During the first half of the class pick three students and watch them for two-minute periods at a time. What percent of the time was each student actively attending to the task at hand during each of the five two-minute periods you observed? During the second part of the class, observe the amount of feedback the teacher gives the students and the reward structure used, if any. Repeat this same observation in another class of the same level, but with a different teacher. What differences did you note, if any?

For further reading on models of mathematics learning you might wish to consult one of the following texts:

Bruner, J. S. *The Process of Education.* New York: Vintage Books, 1960.

_____. *Toward A Theory of Instruction.* New York: Norton, 1966.

Flavell, J. H. *The Developmental Psychology of Jean Piaget.* New York: Van Nostrand Reinhold Company, 1963.

Gagné, R. M. *The Conditions of Learning.* New York: Holt, Rinehart and Winston, 1965.

_____. *Essentials of Learning for Instruction.* Hinsdale, Illinois: The Dryden Press, Inc., 1974.

Meeker, M. N. *The Structure of the Intellect: Its Interpretation and Uses.* Columbus, Ohio: Charles Merrill, 1969.

Sinclair, H. "Piaget's Theory of Development: The Main Stages" in Rosskopf, Myron; Steffe, Leslie; and Taback, Stanley (Eds.) *Piagetian Cognitive-Development Research and Mathematical Education.* Washington, D.C.: NCTM, 1971.

Wadsworth, B. *Piaget's Theory of Cognitive Development.* New York: David McKay, 1971.

5 ODDS AND ENDS THAT MAKE A DIFFERENCE

The evaluation and grading of students is part of every teacher's work. Consideration of how students are to be evaluated will normally affect the depth with which content is developed and the emphasis placed on content. Unfortunately, it is not unusual for a teacher to omit certain content because students' achievement on that content is difficult to evaluate.

Recall from Chapter 3 that analytic testing can be done on three levels: the abstract level (what children write), the concrete level (what children do), and the verbal level (what children say). A child who is less verbal than peers often cannot explain a mathematical idea even though she or he may thoroughly understand that idea. A child may have much difficulty on tests, but almost no trouble with the same content in situations with less pressure. It is essential, if a child's ability and understanding are to be accurately evaluated, that the teacher be ready to evaluate the child at all three levels. This is particularly true when working with children who have special handicaps. Make every effort to choose the evaluation modes that best minimize the effects of the handicap. This is true whether the handicap is physical, mental, emotional, developmental, or social.

It is quite possible to grade a 100-point test and on completion not know any more about the child's knowledge or ability than before the test was given. To gain as much information as possible from an evaluation, each question, problem, or task should be directly related to one of the learning objectives. Figure 5.1 gives this relationship.

Note that this kind of evaluation is a rather negative one that only shows when the child has *not* mastered the objective. Such evaluation does not tell us when the child *has* mastered the objective (see Figure 5.2). This deficiency is a serious problem in the evaluation of student achievement.

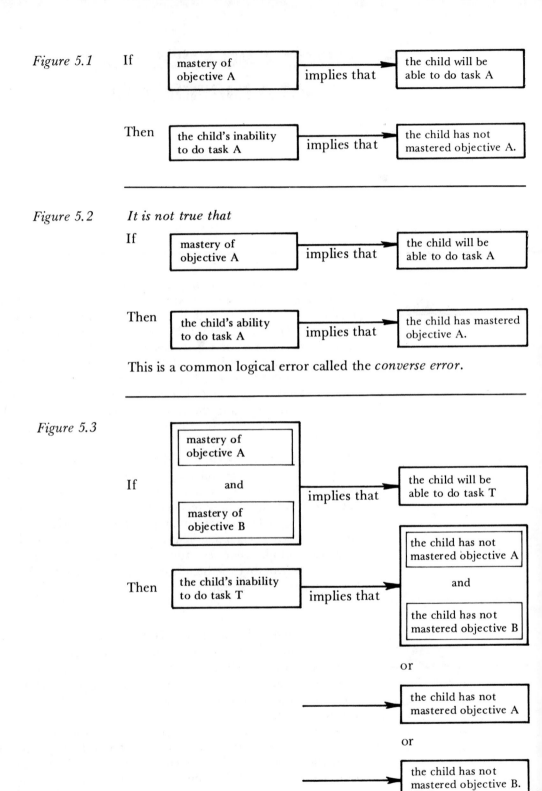

Figure 5.1   If [mastery of objective A] implies that [the child will be able to do task A]

Then [the child's inability to do task A] implies that [the child has not mastered objective A.]

Figure 5.2   *It is not true that*

If [mastery of objective A] implies that [the child will be able to do task A]

Then [the child's ability to do task A] implies that [the child has mastered objective A.]

This is a common logical error called the *converse error.*

Figure 5.3

If [mastery of objective A and mastery of objective B] implies that [the child will be able to do task T]

Then [the child's inability to do task T] implies that [the child has not mastered objective A and the child has not mastered objective B]

or

[the child has not mastered objective A]

or

[the child has not mastered objective B.]

The relationship between objectives and tasks to test the objectives is further complicated when tasks require knowledge of a number of things. Figure 5.3 illustrates this. Note that the student's inability to do the test item (the task) tells us that one of several things is true, but *we do not know which.* This is another serious problem in the evaluation of student achievement. An example of this relationship is illustrated by the following case study.

On a test, Charles, a fourth-grade student, was unable to find the product, 23 x 35. His teacher analyzed the difficulty by considering what Charles needed to know in order to do the exercise. She decided that if Charles had mastered whole-number addition and could multiply two-digit numbers by one-digit numbers, he should be able to find the product in question. This is illustrated in Figure 5.4. (Note that in this case Charles' teacher has greatly oversimplified the component skills for multiplication of two-digit numbers by two-digit numbers.)

The dilemma that Charles' teacher had to face was that his inability to do the exercise meant that (1) he had not mastered whole-number addition, (2) he had not mastered multiplication of two-digit numbers by one-digit numbers, or (3) he had mastered neither of these two skills. She needed more information to determine which of these was true.

In recent years attempts to tie objectives more closely to tasks for testing those objectives has led to the development and use of behavioral objectives. Although behavioral objectives do not solve the problems suggested above, they do provide a nice way to sidestep them.

For example, when our real instructional objective is for the children to know how to add any two whole numbers, we may state our objective behaviorally as follows:

When given five whole-number addition problems the child will add correctly on at least four of those problems.

Figure 5.4

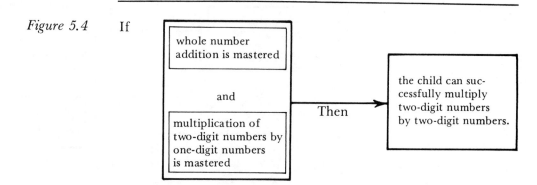

Note that when an objective is stated in this way, you know just how to test the objective (give five whole-number addition problems) and just how to evaluate the test (require at least four correct). A behavioral objective sets forth the student behavior that we want to see displayed as evidence that the real objective has been met. Since the behavioral objective and the evaluation task are essentially the same, the relationship is as illustrated in Figure 5.5.

Now our evaluation can be both positive and negative. If the child cannot do the evaluation task, then we know that the objective has not been met. But, also, if the child *can* do the evaluation task, we know that the objective *has* been met. This is the greatest advantage of behavioral objectives. *The greatest disadvantage is that there is a tendency for the teacher to focus on superficial behaviors rather than on underlying knowledge and skill; to focus on giving the appearance of learning rather than on learning.*

Remember that a behavioral objective is a mere shadow of the real objective. *It may take several behavioral objectives to reflect accurately all the facets of one real objective* (Figure 5.6). Behavioral objectives only have value as they relate to the real objectives. The given behavior(s) should be good indicators that the underlying real objective has been met.

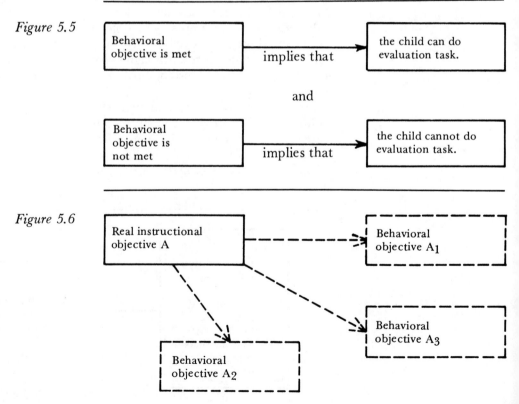

Figure 5.5

Figure 5.6

Teachers teach and children learn. But children don't learn everything that teachers teach, nor do teachers always teach everything that should be taught. Since invariably some objectives are met and some are not, how does one determine if the teaching has been effective? Are the students learning what you as a teacher want them to learn, what their parents want them to learn, and what the school administration wants them to learn? Are they learning what other children are learning? Are they learning what they want to learn?

Most schools have some system for evaluation of student achievement which includes a combination of teacher-made and standardized tests. That combination of testing instruments and techniques should answer, as well as possible, the preceding questions. The answers to those questions will generally indicate whether the teaching has been effective.

Of course, one must also recognize that all objectives are not equally important. Some are more basic than others. Some are more useful than others. Effective teaching is that teaching which focuses on the more important objectives. Teachers often find that they cannot teach everything they would like to teach. Student ability, student background, time constraints, or other factors temper what reasonably can be done. It may then be necessary to assign priorities to objectives in order to decide which to include and which to omit.

The setting of objective priorities is influenced by the needs of society and the needs of the child, as well as by the logical needs of the subject. Will meeting the objective in question help the child take a useful place in society? Will meeting the objective satisfy a personal need of the child? Is the objective a reflection of content that is a logical or pedagogical prerequisite of that which will be studied later? Such considerations can help the teacher to decide whether an objective should be omitted or delayed till another year.

Betty is a fourth-grade child who is a slow learner. Before beginning a measurement chapter, her teacher examined the learning objectives of the chapter and decided which of those objectives should be emphasized most with Betty. The chapter objectives were:

1. Measure lengths to the nearest inch, half-inch, quarter-inch, or centimeter.

2. Measure perimeters of some figures.

3. Make some simple conversions in the English system of length measures (4 ft. = $\square$ in.).

4. Measure areas of figures by using a grid and counting squares.

5. Find the volume of figures pictured as a collection of cubes.

6. Make some simple conversions in the metric system of length measures (□ cm = 1 dm).

7. Measure lengths to the nearest eighth-inch.

8. Use decimals to write some metric lengths.

Betty's teacher decided to help Betty attempt to meet objectives 1, 3, 4, and 6 first, and then, if successful, to work on objectives 2 and 5. Objectives 7 and 8 were assigned the lowest priority and would only be attempted if all the others were successfully met.

In recent years, many school districts have moved away from the use of letter grades in the elementary school. However, regardless of whether students are assigned an A, B, C, D, or F, or a grade of E, S, or U, whether the end of the year results in PASS or FAIL, or whether student progress is reported to parents during a parent/teacher conference, *it is always one job of the teacher to assess pupil progress.* Are the students learning what we—the teachers, the schools, the parents—want them to learn?

Consider the accompanying student papers. Check the work and decide how you would grade the papers. Which student should be given the higher grade? Which student more nearly knows the content? How many examples were missed? What were the causes

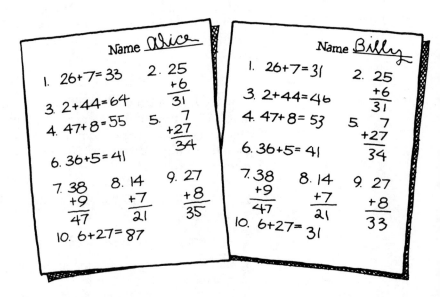

of the errors? What basic facts were not known? What big ideas were not understood? Questions like these should be asked in order to determine a grade that accurately reflects the child's achievement. As you study Chapters 6 through 13, you will be learning about the important facts, ideas, and procedures related to elementary school mathematics content. You will also begin to develop a system of priorities for your own instruction that can help you to assign more accurate and reasonable grades.

---

1. Choose a learning objective from each of three different grade levels of a textbook series and describe how you would evaluate that objective using
   (a) Written exercises (give three items)
   (b) Concrete materials
   (c) Interview (what questions would you ask)

2. Examine the case study (Betty) on page 103 and answer the following questions.
   (a) Why did Betty's teacher set the priorities as she did?
   (b) Would you assign different priorities to the objectives? Why?

3. Examine the case study (Charles) on page 101. It was noted that Charles' teacher had greatly oversimplified the component skills needed for multiplication of two-digit numbers. What additional prerequisite skills might have been listed?

4. Choose a chapter from an elementary textbook series.
   (a) List the learning objectives for the chapter.
   (b) Assume that you cannot cover all the objectives and assign priorities to them.
   (c) Develop a pencil/paper test to evaluate student achievement on the chapter.

5. Choose five objectives from those listed in Appendix A and restate them as behavioral objectives.

Teacher evaluation is a difficult topic to deal with and is often avoided in preservice methods courses. It is important that prospective teachers be aware that they will be evaluated—as they should be. They will be evaluated by their most immediate administrative supervisor and this is expected. *They will also be evaluated by parents and by their students.* This is perhaps the more severe evaluation because *parents and students will evaluate the teacher on the basis of immediate and obvious results.*

All teachers, especially those working with special-needs children, find evaluation on the basis of results a merciless kind of thing. Some students have less ability, poorer backgrounds, or a lack of interest. They will not learn as much. How, then, can a teacher be fairly evaluated on the basis of results? Is such an evaluation unfair? Perhaps, but teachers are hired to produce results.

What do you teach the child who is physically handicapped, learning disabled, poorly prepared, or disinterested? What should the slow learner learn? Perhaps the best, the most reasonable, answer to this question is that, regardless of the child's ability, circumstance, or condition, *when you are finished the child should know more about mathematics than when you began.* There should be affective and cognitive results.

Of all the different ways a teacher is evaluated, the most important of all is the way the teacher evaluates himself or herself. Honest self-evaluation improves results. The teacher who is able to tear down psychological defenses and ask, "How am I really doing?" is the teacher who is most likely to change instructional approaches to improve learning. This is the teacher who is willing to ask a colleague how to teach a difficult topic. This is the teacher who is willing to accept the help that is available all around.

Four approaches to self-evaluation follow in the form of four questions. The most effective form of self-evaluation is the one that seeks satisfactory answers to all four questions.

Question: Are my instructional objectives being met?

Am I teaching those things that I have determined to be important? Am I carrying through the things that I decide to do or do I allow myself to become unnecessarily distracted from my purpose? Am I attempting at least as much as other teachers in similar situations?

Question: Are the student learning objectives being met?

When I teach something, does anyone learn it? Are my students learning at least as well as other similar students? Are my students at least learning the most important things?

Question: Is there satisfactory rapport between my students and me?

Do my students enjoy studying under me? Do my students consider me their friend? Do my students acknowledge my authority? When a visitor enters the room, do my students try to make me look good or do they try to make me look bad? Do I like my students?

Question: Does my stomach hurt?

What does my stomach tell me at the end of the day? Am I subconsciously satisfied with my situation or is there a deep feeling of turmoil and dissatisfaction? Does my stomach tell me that today I did a good job or does it tell me that things didn't go well? Am I happy in my job?

Suppose a teacher's honest self-evaluation is unsatisfactory. Suppose a teacher determines that he or she simply is not doing a satisfactory job. What then? A number of options are available. The teacher might

1. Find ways to hide the ineffectiveness

2. Find ways to improve knowledge and methods

3. Find another line of work

Avoid the first of these options at all costs. If you become trapped in a job that you hate and that provides no real satisfaction, you are almost guaranteed a miserable life. Even more important, when a teacher begins adopting defensive attitudes and spends time covering up basic ineffectiveness, instruction invariably continues to deteriorate and the students suffer.

The third option—to find another line of work—is harsh, but one that may be both realistic and necessary. Many teachers who have invested a large portion of their lives and a not inconsiderable amount of money preparing to be a teacher find themselves trapped in a profession for which they are ill-suited. It is unfortunate that it sometimes takes a year or two of unsatisfactory experience before a teacher discovers this.

Along with the time, energy, and money investment is a personal (and public) commitment that the teacher has made to the profession. There may be a tremendous *perceived* loss of face in leaving teaching. Teachers sometimes feel that a change of professions will be interpreted by others as meaning the teacher "wasn't good enough" or "couldn't handle the pressure."

Although a change of professions is seldom interpreted negatively we may have unrealistic fears of what others may think. Or we may be reluctant to admit that the first choice of professions was not a good one. All of these may seem to be powerful reasons for staying with it, for not giving up, for trying one more year. However, there are more compelling reasons for changing if we find ourselves unsuited for teaching. There are long hours, days, and years of discontent to look forward to. A teacher who doesn't want to be a teacher can come to hate the job, the school, and the children. And the children invariably suffer.

Motivating students and maintaining classroom discipline are teaching endeavors that nearly all teachers recognize as of high importance. Regardless of the teacher's knowledge of mathematics, regardless of the teacher's ability to create lesson plans, regardless of how conversant with acceptable pedagogical practices the teacher may be, motivation must be achieved and discipline maintained if that teacher is going to be effective.

Motivation is important, yet it is often difficult to achieve. Perhaps the reason many students are not motivated to learn is that their teachers think motivation is something you do to your students. Motivation is not something that comes from without. *All motivation comes from within.* We choose to do those things that feel good—things that are pleasing to us physically, emotionally, intellectually, spiritually, and psychologically.

The child who has been successful will strive for more success. The child who has won wants to continue to play as long as he feels there is a chance of winning again. *The child who has had the pleasure of understanding will want to understand more. The best motivation for learning is having learned.* Such intrinsic motivation cannot be rivaled by anything that the teacher can do *to* the child.

However, good teachers do not ignore extrinsic motivation techniques. A pat on the back feels good physically. Praise from the teacher and peer recognition feel good psychologically. Having a gold star next to her name will often make a child feel good if there is no social stigma attached to that star. A piece of candy tastes good. A token that can later be traded for something desirable is indirectly pleasing. But, ultimately, such extrinsic motivators are only useful if the child considers them to be pleasing. Do they make him feel good?

If the child perceives the teacher's reward to be good, the child will tend to repeat the behavior (the homework, the attempt to follow directions, the correct response) that brought about the "good" result. If, however, the child perceives the reward to be bad, he will tend to avoid repetition of the behavior which brought about the "bad" result (Figure 5.7). Again, specifically what we do to reward or motivate the child is not important. Rather, the important thing is whether what we do pleases or displeases the child and whether the child considers what we do to be a result of his behavior.

Figure 5.7

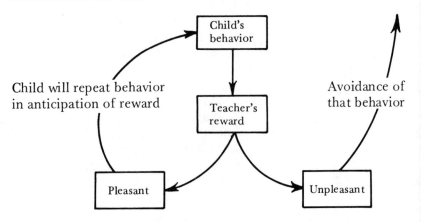

Unfortunately, it is not always easy to predict what the child will like. Some children may want a gold star next to their names; other children may feel that a gold star would mark them as socially unacceptable among their peers. Some children may enjoy a pat on the head; others may only be embarrassed. Some children may want their teacher's approval; others may dislike their teacher enough to avoid any behavior that might result in a show of approval. Sometimes a public show of displeasure by the teacher will cause a child to stop turning in sloppy papers. At other times, that public reaction from the teacher may provide the child with some much-needed attention and, as a result, actually reward the child for sloppy work.

Take care in the use of extrinsic motivational devices. Observe the child's reaction to each motivational attempt to learn which things are really motivational to a particular child and which things are not. All children cannot be motivated the same way.

Jan and Bill are fifth-grade students in the same class. In order to motivate students to work more quickly and accurately, their teacher has developed a math center with recreational math materials. When students have successfully completed their assigned work they are given tasks from the math center to do "just for the fun of it."

The teacher observed that, after a period of time, Bill was consistently finishing assignments more quickly and with greater accuracy. However, after an initial improvement in speed, Jan reverted to her earlier slow rate of work and did not thereafter show any real improvement. The teacher discovered that the math center materials, which were nearly all puzzle-type activities, were fun for Bill but only frustrating for Jan. The two students, quite similar in ability and achievement, were not motivated by the same things.

Discipline is closely related to motivation. If a child is motivated to do *what the teacher wants* that child to do, then discipline problems tend to be minimized. If the child does not want to do what the teacher wants, then discipline is required. *Discipline is doing those things that we know should be done, or those things that must be done, or those things that are good to do, even when we don't feel like doing them.*

Associated with discipline is the word *control*. A disciplined classroom is a classroom under control; a disciplined person is a person under control. Ideally, of course, the child should exercise self-control. Self-control is, in fact, often considered to be a necessary condition of maturity. The primary objective in classroom discipline should be for each child to exercise appropriate self-control.

All actions taken by the teacher to impose control on the child should be toward this end—that the child exercise appropriate self-control.

That self-control may include being quiet when quiet is needed. It may include speaking up when the child would rather withdraw shyly. It may include obeying rules even when it might be more fun to do otherwise, or working to change rules in an orderly, acceptable way. It may include continuing to work on a job when it has become boring or speaking politely to someone who is not very likeable. It may include going along with the crowd when the child would rather do his own thing, or it may include standing up to the crowd.

How can a teacher help a child to become self-disciplined, to exercise self-control? Although no specific technique can be expected to work in every situation, there are a number of approaches that are generally successful.

Establish Behavioral Limits and Don't Waver

Many educators and psychologists feel that children (and adults) are constantly searching for their limits. We all have limits within which we exist. There are physical limits—limits to how high we can reach, how much we can lift, how fast we can run. There are emotional limits—how much we can stand emotionally. There are intellectual limits—how well we can learn. There are skill limits—how well we can do things. There are social limits—limits to what society will allow, limits to what our peers will accept, limits to how far a rule can be bent.

Of course, limitations are not static. They change with growth and maturity. They change as one learns and gains experience. They change as society changes. A four-year-old child has a different set of limits than that same child ten years later. A fifth-grade child is limited differently during a math lesson than during recess. A child is not truly comfortable—emotionally, intellectually, physically—until he has a feeling for his limits. Children need to know what set of limits go with what situation.

Two different reactions are often seen when children do not know where their limits lie. Some children may be reluctant to try new things for fear of exceeding their limits. They don't want to get into trouble, get hurt, or look foolish. As a result, they try only a small amount of what they could safely try (Figure 5.8). Other children who are less cautious may be constantly hurt, may be in trouble socially, or may be irritating to their teachers because they do not know or have not accepted their limits (Figure 5.9).

On the other hand, children who know where their limits are have complete freedom within those limits. They will attempt more with less fear and without getting into trouble physically, socially, or emotionally (Figure 5.10). The wise teacher will help children to see that, by gaining new knowledge, by exercising more self-discipline, by gaining new skills or improving old skills, they can actually extend their limits.

When a teacher wants to establish certain social or behavioral limits on a class, it is not enough that the class be told "You may do this. You may not do that." Certainly children must be told what is expected, but teachers should be aware that such statements will invariably be tested. If children are told that they are not to behave in a certain way, but then are allowed to get away with that behavior, they have learned that the behavioral limit is not where they were told it would be. Since the limit is still unknown, they

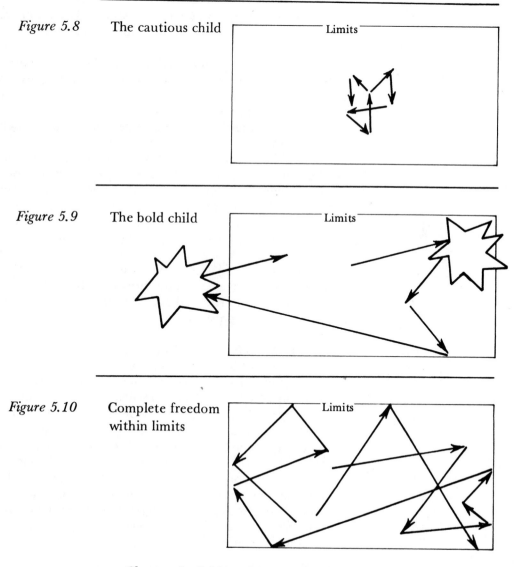

Figure 5.8   The cautious child

Figure 5.9   The bold child

Figure 5.10   Complete freedom within limits

can be expected to push their behavior further and further from the norm until the true limit (what is really allowed) is discovered.

Teachers often find themselves in trouble with classroom control because by giving "one more chance" they are teaching the children that what they say really doesn't mean anything. Empty threats not only do not establish control, but they also undermine the teacher's believability. Tell the child what behavior is expected and what is not allowed, then demonstrate as early as possible that you meant what you said. *Establish behavioral limits and don't waver.*

Closely related to the establishment of behavioral limits is the notion of consistency. The teacher must be fairly consistent in demands placed on the children. If the behavioral limits (what you can get away with) fluctuate in a seemingly random manner, at no time will the child really know where those limits might be.

Dislike Undisciplined Behavior, But Like Undisciplined Children

It is important that children understand that any disciplinary measures reflect disapproval of behavior rather than disapproval of them personally. If children become convinced that the teacher doesn't like them, they will also be convinced that behavior really makes no difference. As a result, the teacher's behavioral expectations will cease to have any impact on them. The child must understand and *feel* that any punishment, regardless of how mild or severe, comes as a direct result of the behavior, not simply because "the teacher doesn't like me."

Don't hold a grudge, or even appear to hold a grudge. *As soon as the child's behavior is acceptable, show approval.* Take care that the approval is sincere and not superficial. (Children are expert at identifying fakes.) Also, take care that the approval does not embarrass the child, having an opposite effect from the one desired. Be sure that the approval takes a form that the child perceives as a reward, not a punishment.

An Idle Mind Is the Devil's Workshop

One goal of instructional activities must be the constant involvement of all the students in learning activities. Students may be physically involved, but physical involvement is seldom what we are really after. Students should be *mentally* involved. They should think along with the teacher during the demonstration of mathematics when doing examples and solving problems. They should think about the related mathematical ideas when manipulating physical materials or playing games. Indeed, physical activities in the mathematics classroom—activities where children are moving about, using materials, playing games—seldom have value unless they serve as a vehicle for achieving greater mental involvement with mathematics.

Unfortunately, what happens in the classroom often involves the student only minimally. For example, during an activity where children are taking turns, it is common for a single child to be "doing" while three or four are paying attention and twenty-one others are not involved. Even though the activity itself may be excellent for those involved, the 84 percent class noninvolvement is a potential source of discipline problems.

Children are not good "waiters." If they are forced to wait for their turn, to wait for materials to be passed out, to wait for the teacher to draw a chart on the board, to wait for whatever, they tend to find all kinds of things to fill the waiting time. It is during such "dead" time that a child might pass a note, throw a spitball, kick Charlie, pull Lucita's hair, or do any number of things that are far more interesting than just waiting.

Teachers need to be constantly alert, recognizing noninvolvement of students, and searching for ways to adjust instructional techniques to increase involvement. Both of the following case studies illustrate changes in classroom activities which increase involvement.

The teacher constructed a cardboard apple tree with numbers on the apples, then made apples with problems such as $2 + 5$, $8 + 1$, etc. She distributed the apples to her students and had them come forward one at a time to paperclip their apples to the correct answer.

The Discipline Problem

Cindy's turn was first. After she finished her turn, she got into an argument with Derek, who was waiting for his turn. The activity was interrupted and not resumed. The teacher had spent over an hour preparing the materials for the activity and was discouraged by the wasted time and effort.

An Alternate Plan

Three days later the teacher identified two children who needed extra practice on addition facts and had them attach all the apples in the correct positions while everyone else was engaged in another activity. She checked their work, discussed their errors with them, then repeated the activity with another group.

After teaching a developmental lesson on addition of unlike fractions, the teacher copied fifteen practice examples on the chalkboard for the students to complete.

The Discipline Problem

While the teacher's back was turned, George threw a pencil at Bill. The pencil hit Bill point first, causing Bill to cry. Several other children began making fun of Bill for his crying. The class was disrupted. The teacher lost his temper and no more mathematics was taught or learned that day.

An Alternate Plan

The teacher decided that the next time a similar practice activity was planned, he would have the problems already written out on an overhead transparency.

Recall that in Chapter 3 it was emphasized that much of the school day is used in transition from one instructional/learning activity to another. It is not a bit unusual for a class to use 15–20 minutes getting from a science lesson to a math lesson or from a reading lesson to an art lesson. But even more important, much time is used getting from one activity to another within a given math lesson. Tasks such as putting away books, distributing rulers, forming teams, writing problems on the board, getting out counting sticks, and a seemingly endless list of similar tasks need to be done. However, they consume time that is, from the child's point of view, dead time—time that is spent waiting, time that is wasted.

It is virtually impossible to eliminate such activities entirely. Many teachers, though, are quite successful in minimizing dead time and keeping the children actively involved in learning. Success in the use of time is almost always the direct result of an awareness of the importance of transition activities and of careful planning.

There are no pat answers that will solve every teacher's transition problems, but the following case studies give some approaches.

Mr. W started his math class with a small-group activity. He used ten minutes of class time forming the groups. Mrs. T also wanted to start her math class with an activity using small groups. When the children came back from music class, she met them at the door and handed out cards with problems on them. Mrs. T told the children to find their groups by finding everyone else with the same answer. As soon as the groups were formed, Mrs. T began the small group activity.

In this case, the transition activity that Mrs. T used was itself a useful practice activity. It provided a smooth flow from one activity to another with a minimum of "dead" time.

Miss D used an activity in her math class that required a variety of manipulative materials. In order to shorten the time needed for distribution of materials, she sorted the materials before school and placed them in paper bags to be passed out to individual children.

Miss S had a relatively long math lesson planned for her second-grade class. The lesson started with a demonstration/discussion activity and ended with children working in their workbooks, which were kept stored on a shelf at the far end of the room. Since the lesson was a long one, she knew the children would become restless. Instead of keeping the children seated while the books were passed out, she concluded the first part of the lesson by leading the children in a game of follow-the-leader. They hopped three hops. They held up seven fingers. They counted objects as they passed. And, among other things, they each picked up a workbook as they passed the shelf where they were kept.

When the children returned to their seats, they were ready to begin working and were better able to concentrate on their work. It should be noted that in this case the transition activity was not intended to save time, but to eliminate possible discipline problems that might result from a lesson that was a bit too long.

The question of how to organize the classroom to produce the best learning environment is a question with no easy answer. In fact, it is a question for which the answer will vary from teacher to teacher and from class to class. The needs, abilities, interests, and limitations of the students must be considered. How well do the students function independently or in groups? How much noise, movement, or other distractions can they tolerate?

Nearly as important are the abilities, interests, and tolerances of the teacher. Under what set of circumstances can the teacher produce the best results? Does the teacher function best leading small-group discussions, activities, and investigations? Is the teacher effective in organizing and initiating independent study? Does the teacher work well in a one-to-one, teacher-student setting? Is the teacher effective in a lecture/demonstration presentation?

Every teacher needs to discover which techniques work best for him or her. The instructional plan chosen should take into consideration both the student and the teacher. The plan should take advantage of strengths and avoid weaknesses—those of the students and those of the teacher. Teachers should not be afraid to try new or different approaches, but neither should they be afraid to discard them when they don't work.

Organizational plans range from those that are totally individualized to those that deal with the entire class as one group. Certainly, children with special needs require a great deal of individual treatment. However, one need of such children and of all children is their need for interaction with other children. They need to interact with other children in their study of mathematics. It follows then, that, as a general rule, teachers should attempt to involve special-needs children in as much group activity as possible.

If the teacher chooses to use as the primary approach to a child an essentially individualized process, it is necessary that the teacher devise ways for that child to work with a group, to use other's contributions and to learn how to make a personal contribution to the group effort. If the teacher chooses to use group instruction, then ways should be devised for the child to get needed individual attention from the teacher and for the child to have opportunities to make personal discoveries, to "do his own thing."

Organizational plans may call for a variety of teaching techniques: telling, showing, asking, listening, and observing. There are times when the most appropriate thing a teacher can do is to demonstrate, using a physical model, how a mathematical idea works in the real world. At other times it may be most appropriate to simply tell the child what we want the child to know. Sometimes it is best to ask leading or open-ended questions and to let the children arrive at their own versions of the mathematical idea.

It is always appropriate to listen. Teachers should listen when questions are answered. Teachers should listen when apparently irrelevant comments are made. Teachers should listen as children talk to them and as children talk to other children. We can learn much about a student's needs—about what the student knows and doesn't know—by listening to what the student says. And, it may be equally important to listen to what the student doesn't say, to listen for those things that the child avoids.

It is also always appropriate to observe. Observe what the child enjoys and what the child dislikes. Observe how the child reacts to what you say or what you do. Observe what the child does with physical models. Observe what the child writes. But be sure to observe the child's eyes. A good teacher can often learn to judge the degree of understanding and interest in the child's eyes.

Organizational plans should also take into consideration the physical setting of the classroom, the laboratory materials that are available or not available, and the printed curriculum materials that are provided. The teacher should always ask, "Will my classroom—its shape, size, and arrangement—allow successful implementation of the instruction that I have planned?" Sometimes the physical classroom is not consistent with the needs of a given instructional procedure and that procedure should be discarded. More often, the physical classroom will need to be rearranged or changed in some way to encourage success. In either case, it is important that the teacher think ahead. It is important that problems that might be caused by instruction that is inconsistent with the physical space be anticipated and avoided.

The teacher should always ask, "Do the procedures I plan to use require the use of any laboratory materials or equipment? Are the necessary items available? Will I need to locate, purchase, borrow, or construct these items before my instruction can be successful?"

Of course, sometimes it is wiser to adapt the instructional plan to avoid the need for unavailable equipment than to attempt to secure it in time for the lesson. In any case, the teacher must plan ahead. And finally, the teacher needs to consider whether the planned instructional approach is consistent with the printed curriculum materials that are provided. Will additional materials need to be prepared? Will the use of certain text pages need to be avoided to eliminate possible confusion? Will gaps result? If too many adaptations to the existing printed materials seem to be called for, the teacher should reevaluate whether the instructional plan is consistent with sound curriculum principles.

In summary, effective organization of the learning environment depends on the teacher's ability to achieve balance between student strengths and weaknesses and teacher strengths and weaknesses within the constraints of available physical resources. Every situation has its limitations. Seldom is it possible to choose and use the best conceivable plan. Rather, the teacher must strive to do the best job he or she can possibly do within the limitations that exist.

---

1. Think of the worst teacher you can remember having. List the characteristics possessed by that teacher that you feel contributed to his or her ineffectiveness.

2. Think of the best teacher you can remember having. List the characteristics possessed by that teacher that you feel contributed to his or her effectiveness.

3. Talk with at least two public-school teachers about teacher evaluation.
   (a) What factors do they consider to be important in teacher evaluation?
   (b) What factors do they think their principal considers to be important in teacher evaluations?
   (c) What characteristics do they consider to be desirable in the other teachers with whom they work?

4. Ask at least two parents to describe the characteristics they expect to find in a good teacher (the kind of teacher that they would like to have for their own child).

5. Ask four elementary-age children whether they like math. Then ask them why.

6. Ask some teachers what they do to motivate children to study mathematics.

7. List at least four of your own personal characteristics that will limit your effectiveness as a teacher of mathematics. Which of these personal limitations can you eliminate or at least reduce and which will you need to learn to work around or live with?

8. Develop a mathematics lesson plan for any grade level. Pay special attention to transition activities. (How will you form groups, pass out materials, get children where you want them, etc.?)

6 PRENUMBER EXPERIENCES

The development of prenumber concepts and skills is a very important component of the early school curriculum for children with special needs. The work of child development and mathematics education has indicated the vital role played by these topics in the later development of basic mathematical concepts, skills, and principles in number, operation, measurement, and geometry. In addition, many of these skills play vital roles in early language development and reading skills.

This chapter focuses on various ways in which children group order, and structure the objects they encounter or think about in their daily lives. Activities that develop skills dealing with form recognition and aspects of perception are also considered. Many of the ideas from this chapter suggest a natural sequence for mathematical beginnings at the early primary level and also for children with developmental lags due to mental, sensory, or physical handicaps.

CLASSIFICATION EXPERIENCES

Teacher Background

Classification, or the grouping of objects having a similar property or properties, plays an important role in the development of mathematics and other school subjects. In mathematics, classification activities provide a foundation for the growth of number and measurement concepts, as well as for the development and growth of logical thinking. Such activities are particularly useful for establishing eye-hand coordination and other motor skills used in the writing of numerals. In reading and language learning, classification provides an avenue to the recognition and discrimination of letters, sounds, and words. In social studies, science, and many other areas of study the impact of classification is just as great.

For most children, classification activities and concepts are a natural outgrowth of maturation and other factors in their environment, such as parents, peers, and learning opportunities. For others, these skills and concepts are almost totally nonexistent. It is important that preschool mathematics curricula for special needs children help them acquire or develop classification capabilities before entering the study of number and more formal work in measurement and geometry. The thinking and oral language involved in successfully carrying out classification tasks play a subsequent role in helping children bridge many of the factors that inhibit learning: lack of recognition, discrimination, and memory skills; faulty organizational and language skills; and comparison and contrast skills.

An important outgrowth of an early mathematics curriculum which emphasizes classification activites is the formation and use of logical language. While first attempts at classification normally involve only one grouping property, it is not long before children start making more involved grouping rules, such as those involving the terms *and*, *or*, and *not*. These situations may arise from attempts to group things on the basis of size, shape, and color. For example, one child might attempt to sort out all large, round, blue buttons. Such activities, if started from a verbal suggestion, will lead to an increased awareness of the meanings associated with many of the common logical connectives. These activities also lay the foundation for further work in classification.

In everyday life people are classified with respect to sex, appearance, personality, physical build, intelligence, and a myriad of other characteristics. Containers are grouped on shelves of a grocery store by their contents. Within similar types of foods, brands and different sizes of containers are further grouped. Automobiles are classified according to their make, color, body style, and horsepower. Such a listing of the ways we classify, or group, objects could be extended as a classroom activity.

Classification plays an important role in mathematics. The ability to describe ideas of an abstract nature requires many classification terms and skills. Stating a definition involves grouping together ideas and terms with the hope of simplifying a given situation from a linguistic point of view. For example, if one wishes to talk about numbers that have exactly two distinct whole-number factors, a mathematician finds it much easier to talk about *prime numbers*. This process of using classification properties, or attributes, of objects to create definitions is common both in mathematics and everyday life.

In mathematics small groups of objects are commonly used at the preschool level and in the primary grades to initiate the questions of "How many?" and "How much?" The sets of objects, or groups, used in these situations are usually the result of some grouping activity. These situations serve as springboards to the study of number, operation, and measurement. Students start to classify things according to their size, mass, appearance, and other attributes. Many of these activities evolve into the study of orderings (seriation) and into measurement. Hence, having knowledge of classification skills is an important prerequisite for the special child before entering later phases of the elementary mathematics curriculum.

Classification activities also provide an avenue to using the senses and building sense-using skills for children whose skills in these areas have been impaired in one way of another. These information-gathering skills and senses can be built or improved through activities which require children to collect data, organize it, and make decisions accordingly.

Visually handicapped students need experiences that focus on hearing, touching, and smelling. It is important to design activities that expose children to a wide range of attributes in each of these areas. The deaf student needs work with the visual classification of various shapes and forms as he prepares to begin the mastery of sign language and speech reading. Learning-disabled children need classification activities which focus on a wide range of learning skills: organization skills, discrimination skills, long- and short-term memory skills, and revisualization skills.

With these factors in mind, and remembering the role that classification plays both in mathematics and language development, let us turn to the psychology of classification and the stages and activities of classification learning at the classroom level. This will be followed by an overview and introduction to activities related to the seriation, or ordering, of objects according to some property they possess.

PIAGET'S ANALYSIS OF THE DEVELOPMENT OF CLASSIFICATION ABILITIES

The mental processes from which classification abilities develop are fairly well-known through the work of the Swiss scholar Jean Piaget (1896–1980). Piaget, along with his co-worker Barbel Inholder, outlined much of this structure of the development of a child's classification abilities in the book *La Genèse des Structures Logiques Élémentaries: Classifications et Sériations* (1959).

In it they describe the development of classificatory abilities through three stages. The first of these stages, lasting from about two and one-half to five years of age is noted by the child's inability to sort through a collection of objects according to a fixed logical rule, e.g., "Find all of the square blocks." Rather, the logical criterion used by the child seems to be in a constant state of change during the sorting process. It often happens that the end of the sorting process is viewed by the child as some sort of figural pattern—a plane, a car, or some other familiar object (see Figure 6.1)— rather than some class of objects having some related property.

While the sorting process may seem to have some logical basis at points in the sort, the resulting class does not fit any one of the intermediate rules used by the child. For this reason and the fact that the child often ascribes the shape of the pile as being some object, Piaget calls this first stage of classification abilities the stage of *figural collections*.

The child in the age span from two and one-half to five is logically unable to handle the problem of clearly defining the properties that characterize membership in a given class. Rather, the child's definition constantly changes as each new block, or small set of blocks, is considered. The child often is unable to sort out one critical attribute from among several more attractive noncritical attributes. In this process the child may be sorting squares and actually start out with a square. It may be a red square, so the next block may be a red triangle. At this point, the child remembers the attribute of triangularity and selects a blue triangle for the third block. This process of change of rule continues until all the blocks are used or the child tires of the activity. This type of sorting process is typical of the child in the stage of figural classifications.

The second stage of classification abilities develops somewhere around the age of five and lasts until about the age of eight. This is the stage of *nonfigural collections*. The child in this stage has the ability to form groups of objects based on a fixed rule, but still has difficulty in handling the relationship between a class and one of its subclasses, e.g., the relationship between the set of all quadrilaterals and the set of squares. Here the child becomes confused about whether the squares or the nonsquares are part of the set of quadrilaterals and whether or not the set of quadrilaterals may be identical to the set of squares. For example, in the case illustrated in Figure 6.2, the child would note that all quadrilaterals are not

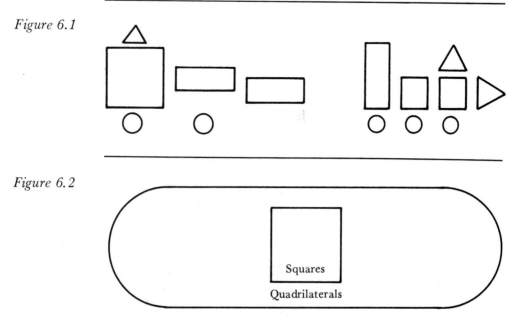

Figure 6.1

Figure 6.2

necessarily square, but would have difficulty in handling the question "Are all squares quadrilaterals?" At this stage the child cannot distinguish between a subclass and the entire class itself. Once the child masters the ability to handle the inclusion relationship for classes, the child moves to the third and final stage, the stage of *classifications*. At this point, the child is no longer confused about the sorting rule or the relationship between a given group and either subgroups or supragroups of that group.

What Do *You* Say?

1. Why should nondirective classification activities precede directive ones at the preschool and primary grade levels?

2. What is the purpose of having children focus on the verbal aspects of classification?

3. Why should smaller rather than larger objects be used with special needs students for classification tasks?

4. How do multiple classification tasks differ from the more basic type of classification activities posed for young children?

DEVELOPING BASIC CLASSIFICATION SKILLS

As mentioned earlier, classification is basically a sorting or grouping activity. The following sequence of classroom experiences is designed to help special students acquire classification skills necessary for success in school and in life. The sequence includes a continuum of experiences ranging from those involving work with the basic senses to those dealing with abstract reasoning. In addition, the experiences deal with both quantitative (numerical) and qualitative (judgmental) aspects of classification.

Early Classification Experiences

First classification experiences for children should involve the free, nondirective sorting and examination of familiar objects. These objects may be simple things such as rocks, spools, pencils, coins, toy cars, or other small items. Particularly for children who have visual perception problems, the objects should be small enough to fall within the central vision area. This allows for total vision of the object and the ability to see all of its major attributes at once.

By allowing opportunities for children to participate in nondirective classification activities, teachers can more readily recognize a child's level of logical thinking and expressive language abilities. The following examples illustrate how such activities might be carried out. Note particularly the observational guidelines that are presented and the emphasis on the language development that accompanies each classification activity.

Activity 1: Free Sort.

In Free Sort children are asked to sort articles into piles using any rule they wish. The teacher's role during the activity is to observe each child's choice of sorting properties and to note whether a child is consistent in applying the chosen rule. As children finish, ask them to tell why they grouped the articles as they did. This encourages children to grow in their ability to express their ideas and to review their work. Besides cueing the teacher to a child's progress in oral communication skills related to classification, the verbal description of the classification activity will give further insight into the reasoning used by the child while sorting.

Early classification activities such as Free Sort deal with the sorting of materials on one basic attribute. Usually the attribute will reflect some gross difference in the objects available for classification. Research studies have shown that children often focus on a ceptual difference (color, size, shape, or position); a functional difference (use or function of a group of objects); a nominal category (what the object's name is); or a fiat equivalence (child says the objects are alike, but is unable to tell how even when asked).

Activity 2: Sorting Box.

In this activity, children are given a box, like the one pictured in Figure 6.3, into which they are to sort the objects. Such a box can be easily made by cutting down a box from the local grocery store and making partitions out of the cardboard pieces removed. This box can be used to hold various categories determined by a sort. Collections of buttons, seashells, unshelled nuts, nails, small toys, corks, and other objects are appropriate. Give children a mixture of the items and ask them to sort them into piles. As before, children should be asked to describe their sorting (classification) scheme. Older children might be asked to label the categories formed. Alternately, to help develop verbal expression, have children make up a riddle about the objects in one of the box partitions, to see if others can guess what is being described.

Both of these examples are free-form classification situations, in that they do not impose a sorting scheme on the child. The Sorting Box activity is somewhat more restrictive in that it provides a position for depositing the items. In designing activities of this type, the teacher should provide the child with experiences using a variety of possible sorting attributes. Materials might be provided that would suggest to the children the various sorts listed in the research findings. In addition, other materials could be made available that would suggest groupings by mass, taste, texture, smell, or transparency. If children already have a firm grasp of number meaning, materials might even be used that suggest grouping on the basis of number. For example, given a box of wooden polygons, a child might sort by the number of sides. The wider the variety of the materials used in the various experiences, the wider the range of the child's classification skills will be. The broader the experiences, the greater will be the exposure to language—so important in the early development of the young child.

Figure 6.3

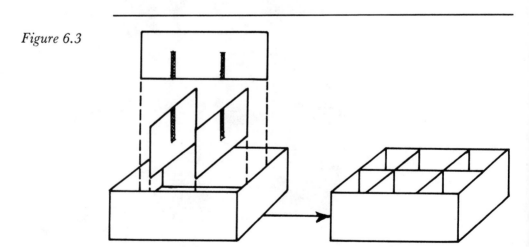

The classification activities that follow exemplify two different types of open-ended grouping experiences. Through opportunities such as these, teachers can encourage creativity and still cue in to the level of logical thinking and language skills possessed by the child.

Activity 3: Pick A Pair.
In this early classification experience, the child selects a pair of objects from the classification materials provided, or from the objects in the room. The group of children with whom the child is working then try to guess the properties the objects have in common. This activity is especially effective when the classification property chosen is one which involves the children themselves.

Activity 4: Ten Questions.
In this activity, a child picks an object from a fairly small set of items. The other children try to identify the object by asking about the attributes of the object. They can only use questions which can be answered by yes or no. This game provides preparation for later classification experiences having more than one important attribute involved.

Directed Classification Experiences
As the child grows in the ability to classify on the basis of internally perceived relationships, it is time for the teacher to begin to direct the child into experiences where the critical attributes or sorting rules are stipulated by someone else. By this time teachers are aware of each child's strengths and weaknesses with respect to general classification tasks, and they can now structure activities to help children grow and develop more mature thinking and language patterns. The following activities indicate some classroom experiences that involve the forming or the recognition of classification rules.

Figure 6.4

Activity 1: I Know.

Tell children you will name objects that belong together in some way. When they think they know *why* the items are grouped together, they should say "I know!" and describe the set named. For example, oranges, hamburgers, carrots, and cereal can be grouped together. They are foods people eat.

Activity 2: Go For.

This is an activity in which a child is requested to go to a collection of objects and bring back a particular type of object. For example, a child might be instructed to go to a box of buttons and bring back "a round button," "a metal button," or "a button with holes in it." It is easy to see how these activities help bring together visual and auditory skill aspects of classification.

Activity 3: Odd One Out.

Arrange objects on a table or chalk ledge in groups, four objects alike in some way and one different (Figure 6.5). Have children take turns finding the one that does not belong.

Activity 4: Simon Says.

The usual rules for Simon Says prevail, except a child or a teacher calls out commands involving common classroom attributes that are specific to the children, their features, or their school supplies. For example, one might say: "Simon says: All children having brown hair stand on one foot" or "Simon says: All students who walked to school today clap their hands." This activity strengthens auditory listening skills related to classification.

Activity 5: Comparison Boards.

An effective activity, one which combines classification and the reinforcing of visual discrimination skills, is Comparison Boards. This activity requires the use of a board such as that shown in Figure 6.6, with a pair of words written and illustrated at the top. You will also need a wide variety of cards that have objects or pictures of objects on them. The students need only look at the heading on the board and then pick a pair of cards and place them in the appropriate columns of the Comparison Board. Comparison classifications which might be used to head the board are: same, different; rough, smooth; hot, cold; heavy, light; smells bad, smells good; sharp, blunt; long, short; high, low; and so on.

Figure 6.5

Figure 6.6

Activity 6: Form Factory.

The Form Factory involves the use of cutouts and stencil forms into which they can be placed. Make the cutouts by cutting them from heavy posterboard, or cut them from paneling of stiff foam rubber. To do the activity, the child picks out the forms and places them in the correct cutout hole. A storyline could be made for the activity that would be based on the idea that the shapes need to be packed for shipping. The students would then play the role of shipping clerks.

Activity 7: Shape Bingo.

Each child is provided with a Shape Bingo board like that in Figure 6.8. The items in the first column are all colored red, the items in the second column are all colored blue, and those in the third column are all colored yellow. The possible shapes one might use are squares, triangles, circles, hexagons, rectangles, ovals, and possibly trapezoids. The cards should all be different and covered with contact adhering paper, so that students can mark on or cover the items as they are called out. A master sheet of possible combinations of the shapes and colors can be made for the caller. This paper might be covered with the contact adhering paper so that items can be marked off as they are called. An illustration of

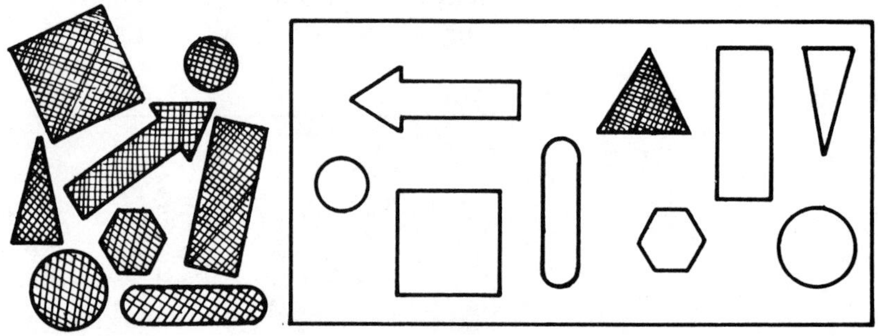

Figure 6.7

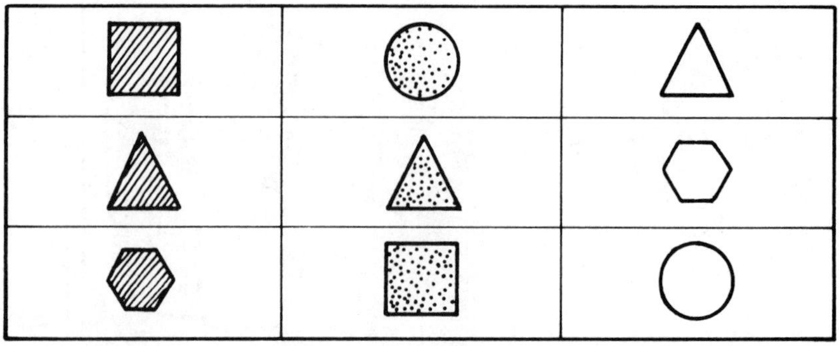

Figure 6.8

such a master board is shown in Figure 6.9. Various types of game boards might be used and the method of determining the winner may vary from one game to another (three in a row, three in a column, three on a diagonal, or all four corners).

Activity 8: Pattern Sort.
    This is an activity that can be done by a child working alone. The basic format of the activity allows for a myriad of variations. One version involves the sorting of pictures of cars and trucks that have been glued to index cards. Other variations might use wallpaper scraps, cloth, pictures of objects, or pieces of sandpaper in the sorting. A sorting box such as that shown in Figure 6.10 could be constructed with six to nine partitions. For the car-truck sort, categories might include sports cars, race cars, sedans, station wagons, pickups, dump trucks, and semitrailer trucks. When the child finishes the sort, all the teacher need do is check through the cards in each partition to see if the sort has been correctly carried out. When variations arise in the child's sort, the teacher might ask why that sort was made the way it was.

Figure 6.9

RED	BLUE	YELLOW
⬤	⬤	◯
▪	▫	▢
⬡	⬡	⬡
▲	▲	△
▬	▭	▭

Figure 6.10

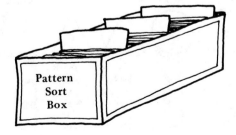

Pattern Sort Box

Many other ideas for classification activities for the child involved in making simple discriminations can be found in *Workjobs for Parents* and *Workjobs* by Mary Baratta-Lorton (Menlo Park, California: Addison-Wesley) and in *Mathematics Learning in Early Childhood* (Reston, Virginia: National Council of Teachers of Mathematics).

Multiple Classification Activities

The basic classification experiences focus on a single attribute that defines the groupings. In real life, classifications are usually not so simple. As teachers expand the study to include situations involving multiple criteria, they must also help children enlarge their language capabilities. In the preceding work, attributes such as shape, sound, and texture were considered together with the development of the associated language.

Activities should now focus on deepening children's analyses of the characteristics of objects. While an object may possess the attribute of shape, the focus must now narrow to *what kind* of shape. That is, is it a square, a triangle, a circle, a trapezoid, or a hexagon? Is the shape colored red, yellow, blue, green, or some other color?

Logical blocks are among the simplest sets of materials available for classification activities that involve multiple-attribute identifications. These materials suggest a wide range of classification activities that focus on the attributes of color, shape, size, or thickness.

A set of logical blocks consists of squares, triangles, circles, and hexagons. Each of these blocks comes in one of three colors: red, yellow, or blue. Further, each of these combinations of blocks is either large or small in size and is either thick or thin. This set of forty-eight blocks exemplifies all possible combinations of the attributes of shape, color, size, and thickness. (See Figure 6.11.)

Figure 6.11

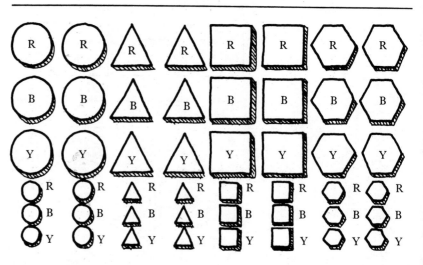

1. What classification activities can you think of for using logical blocks with children having special learning needs?

2. What adaptations (of the blocks, special activities, or instructions) might be necessary for blind or visually handicapped children?

3. What uses of the blocks could be made with children having memory problems?

4. What modifications might be made to deal with the blocks in a problem where the teacher is dealing with a child having expressive language problems?

5. How might the blocks be used to develop motor and perceptual skills within the area of classification?

6. How can the blocks be used to promote the development of logical thinking and the associated language skills?

The accompanying questions should serve as a guide for the thoughtful analysis of each of the following multiple classification activities. Examine each of these activities carefully, thinking of their use with the child who has special learning needs.

Activity: The One Difference Game.

This is an excellent way to introduce children to the logical blocks following a time of free play. (*Free play* refers to the time children should have to play with and explore the blocks as they like. Before any new manipulative is used in math class, it is quite necessary to allow children free time to work with the materials to satisfy their natural curiosity. Only then will they be ready to go ahead with a structured learning activity. During the free time, observe the children's play patterns.)

The basic rules governing The One Difference Game can be modified to match the developmental level and capabilities of the children involved. These rules are:

1. Use some means to decide which child goes first. Perhaps allow the child who first draws a red block from the bag to go first.

2. The first player selects a block from a mixed collection of blocks and places it in the middle of the working area.

3. The next player, the one to the left of the first player, must select a block which is different from the one just played in *exactly one* attribute. This piece is then placed next to the first piece in the playing area.

4. The next person then selects a block from those not yet played

that differs from the last block in exactly one attribute. This piece is then played next to the last piece played.

5. Play continues in this manner with each person getting one point for each block they correctly play. If a person should play incorrectly, anyone can challenge the play. If a challenge holds, the person playing the piece loses their point and the challenger receives it instead. If the challenge is not correct, the person challenging loses their next turn.

6. Play continues until no one can play another block. At this point, the player with the largest total of points is declared the winner.

Variations of The One Difference Game are easily constructed. More advanced students who can recognize and write letters of the alphabet can be encouraged to record game moves on paper. This helps them focus on the attributes and gives them a written record of the attributes used in each play. One appropriate code for recording the moves is as follows: R—Red, Y—Yellow, B—Blue, L—Large, S—Small, Tk—Thick, Tn—Thin, □—Square, ○—Circle, △—Triangle, ⬡—Hexagon. A large, thick, blue circle might then be coded as L,Tk,B, ○ . If the play L,Tk,R,○ is followed by the play S,Tk,R, ○ , then a correct play has been made as the pieces differ in only one attribute. The play L,Tk,R,△ followed by S,Tn,R,△ is not correct as the two blocks differ in two attributes.

Obvious spinoffs of this activity for more advanced students are The Two Difference Game, The Three Difference Game, and The Four Difference Game. Another variation is The One-Two-Three Difference Game, where the successive blocks must differ by one, two, and three attributes from the preceding block. Another possibility is to roll a die with the numerals 1, 2, and 3 on its faces to determine the number of differences a given piece must have from the prior piece.

Modifications to tone down the sophistication of the games for less mature learners are also possible. A basic suggestion is to limit the number of attributes involved in the block set. For example, one might eliminate all of the thin or all of the small blocks. If more than one attribute is omitted in a round of play, it may be necessary to provide multiple copies of some of the remaining blocks. Alternately, one might eliminate one attribute (size, thickness, or color) and limit the number of different shapes used.

The logical blocks can also be used to simultaneously extend logical thinking and multiple classification skills that involve the words *and*, *or*, and *not*. These connectives play a major role in our everyday language. They also serve an important function in mathematics. Elementary school children meet them in the definition of mathematical concepts and in the description of basic properties.

Basic meanings for these words can be informally developed

using the blocks. The word *not* can be dramatized by asking a child to find all of the blocks that are *not red*. Children understand the meaning of the term *not* associated with red if they can correctly sort out all of the blue and yellow blocks. An illustration of what happens in this sort is shown in Figure 6-12. Mathematicians would say that the set of blue or yellow blocks is the *complement* of the set of logical blocks.

Another type of multiple classification rule is one that employs the connective *and*. When applying this type of rule, both of the linked conditions, or attributes, must be present. If, for example, one wished to sort out all of the blocks that were red *and* square, one would probably just say, "Find the red squares." The use of the word *and* is sometimes just understood from the context. To find these blocks, children would have to look at the set of logical blocks and identify those blocks which were both red and square at the same time. These would be the blocks that are in the shaded region of Figure 6.13. This set of blocks, known as the intersection of the set of red blocks and the set of squares, is common to both sets. The remaining blocks are those that are square, but not red; those that are red, but not square; and those that are neither red nor square. Mathematicians call the rule using *and* to connect the critical attributes a *conjunctive rule* and the resulting set of objects the *intersection of the sets* having the critical attributes as their labels.

A third multiple sorting rule is one which uses the connective *or*. Suppose, for example, a child were asked to find the blocks which were yellow *or* red. This set of blocks would be made up of blocks which are yellow as well as all blocks which are red. If a

Figure 6.12

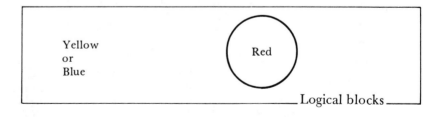

Figure 6.13

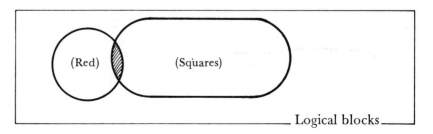

block is *either one* of the two colors mentioned, it belongs to the set of *yellow or red blocks*. This sorting rule leads to the situation illustrated in Figure 6.14.

It is obvious that to get into the final set, it is sufficient that a block be only one of the two colors mentioned. This often leads to problems in a child's sorting of the blocks. In mathematics, the term *or* is interpreted to mean *one or the other or possibly both*. Hence the resulting set consists of blocks that may or may not share common properties. The confusion may result from a lack of concrete experiences that help students differentiate between the mathematical meanings of the terms *and* and *or*. In the illustration in Figure 6.14, no blocks are both red and yellow at the same time. Mathematicians refer to the shaded portion of the diagram as the *union* of the set of red blocks and the set of yellow blocks. The sorting rule using the term *or* is called a *disjunctive rule*.

There are several situations in everyday life which involve complex sorting rules. Many are found in applications of law and governmental regulations. When one describes a person, a conjunctive rule is often used: "John is tall, dark, *and* handsome." In reading over a contract or in applying a taxonomy in a biology class, many disjunctive rules are encountered. There are other instances, such as that typified by the "Do not walk" traffic signal and by certain park signs, which embody negative sorting rules to tell you which activities you are not allowed to pursue. The following activities give ideas that may be carried out with children. They involve using the blocks and different types of sorting rules.

Matrix activities involve compound, or multiple, sorting rules using the word *and*. The easier ones involve the completion of a simple array of blocks such as that shown in Figure 6.15. Children can use any block in the set of logical blocks to complete the pattern that has been started.

Figure 6.14

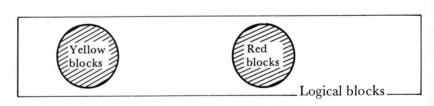

Figure 6.15

A little reflection on the pattern indicates that the first row contains large, thick, red pieces. The first column indicates that the blocks in it must have the same shape and thickness, but be of different sizes. The block in the top is red and the one below is blue, so one might guess that the second row consists of blue blocks. Hence, a likely candidate for the completing square is the S, B, Tk, □. A different person looking at the same array might see, on occasion, a different sorting rule and pick a different block; but for the most part there would be agreement.

Special sets of matrix cards could be made with the answers printed on the back sides for student use. Cards could also be made picturing objects other than logical blocks. Some children are more successful with the cards when the blocks or the objects on them are available for students to use when solving the problems.

Activity: The Detective Game.

This game, designed for two players and a banker, focuses on the use of the word *not*. To begin, each player in turn draws blocks from the bank until all are divided between them. They each lay their blocks on a table, on their own side of a 30–50 cm high partition that separates them. This arrangement is shown in Figure 6.16. In turn, players ask each other for a block that is *not* on their side—one that they think the other player has. If the opponent has the block, it is given to the player who requested it. Then the block is given to the banker who removes it from sight (banks it) and gives the player who requested it one point. The game continues in the same fashion until each player has had ten requests or an allotted amount of time has been used. The player with the most points is declared the winner. A variation to the scoring procedure, one illustrated in Figure 6.17, requires that the banker keep the blocks won

Figure 6.16

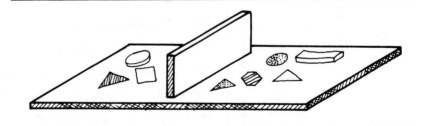

Figure 6.17

by each player in separate piles (but out of sight). At the end of the game the players compare, by matching one-to-one, to see who has more blocks. In either case, the game builds both short- and long-term memory skills while assisting in developing language skills.

The Detective Game involves children in thinking about what blocks have already been played, and then asking for a block they believe is still in play, but *not* on their side of the table. The game reinforces both short-term and long-term memories. It is suggested that one limit the number of shapes or other attributes for beginners, and only gradually increase the number of blocks in play.

Various hoop games can be played with children to help them learn the meanings of the intersection (*and*) and union (*or*) of sets. The circles representing the sets can be constructed out of hula hoops or out of yarn.

---

1. Be prepared, as your instructor directs, to answer the major questions on pages 122 and 131 of this chapter.

2. Make a list of instances in which you have used classification skills today.

3. Describe a classification scheme for use with sandpaper. Be explicit as possible in the categories you would use in setting up your sandpaper classification scheme.

4. Think of all of the buttons you have seen in your life. List the attributes which could be used to sort these buttons. What are some of the common multiple attribute combinations?

5. Look at the registration form for a common credit card below. What attributes does the card ask about? Are all of these applicable to all people?

Name _____

Address _____

Town _____ State _____ Zip _____

Bank _____

Savings Acct. No. _____ Checking No. _____

SS No. _____ Age ____ Marital Status _____

Occupation _____ Salary _____

Credit References—1. _____

2. _____

---

6. What are the defining attributes of the concept of a trapezoid? (If you don't remember what a trapezoid is, look it up in a geometry book.)

Use a Set of Logical Blocks in Completing Items 7–10

7. Complete the matrix shown in Figure 6.18 using only the thick blocks from the set of logical blocks. A letter within the shape indicates a large block and the letter of the color on the outside indicates a small block. Each block in the set of thick blocks can be used only once.

8. Find the three large triangles, the three large squares, and the three large circles (all thin) and arrange them in the matrix shown in Figure 6.19 so that all four of the following properties are satisfied at once:
   (a) Each row (across) has each one of the shapes in it.
   (b) Each row (across) has each one of the colors in it.
   (c) Each column (down) has each one of the shapes in it.
   (d) Each column (down) has each one of the colors in it.
   When you have filled in the matrix, look carefully at your pieces on the diagonals. What do you notice? Do you think that this will always happen?

9. Place two string loops on the table to represent the boundaries of a set. Work only with the large squares and the large triangles in completing this exercise. Place all of the blue pieces in one of the loops. Place all of the squares in the other circle. Do these sets have any pieces in common? Draw a picture showing your final sorting. Label all of the regions formed, telling the type of block to be found in each.

10. Use all of the blocks for this problem. Place a string loop on the table. Place all of the triangles in the string. Using a second string, enclose all of the thin pieces. Draw the resulting configuration and label each of the regions formed. Also show the number of blocks contained in each of the regions. Don't forget to label the type of blocks found in the area outside the string loops, but still in the set of logical blocks. Use only the connectives and the attribute names to label the regions.

11. Create a card file of classification experiences of the simple (single-classification) type to be used with young preschool children. Create activities that use many senses. In the upper left corner of each card indicate (a) whether the activity is a directive or nondirective one and (b) the specific senses or thinking skills emphasized by the activity.

12. Describe a classification experience for a young child using logical blocks. Make a list of questions you would like to have the child respond to. Relate the activities and replies to Piaget's stages of classification development by describing how the performance on the activities and questions relates to Piaget's stages.

13. Conduct an interview with a primary-aged child using the classification activities from question 12. Prior to your interview construct an observation checklist based on the performance levels you identified in 12. Check your sheet with your instructor. After the interview write up your observations and submit them to your instructor.

14. Describe a classification game for a visually impaired child involving the word *not* and materials. Explain how the activity relates to the child's impairment and needs.

15. Develop and make a classification game which can be played by two teams of two players each at the fourth-grade level. Describe the particular class of students for whom the game is specifically designed and how it helps their growth in classification skills.

16. Outline a classification activity for a junior high learning-disabled student who is having trouble discriminating between even and odd numbers.

Figure 6.18

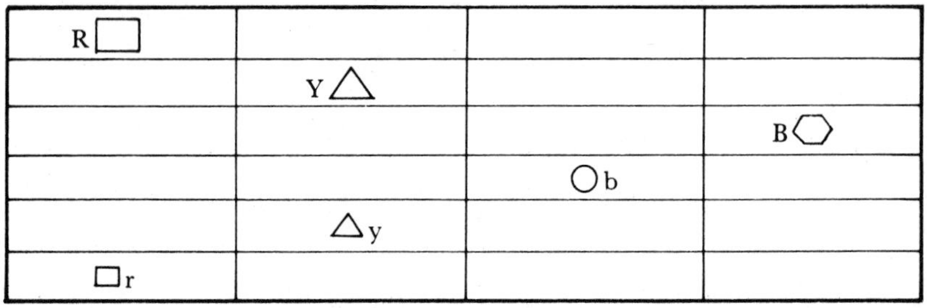

R ☐			
	Y △		
			B ⬠
		○ b	
	△ y		
☐ r			

Figure 6.19

SERIATION EXPERIENCES

Teacher Background

Following closely on the development of basic classification skills in children is the development of seriation skills. Seriation is the process by which one compares and orders objects on the basis of some attribute they possess. This ordering is made in reference to the degree of the attribute each of the individual objects possesses. The attribute selected may be one of length, height, thickness, mass, color, or something else.

Seriation skills developed by children are important prerequisites for the study of number and the study of measurement. Before children are asked to order numbers or interpret measurements at the symbolic level it is important that they have had many experiences ordering objects on various physical levels. Then, when children have an intuitive feel for what it means to seriate, applying that notion to number and measurement is easier and more meaningful.

The impact of seriation concepts and skills extends far beyond the elementary school level. Seriation relates to common daily situations, vocational applications, and the study of relationships in advanced mathematics and other areas. Developing ability to see relationships overrides the whole process of seriation. At an early age children encounter situational needs to understand common relationships such as *same*, *more than*, and *less than*. A simple example is the purchasing of items at a local store. "This costs 64 cents. I have 48 cents. I need *more* money." It is therefore imperative that all children, including those with special needs, develop a good foundation for seriation in the early stages of their mathematical education. This foundation, like that for classification, should consist primarily of activities that allow children to develop necessary concepts and skills in active, hands-on experiences.

Piaget's Analysis of the Development of Seriation Abilities

Piaget has also studied the developmental process through which children acquire relational skills which are called seriation skills. Like the developmental pattern for classification, the taxonomy of seriation skills contains three levels. These three stages show the development of the skills spanning a continuum from noncomprehension of ordering to ordering according to a well-developed systematic process.

The development of the various seriation skills differs, depending on the particular materials used and the methods by which children are asked to order the materials. For example, the ability to order a set of objects on the attribute of weight develops two years later than the ability to order objects on the attribute of length. Other factors such as the amount of difference in the materials (gross to fine), number of objects to be ordered, and ordering instructions affect the rate of a child's developmental progress. Once a child has developed a systematic operational ordering process, the number of objects to be sorted plays a smaller role in the student's performance levels.

Using the ordering task of sorting ten small pieces of wood on the basis of length, one can consider the three stages of Piaget's model for the growth of seriation skills. The first stage is *uncoordinated series*. During this stage the child progresses from no attempts at seriation to making several small sets or series of two or three sticks. However, the small series may not contain the elements in a correct serial ordering (Figure 6.20). In addition, the child does not attempt

Figure 6.20

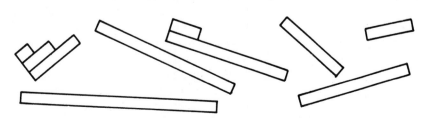

to relate the subseries in any way to one another. Most children pass through the stage of uncoordinated series while they are four to five years of age.

The second stage is *trial and error series*, which occurs at about six years of age. In this stage students are capable of serial ordering a series of ten sticks on the basis of length, but they can only accomplish this ordering through trial and error. Students in this stage of development accomplish the orderings through careful consideration of the relationship of adjacent objects and a multitude of tests. They use little or no knowledge of how the objects they are comparing are related to other objects contained in the set. That is, they order only through an extensive set of comparisons of two objects at a time.

The third and final stage of seriation development for length orderings occurs when the child is between seven and eight years of age. This is the stage of *operational seriation.* In this stage, the child orders the sticks by making use of the knowledge of relations. That is, if the child knows that rod c is longer than rods a and b and at the same time rod c is shorter than rods f and k, then the child knows that rods a and b are shorter than f and k. A child at this stage does not have to make an extensive number of two-way comparisons to order a group of objects serially. This ability to develop the serial orderings in a systematic fashion employing the theory of relations characterizes a child who is at the stage of operational seriation.

These stages and the developmental time frame varies from child to child. The *order* of the stages will always be the same, but there may be some differences related to the attribute used in the serial ordering and the observational situation.

What Do *You* Say?

1. In what way do seriation skills build on previously acquired classification skills?

2. How important is the child's oral language in early comparison experiences?

3. What major difficulties do children face in successfully completing multiple-comparison seriation activities?

4. How can teachers basically modify early comparison/multiple-comparison seriation activities to accommodate highly visual learners? Auditory learners?

Developing Seriation Skills with Special Needs Children

Seriation, or ordering, is an outgrowth of the focus on attributes during the study of classification. When one begins to sort objects on the basis of an attribute, one also begins to notice that some objects have more of the attribute than others. It makes sense to think about which objects have more of the attribute and which have less. There is a normal tendency to examine a pair of objects and compare the amount of the attribute they have. This leads to the early comparison stages of seriation. Following this, the child encounters larger sets of materials, where a simple comparison will no longer suffice. Multiple comparisons are called for and the task becomes one of making use of relationships to determine a correct ordering more efficiently.

Early Comparison Experiences

The first type of comparison activities that children need is that which encourages the development of language associated with the comparison of *two* objects. In such activities children learn to do the comparative form of the adjectives with respect to the various attributes. There are many such terms: longer, shorter, higher, lower, heavier, lighter, wider, thinner, more, less, and many others. These terms, many of which will be subsequently used in measurement and number work, can be carefully used through individual and small-group work. Both formal activities, such as those described in the following activities, and informal situations can be used to monitor children's growth in the understanding and use of comparisons.

Activity 1: Sort Board.

This activity is similar to the Comparison Board activity used in classification. In this activity the focus is on giving children objects with the same attribute, say length, and then having the children compare the amounts of the attribute present in each member of the pair. The child shows her or his decision by placing the members of the pair on the appropriate side of the Sort Board shown in Figure 6.21. (Note that both the word and pictogram are used in

Figure 6.21

labeling the sides of the board.) The items to be sorted can be pictured on index cards and paper clipped together to establish which objects are to be paired in a given comparison. A code on the back of the cards can provide the children with immediate feedback on how they are doing.

Activity 2: Pick The Right One.

This activity calls for a host of pairs of similar objects where the child is given the command to "Pick the longer one" or "Pick the lighter one" (Figure 6.22).

Both of these activities focus on the development of comparison skills in dealing with pairs of objects. Appropriate materials for the activity are common classroom or junk box items: pencils, spoons, measuring cups, jar lids, and so on. In general, objects should be things the child has had experience in sorting earlier, so the attributes of interest are quite familiar.

One point that must be given special attention is the careful development of the appropriate language for given comparison situations. Consider, for example, the following situation.

Teacher: Which of these pencils is the larger one?

Child: What do you mean, "larger one"? (See Figure 6.23.)

The child has made a more astute analysis of the situation than has the teacher. More appropriate language might have been: "Pick

Figure 6.22

Figure 6.23

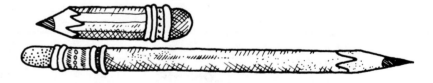

Figure 6.24

the longer one," ". . . the thicker one," or ". . . the heavier one." Teachers should be careful not to use sloppy language in comparison situations.

Activity 3: Pick A Pair.

This activity, similar to that in classification, involves the child in finding a pair of objects which have a given relation to one another. The activity can be modified by requiring children to make (draw, cut, form, model) a pair of objects that have the given relationship.

Note that, while the first example of activity relies heavily on seeing skills, the last two rely heavily on children's ability to deal with comparative language as they hear it. All three examples focus on the comparison of two objects with respect to the degree a particular attribute is possessed, a task most children can master quickly even with trial-and-error approaches. Such immature approaches, however, will not be adequate for handling higher-level or multiple comparisons.

Multiple Comparison Seriation Activities

Once the number of elements involved goes beyond two, the child is faced with a quickly growing number of comparison combinations which must be made if a serial ordering is to be completed using trial-and-error methods. It is at this point that many children (who are developmentally ready) start to look for other more systematic methods of ordering materials. The use of relations such as *transitivity* is frequently brought into play. Suppose a child is faced with ordering three objects, A, B, and C, on the basis of length. To make all possible comparisons requires that the child check out AB, BA, AC, CA, BC, and CB. The task can be shortened considerably if the transitivity idea is used. Once the child finds that A is shorter than B, and B is shorter than C, it is an easy step to deduce that A is shorter than C. That is, it is an easy step if the child has developed the idea of transitive thinking—a product, says Piaget, of maturation *and* experience. Seeing that A is shorter than B and B is shorter than C provides children with all of the useful information they need. Other children fail to benefit from this information since they fail to recognize the transitive pattern or when to apply it.

Another stage in the child's development of multiple-comparison skills is the recognition of *reversability of relations.* Teachers can help by reinforcing children's growth in understanding opposites, especially when these deal with comparative terms. Consider the two relations *is longer than* and *is shorter than.* Suppose a child correctly recognizes that stick A is longer than stick B. The child familiar with opposites can reverse the relation to say "stick B is shorter than stick A."

The ability to use transitivity and the ability to reverse relations are two important stages in a child's development of the ability to perform multiple seriation tasks. Carefully structured activities, such as those that follow, form an important part of the experimental background needed by children to master seriation skills. First three, then four, then five or more objects can be embedded in the seriation tasks proposed. At the same time, through a careful selection of activities, a wide range of tactile, visual, and auditory skills can be reinforced.

Activity 1: Tower Puzzle.

The Tower Puzzle consists of a set of rings (of wood, plastic, rubber) and a base with a peg extending upward from it (Figure 6.25). The object is to have the child sort the rings and then place them on the peg starting with the disk of greatest diameter and working upward with decreasing diameters. This activity can start with two rings, then three rings, and then a larger number of rings.

Activity 2: Nuts And Bolts.

This activity involves giving a child a board with a number of bolts protruding through it and a corresponding number of matching nuts to arrange by size of opening. When the nuts have been arranged by size (Figure 6.26) they can be screwed onto the bolts, as the bolts are on the board in order of decreasing diameters. This gives the child a quick nonverbal response to the correctness of the ordering.

Figure 6.25

Figure 6.26

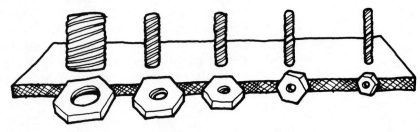

Activity 3: Shoe Size.

The children can each trace the outline of their foot on a portion of a ditto master. Several of these masters can be run off, making sure each child's name is within the outline of the appropriate foot. The outlines can then be used for various size orderings following careful directions. Similar activities can be done with strands of rope, pieces of scrap lumber, and other like materials.

Activity 4: Jars And Lids.

This activity requires two types of sortings. A collection of jars is presented to the children and they are asked to put them in order according to the size of their top openings. (A wide variance should be presented in the other attributes of the jars.) When the child has correctly completed this ordering, the child can then sort the caps from smallest to largest. A final check (Figure 6.28) comes when the matches of lids with jars is attempted. This again provides a quick form of nonverbal feedback that does not require the teacher's attention.

Figure 6.27

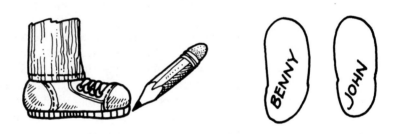

Figure 6.28

Activities such as these help children understand the various relations, their properties, and their applications in common situations. The activities also help children realize that each ordering has a beginning, or origin, and that a specific relationship exists between successive members of the ordered set. In moving to multiple comparisons, the proper verbal form of comparison is the superlative. There is now a heaviest or a lightest in each set of objects compared on the attribute of mass. A similar language development occurs in other multiple comparison seriation situations. The child learns to first focus on the object that begins the comparison, but also to be aware of the direction of the comparison (longest to shortest, heaviest to lightest). While the direction may be fairly obvious from the language, the child needs to be able to *visualize mentally* how the direction determines the ordering. This imagery assists children in applying the properties of the relation to efficiently complete the ordering.

Fred is a first-grade student in a suburban school. His work at the beginning of the year showed a marked aptitude for mathematics, but this initial indication seemed to be off as the year progressed. Fred's teacher was quite concerned, so she administered the *Key-Math* test. His performance on the test gave some ideas of what his problems might be. Overall, he scored a 39 on the test, a score that equates to a grade placement of between 1.1 and 1.2. At the time of the test, he was in the fourth month of the first grade.

On the first scale, Numeration, Fred missed the items in which he had to identify kittens. On the geometry scale, he had difficulty in identifying the basic geometric figures of circle, square, and triangle. Following this initial diagnosis, Fred's teacher followed up with some informal questions using the logical blocks. She was able to determine that Fred knew little about shapes and only a few of the basic colors.

Based on this information, plus that which had caused her initial concern, Fred's teacher developed a series of classification-oriented activities for Fred. It seems that Fred had grown up in an adult-centered environment with elderly parents. He had not had the needed developmental exercises in comparing and classifying a wide variety of objects. As his state did not have mandatory kindergarten, he was far behind his peers in entering mathematical developmental skills which provide the basis for discussing situations concerning operations and the basic number facts.

Following the completion of a series of carefully structured activities like those mentioned on the previous pages, Fred started to make progress consonant with his initial showing. It took over two years for Fred to make up the effects due to his early educational lag in sorting and language skills.

1. Be prepared, as your instructor directs, to answer the major questions on page 140 of this chapter.

2. Think of an ordering situation you have had to carry out during the past week. What attributes did you have to focus on in the completion of your ordering? What degrees of the attributes were present?

3. Which of the following sorting activities would you expect a child to be able to do first, second, and third? Why?
   (a) Order eight jars according to volume
   (b) Order eight sticks according to length
   (c) Order eight rocks according to their mass

4. Answer the same question for the following:
   (a) Order two sticks on the basis of length
   (b) Order ten sticks on the basis of length
   (c) Order ten sticks and ten ropes on the basis of length

5. Review Piaget's analysis of the development of seriation and then analyze a standard textbook's treatment of seriation at the prenumber level. Base your analysis on the following questions:
   (a) How much work is done with seriation at the prenumber level?
   (b) How much does seriation work agree with the analysis Piaget makes of children's development in the ability to seriate?
   (c) What sensory modes does the text emphasize in its activities?
   (d) How extensive is the vocabulary development in relation to single- and multiple-comparison situations?
   (e) How might you improve the activities presented?

6. Develop a seriation task for a visually impaired child that focuses on taste or touch as the main attribute.

7. Develop a seriation task for a hearing-impaired child which focuses on the handling of tactile materials.

8. Develop a seriation activity in game form which can be used by two children working together. Make it a self-checking activity.

9. Develop a seriation task file which includes a number of simple and multiple seriation tasks. Attempt to have activities covering a wide range of attributes.

10. How do seriation activities interact with other prenumber work in a basal mathematics series? Cite specific examples you find in your analysis of a text series.

7 NUMBER AND NUMERATION

EARLY NUMBER CONCEPTS AND MODELS: NUMBERS TO TEN

Teacher Background

As children live and move, they meet numbers: on the clock, on a telephone dial, on the mailbox, on the license plate of the family car, even on the price tag of a toy they'd like to have. Numbers appear in many common daily situations.

How does one help children grow in their understanding and use of numbers? How does one build on early work with objects—classifying, comparing, and ordering them—a more formal study which includes recognizing, naming, writing, and ordering numbers to ten? The first part of this chapter is devoted to answering these questions. The remainder of the chapter discusses ideas for extending children's understandings and skills to two-digit and larger numbers.

One-to-one matching. "First beginnings" of number often focus on likenesses and differences in quantity. Children may be asked to match the objects in two groups, one by one, and describe what they find (Figure 7.1). Sometimes, when one group has as many objects as the other, the matching comes out even. Otherwise a group may be described as having more or less than another.

Figure 7.1

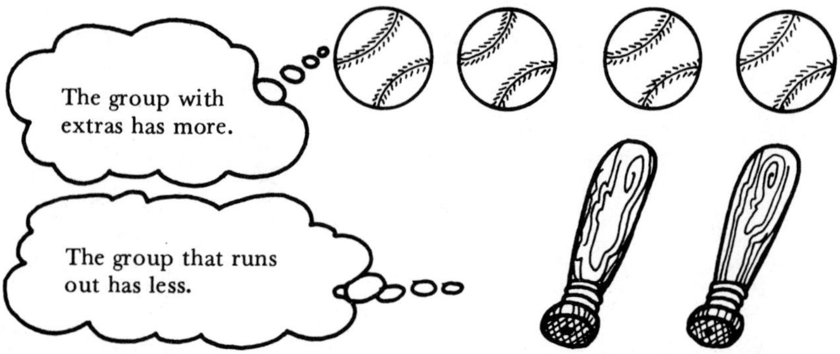
The group with extras has more.
The group that runs out has less.

Occupational development centers and sheltered workshops employing the handicapped often use this matching idea. For example, an employee may be required to pack twenty-four washers to a box. Using a tray with twenty-four spots, the employee puts *one* washer on *each* spot. When all spots are filled, the washers are placed in a box.

Rote counting not enough. Preliminary experiences with one-to-one matching are usually followed by counting activities. Many pre-school and very young primary children can count rotely to 10. That is, they can name the numbers in correct sequence, but may not really understand what those numbers mean. They may be able to dramatize fingerplays and sing number songs but cannot, for example, take five beads from a box upon request. It is not unusual for young children to guess and give the teacher a "bunch." Or they may count rotely as they pick up the objects, saying two or even three number names for each object touched (Figure 7.2). These children do not have the "one number, one object" idea which is a component of rational or meaningful counting.

Conservation needed, too. Even children who can select the correct number of objects for a given number may not have a full understanding of number. An adequate grasp of number means that children also recognize that rearranging a group of objects does not affect its number. In *The Child's Conception of Number* Jean Piaget describes three phases through which children pass. In the first, immature phase, a child believes the number changes when the

Figure 7.2

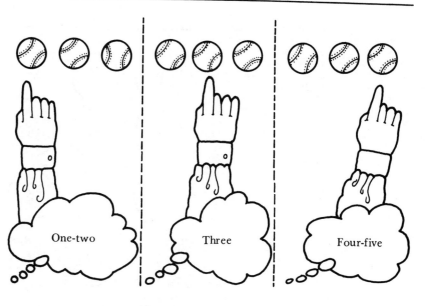

One-two   Three   Four-five

shape or size of an arrangement of objects is changed. Some children describing situations like that of Figure 7.3b will say that there are three chips in each row, but also firmly state that there are more chips in the second row!

A child in the second phase knows the quantity should be the same in each row but, since the length is much different, feels the number has changed. When conservation of number is achieved at about age seven, a child knows the number is the same after as before the transformation, no matter how changed the set may appear. Children in this final phase can reverse their thinking. They realize that the number doesn't change when objects are moved because they can be returned to their original places again. They can show you, by one-to-one matching, that there are just as many in one row as in the other. They are now at a stage where counting is truly meaningful or rational.

The growth which leads to conservation of number is a dual product of both maturation and experience. In the classroom it is best nurtured not only by opportunities that allow children to rearrange and rearrange, to count and recount, but also by experiences that encourage children to check or prove their thinking by matching objects, one to one.

Behind it all: ordinality and cardinality. The mathematical ideas of ordinality and cardinality are embedded in the early development of numbers to ten. Ordinality refers to the relative position or order of an object within a set with respect to the others—first, second, and so on. Cardinality answers the question "How many?" in reference to the total number of objects in a group.

While counting, children make an ordinal use of number. They say *one* for the first object, *two* for the second, and so on (Figure

Figure 7.3

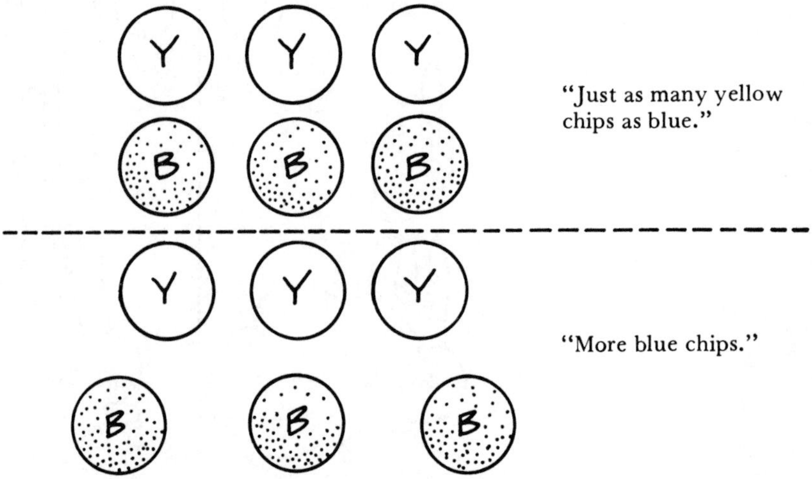

"Just as many yellow chips as blue."

"More blue chips."

7.4). Moving each object as it is counted helps children to count objects only once and not to miss any in the counting. When the counting is complete and children can tell that, in all, there are three balls, they are giving the cardinal number of the group (three). It is important that children realize that the number 3 is associated with all three objects, not with just the last one counted.

Sequencing numbers to ten. Earlier work comparing and ordering quantities is gradually extended as part of this early work with number. Children learn that five, which is more than two, comes after two in the counting sequence. These ideas help in completing short number sequences and in ordering all numbers, zero to ten.

About zero. Developmental work for zero is inserted within the 1 to 10 sequence rather than first. Some textbooks present it after "two" or "four." Others wait until it's needed, for use with "10." The meaning of zero as *none at all* is easier for children to grasp when they can relate it to known quantities. For example: "Four cookies on the plate. Mother, Dad, baby, and I each eat one. Now they are all gone. There are none left. The plate is empty."

A three-phase development. Instruction for numbers through ten emphasizes ideas outlined above and moves through three major phases in elementary school classrooms. These three phases are (1) numbers to 4, (2) early sequencing tasks, and (3) numbers 5 to 10 and are presented below. Samples activities for each phase are also described.

Instruction within each of these phases begins with objects, counting them or physically comparing the number in given groups

Figure 7.4

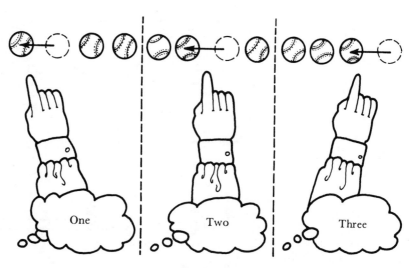

(Figure 7.5). Since children's thinking is typically concrete, real meaning will come from their experiences with physical models of a number.

Instruction next moves to using symbols to describe what is seen (Figure 7.6). Children now learn to recognize and write numerals meaningfully. Eventually most children also learn the meaning of the greater than (>) and less than (<) symbols, and to use these when needed.

Finally, children are able to name and use numeral or symbols for comparison, even when objects are no longer present (Figure 7.7). Taken together, these relationships form a learning triangle which guides instruction and the evaluation of learning during early number work (Figure 7.8). The flow of instruction from work with objects to written work without counters is illustrated in the sample activities described for each phase below.

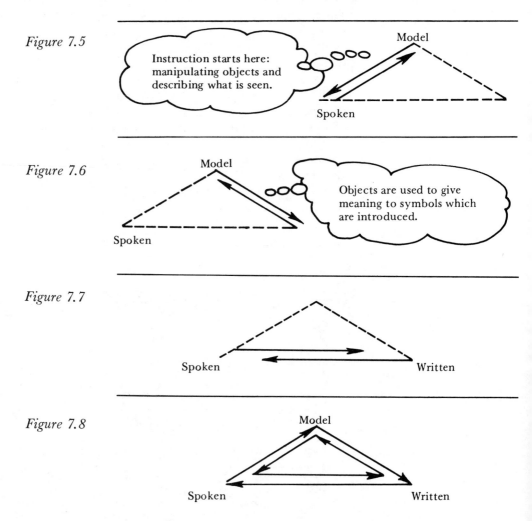

Figure 7.5

Figure 7.6

Figure 7.7

Figure 7.8

DEVELOPING BASIC NUMBER MEANINGS AND SKILLS WITH SPECIAL NEEDS STUDENTS

What Do *You* Say?

1. When is a child ready to begin a formal study of number?

2. How does rational counting differ from rote counting?

3. How can a teacher determine whether a child is conserving on number?

4. When and how should zero be introduced?

5. At what point during early number work should the writing of numerals be introduced? What is the role and nature of gross motor activities in this instruction?

Basic Number Meanings and Skills: Phase 1

Numbers to Four: Sight Groups

Because there are so many examples of "two" in the child's own body—two eyes, two ears, two hands—some textbooks start the early number sequence with this number. Children then add to or take from sets of two to form groups having one, three, four, or even no objects.

Throughout this sequence children will at first need to count from one to determine the number of objects in a group. A major goal for sighted children, however, is that they learn to quickly recognize, *without counting,* the number of objects in groups having four or less objects. Then as the other numbers through ten are studied, children can be encouraged to sight subgroups they recognize and to count on from there (Figure 7.9). This counting on idea is an important one which will recur as a strategy for finding addition sums (Figure 7.10), and for making change (Figure 7.11).

Figure 7.9

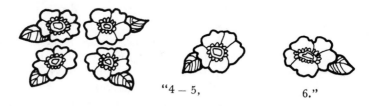

"4 — 5,   6."

Figure 7.10

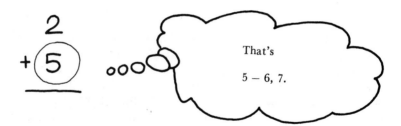

That's

5 — 6, 7.

While working with numbers to four, teachers will want to encourage sight recognition of groups whenever possible. It is a skill with big payoffs. Opportunities to reinforce this skill recur in the sample activities which follow.

Figure 7.11

"13,   14,   15,   25¢.

Paid 25¢

**Activity File:
Numbers 1 to 4**

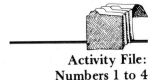

Sample Activities: Numbers to Four
Many children can relate to situations like *one* candle on a cake (baby is one year old); *one* nose, *one* mouth; *two* feet; *three* wheels on a tricycle; *four* legs on a chair or dog. Whenever possible in developmental work, teachers should relate number to its occurrence in a child's surroundings. The activity, Two It Is, drawn from early number work with an autistic child, illustrates this idea. The activity is based on the fact that many children, including autistic and hypoactive children, need to be physically drawn into a situation in order to learn from it.

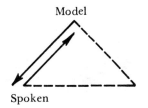

Model

Spoken

Activity 1: Two It Is.
   Need: Two mittens taped to a mirror.
   Procedure: Child is asked to put one hand on each mitten, then tell how many hands were placed. If necessary, teacher can count with the child: "One, two. *Two* hands."
   Follow-up: Find other "twos" in the mirror: two eyes, two cheeks, two ears, two elbows, and so on. Play Simon Says, having children dramatize understanding of "two" by body movements such as clapping or tapping two times.

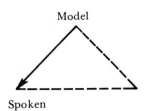

Model

Spoken

Activity 2: Move And Count. (Tactile counting)
   Need: One to four beads, blocks, or other counters.
   Procedure: Children move each object to another part of the table or desk as it is counted. This way they are less likely to count objects twice or skip an object while counting.
   Note: If a child is visually impaired, objects may be placed in commercial or homemade work-play trays. Work-play trays (in two

sizes, 22½" x 13¼" x 1¼" and 18" x 12" x 1¼", wood, with rubber nailheads fastened to the bottom to prevent scarring of desks and to eliminate sliding) are available from the American Printing House for the Blind, 1839 Frankfort Avenue, Louisville, Kentucky 40206. Cookie trays with edges may also be used.

Visually impaired children should be encouraged, even taught, to use whatever residual vision is available. Children with residual color perception, for example, may respond to distinct color differences of objects. Others having even partial light or object perception may be able to discriminate objects if they take them close to their face or to a stronger light source, such as the writing surface of a lighted overhead or a nearby window.

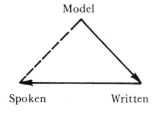

Activity 3: Numeral Recognition.

Need: Large textured numerals. (Use sandpaper or felt numerals; or those drawn with a broad felt-tipped pen and retraced with glue. The dried glue will give a 3-D effect to the numerals.)

Procedure: Child is asked to place objects (to four) on a table or desk. Teacher introduces the matching numeral by asking the child to finger-trace its shape and say its name. From the beginning, correct formation is stressed. Verbal cues are used to help the child remember its shape.

Examples: The 1 is tall and straight. The 2 is a candy cane with a stick on the bottom. The 3 goes round and round like a busy bee. The 4 is all sticks: down, over, down.

Children should be actively involved in the mathematics they are learning. However, hyperactive children, who are characterized by unending motion, require carefully structured activities in order to learn. The following two activities have proved helpful for such cases. The focus is meaningfully interpreting written numerals.

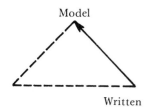

Activity 4: Can It.

Need: Can covered with self-adhesive or construction paper, a numeral written on each; counters; number book for checking work (Figure 7.12).

Procedure: Children place counters in each can to match the numeral on the can. When all cans are filled, they use the number book to check. For example, each counter in the "4" can would be placed over a dot on the "4" page of the number book. There should be just enough with no leftovers.

Children often need to locate numerals from oral directions: finding page numbers; doing selected problems of an assignment; setting an alarm clock upon request; going to a specified room in the building; and so on. The next activity, Mailman, reinforces this skill.

Activity 5: Mailman. (Auditory numeral recognition)

Procedure: Children are given "letters" (unmarked envelopes)

and are orally instructed to deliver them to specific mailboxes. Upon delivery the child can open each envelope and check if the numeral card inside matches that on the mailbox (Figure 7.13).

Activity 6: Logging Along. (Visual numeral recognition)
Note: This activity is particularly good for strengthening eye-hand coordination.
Need: Number logs, as shown in Figure 7.14. They are made from two milk cartons attached end to end and covered with brown paper. A numeral card is glued to each of the four sides of the log.
Procedure: Children take turns trying to catch the number log thrown them. If they can say the numeral that lands "up" on the log, they walk that many steps toward a finish line.

Activity 7: Look Fast. (Sight group recognition)
Need: Picture cards showing one to four objects.
Procedure: Children tell how many they see on each card as the cards are flashed. If necessary, a card may be recalled so a child can count from one to check.

Zero

Conceptually zero is introduced as *none* or *not any at all.* This idea is best taught by contrasting to quantities children have already studied.

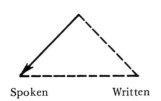

Spoken   Written

Spoken   Written

Figure 7.12

Figure 7.13

Figure 7.14

**Activity File:
Zero Activities**

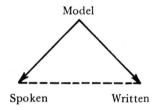

Model

Spoken   Written

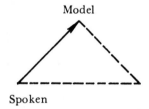

Model

Spoken

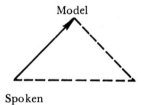

Model

Spoken

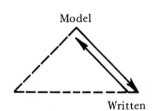

Model

Written

Activity 1: All Gone. (Introductory activity)

Give each child two M and M's. Ask them "How many?" (Two.) "Eat one. How many now?" (One.) "Now eat the last one. How many are left?" (None—not any at all.) Show the children a textured numeral card for zero, and explain that it tells the number of pieces of candy that are left: "Zero. That's the number name for none." Have them repeat *zero* after you as they trace the numeral with a finger.

Activity 2: Fingerplays.

Note: Fingerplay chants often dramatize the zero idea by getting children to count backwards from known quantities.

Example:

Five balloons—see how they soar (hold arm high, five fingers extended). Pop went one! Then there were four.

Four balloons (four fingers high)—way up in the tree. Pop went one! Then there were three.

Three balloons (three fingers high)—one for you, you, you (child points). Pop went another! Then there were two.

Two balloons (two fingers high)—taking a run. Pop went one! Then there was one.

One lonely balloon (one finger high). That's no fun. Pop it went! Then there were none.

None—none—none—that's ZERO! (hold thumb and forefinger together to form numeral "0").

Activity 3: Look Around You.

Children look around the room to answer questions like the following: "How many tigers do you see?" (None.) Children can select and hold up the numeral card (from those on their desks) that means none at all. ("0") Trains, ghosts or other real or make-believe objects that children enjoy can also be used for the activity.

Activity 4: Match. (Mixed review)

Need: Numeral and matching picture cards for numbers zero to four.

Procedure: Children form matches by pairing a numeral card with a picture card illustrating its number value.

Writing Numerals

Writing is begun after children have learned to recognize a numeral and associate it with the correct number of objects. Even when the focus is on writing, however, continued reference to the quantities named is made. The following activities, which move from gross to fine motor involvement, make suggestions along these lines. From the beginning, correct formation is stressed. The verbal cues of Activity 3, page 155, would be used as needed.

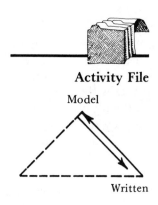

Activity File

Model

Written

Sample Activities: Writing Numerals

Gross Motor Activities

Activity 1: Shadow Play. Place a prepared transparency of a numeral on the overhead so children can trace its image on the wall. After a numeral is traced, children clap as many times as is necessary to pantomime its meaning.

Activity 2: Chalk Talk. Write large numerals on the board and let children use a sponge or their fingers to trace them. Children can draw chalk prints of objects beside each to illustrate the number meaning.

Activity 3: Finger Play. Children use sand trays or finger paint to sketch the numerals. Wheelchair pupils can do this activity on lapboards.

Sample Transition Activities

Activity 1: Stencil In. Children use wood stencils as a guide for forming the numerals.
   Note: This technique is often necessary for children with severely impaired or underdeveloped motor skills. Stencil boards such as those illustrated in Figure 7.15 can be used. They are made from inexpensive wood scraps. Each board should provide a stencil for the numeral and removable blocks on small dowels to represent the quantity. Children are given a board, a box of blocks, paper, and a felt-tip pen. They place blocks on the board to illustrate the numeral and then use the stencil part of the board to write the numeral. Later they can try to write the numeral without the stencil, using the stenciled numeral as a pattern. Tagboard stencils can also be used.

Figure 7.15

Activity 2: Marks-A-Lot. Children use grease pencils to trace over or finish numerals on laminated boards like that shown in Figure 7.16. A green dot shows where to start. When finished, children could cover each numeral square with counters to show how many are represented by each numeral.

Activity 3: Feel And Say. Children place manila paper over window screen and write numerals with a crayon (Press hard!). They then find groups of objects in the classroom which illustrate the number value of each textured numeral they write.

Sample Fine Motor Activities

Activity 1: Trace 'N Write. Prepare a sheet similar to that shown in Figure 7.17. Children trace the sample numeral with a finger and then draw over it with a primary pencil. Using this as a pattern, they finish the row with 4's. Children then color the set of four balls.

Activity 2: Count Down. Children count the dots and write the number counted at the bottom of the card (Figure 7.18). Later,

Figure 7.16

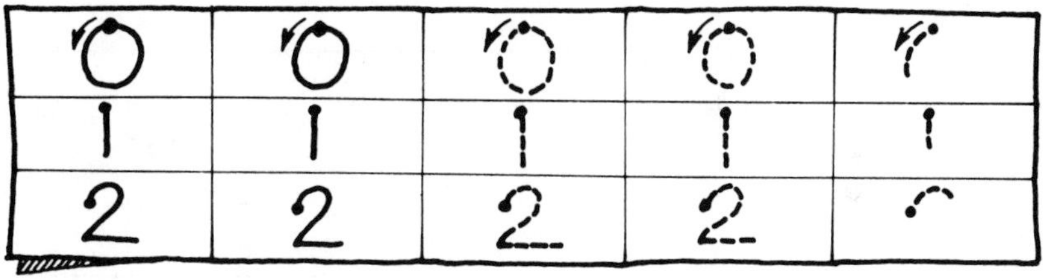

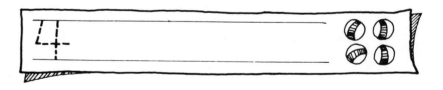

Figure 7.17

Figure 7.18

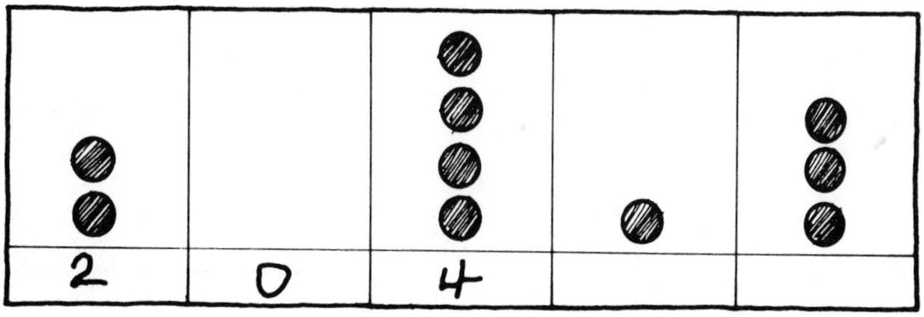

children can cut the columns apart and arrange them in counting sequence.

Note: For special cases, textured dots can be used.

Discussion

The preceding activities suggest ideas for developing basic concepts and skills for numbers to four. Instruction is guided by three major goals:

1. Hearing a number spoken, children will be able to use counters to represent that number and write the corresponding numeral (Figure 7.19).

2. Shown a cluster of four or less objects, children will be able orally to name the number represented and write the corresponding numeral. Ideally, they would recognize the number in the cluster quickly, without counting from one (Figure 7.20).

3. Shown a numeral, children will be able to correctly name it and use counters to illustrate its number value (Figure 7.21).

Children who can successfully perform each of these tasks demonstrate their grasp of basic number meanings and skills for numbers to four.

Figure 7.19

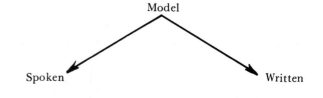

Figure 7.20

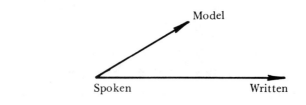

Figure 7.21

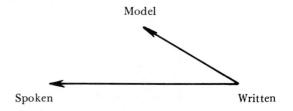

Basic Number Meanings and Skills: Phase 2

> **What Do *You* Say?**
>
> 1. When, in the teaching sequence for numbers to ten, should number seriation tasks be introduced?
>
> 2. How can the ideas of *more* and *less* be gradually extended and related to completing number sequences?
>
> 3. How can the ideas of *more* and *less* be extended to identifying and writing numbers before, after, or between other given numbers?

When children can demonstrate their understanding of several numbers, such as those to four or five, instruction can focus on comparing and ordering those numbers. Earlier work in which children used one-to-one matching to tell whether a group had more, less, or as many objects as another is extended as the basis for this instruction. After representing two numbers with objects, children match one-to-one and compare. At first numbers are chosen so that obvious perceptual differences in quantity cue the comparison. For example, four objects are rather clearly recognized as more than one or two.

Activity File: More, One More, Number After

The following sample activities outline a sequence for helping children compare and order numbers. The activities described focus on finding the number that is *more* or *one more* than another. Some of the activities are designed to help children realize that when a number means more, it comes after another when you count. Once these ideas are grasped, a similar sequence for *less* and *number before* would be introduced.

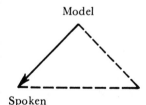

Model

Spoken

Activity 1: Stack And See. (Comparing numbers, obvious perceptual differences)

Child is given a "4" and a "2" card. "Stack four red blocks on the red paper. Stack two blue blocks on the blue paper. Use your cards to label each stack. Which number means more? How do you know?"

Note: Children can use one-to-one matching as necessary to find or "prove" answers. The group with extras has more. Followup worksheets on which children circle the greater number reinforce activities such as this. Lines can be drawn to match pictures one-to-one.

It is not necessary that all children learn the greater than (>) and less than (<) symbols. Some children, such as those in a nonalgebra track, may never need them. If presented at this time, they may be likened to an alligator's mouth—always open to grab the greatest number treat.

Example: $4 > 2$: Four is greater than two.

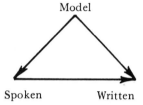

Activity 2: Step And Count. (*More* means *after* in the counting sequence)

Let one child be "4" and another "2." Each takes a dot card to represent the number. Both start at the GO or "0" mark on a large walk-on number line (Figure 7.22). Begin counting slowly to four. The two children step along the number line to keep up with the count, stopping on the numeral that matches their cards.

Follow-up discussion: "Yes, 4 took more steps, went further than 2. 4 means more, so it comes after 2 when you count."

Note: Followup worksheets should picture objects to aid number comparison before presenting exercises with numbers alone (Figure 7.23).

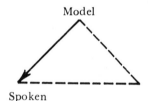

Activity Sequence: One More. (Comes right after when you count)

Note how work with objects and discussion precede paper and pencil exercises in the following sequence.

Activity A. Build.

Have children put down one block, two blocks, three blocks and four blocks as shown in Figure 7.24. Discuss how each time they added one more, the structure became bigger.

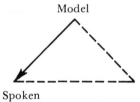

Activity B. Match Up.

Hold up four fingers on one hand and three on the other. Children tell you which hand has more fingers up. "How many more?" Show that by matching up the fingers one-to-one, they can check and *see* the one left over.

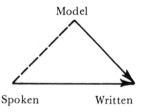

Activity C. Kerplunk!

Children drop a certain number of washers into a pie pan and write the number that tells how many they dropped. Then they drop one more washer into the pan and write the numeral to tell the number of washers now. Encourage children to count on in the number sequence from the first number, rather than starting from one. Prompt if necessary. Then count the washers to check. Help children realize that when a number means one more than another, it comes right after it when you count.

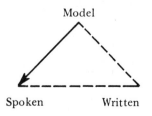

Activity D. Bead Tally.

Prepare sets of beads strung on wire, pipe cleaners, or string, so that all but one bead, the last, is the same color. Use masking tape to label each string. (Use alphabet letters if the children know them or code them with colors or simple line drawings.) The bead sets

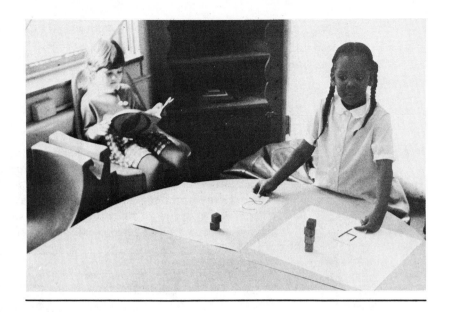

Figure 7.22

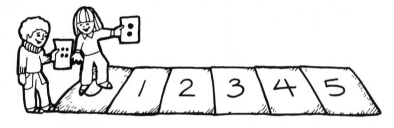

Figure 7.23

Picture
model

Written

Example
Exercises

Tell how many. Then circle the
number that means more.

Figure 7.24

can be used during class sessions to emphasize, for example, that four is one more than three. Four comes after three when you count.

Follow-up Worksheet (Figure 7.25). (Requires the use of bead strings)

Model

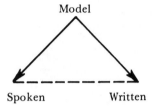

Spoken   Written

Activity E. You Tell It.
Prepare a deck of large numeral cards (textured for visually impaired readers). Lay out a walk-on number line on the floor with masking tape. Ask children to study it so they can remember where the numbers are. Then, turning away from the number line, each child draws a card from the deck and tells which number is one more than that on the card. To check, children find the first

Figure 7.25

Figure 7.26

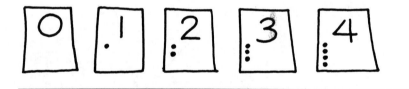

Figure 7.27

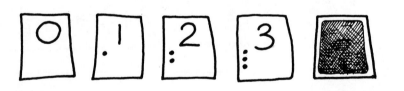

Figure 7.28

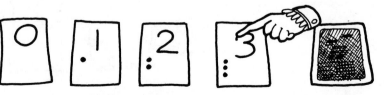

numeral on the number line and show how just one more move locates the second. Then count from one to stress that when a number is one more than another, it comes right after it.

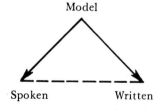

Model

Spoken   Written

Activity F. Peek! (To help children name the number after that given, without counting from one each time)

Lay out numeral cards (Figure 7.26). Let the children see you turn one card over (Figure 7.27). Point to the card preceding it, in this case, "3" (Figure 7.28). "What's this numeral?" (Three.) "What comes after three?" (Four.) Turn card over to check.

Variation A: As above, but children do not watch as you turn over the card.

Variation B: Use a ruler and tagboard arrow card. Cover numerals to the right of the given numeral. Children say (guess) the number after, then peek to check.

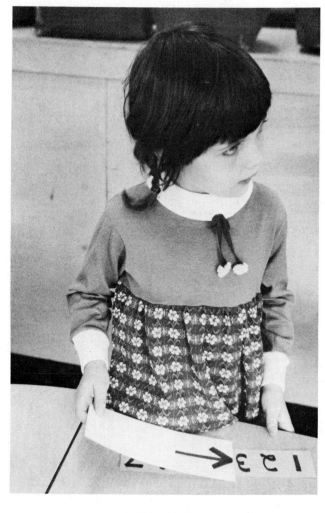

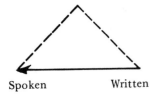

Spoken   Written

Spoken   Written

Model

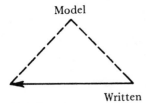

Written

Follow-up game: You Name It.

Provide a deck of numeral cards for numbers children have studied. In turn, children turn over a card and tell the number after. A ruler can be used, if necessary, to check. If correct, card is kept. Otherwise card goes to discard pile for reuse. Winner: Child with most cards when deck is used up.

Variation: Children play a prerecorded tape which gives a number, pauses, then tells the number after. For each number given children, in turn, try to *say the number after* before it is played on the tape. Children make a tally mark each time they "beat the tape." Winner: Child with most tallies when tape has run out.

Activity G. Washday. (Ordering all numbers, one to four)

String a rope between two chairs and give children clothespins and paper shirts on hangers (Figure 7.29). Children pin clothespins to the top of each hanger to match the number on the shirt. They then place the hangers on the clothesline in order, so that the numeral on each shirt means one more than that to the left of it on the line. If necessary, children can match the clothespins for each one-to-one, or count from 1 to help them decide.

Activity H. War.

At each turn, the child who turns over the card that means more wins both cards. Introduce the variation that when a child turns over a card that is *one* more than the opponent's, two cards (the card the opponent is playing and that next in line) are taken. Show children how to use markers and one-to-one matching, or even a ruler, to help them if they are not sure which means more.

Figure 7.29

Figure 7.30

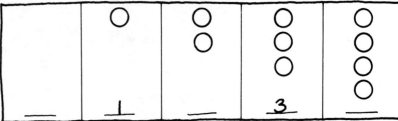

Ideas for "less."

Similar activities for *less* and for helping children recognize that a number one less than another comes right before it in the counting sequence are introduced next. Then children would be ready to complete short number sequences, filling in numbers before, after, or between those given (see Figure 7.30). Eventually picture cues are dropped, but children should know how to use objects or a ruler to help themselves out when necessary.

Basic Number Meanings and Skills: Phase 3

Numbers Five to Ten: Illustrating, Reading, Writing, and Ordering Them

---

What Do *You* Say?

1. How does developmental work for numbers to four differ from the development of other numbers through ten?

2. How are pre-addition and pre-subtraction activities related to early number work?

3. When and how should number words for numbers to ten be introduced?

4. How can instruction for money be correlated with early number work?

The numbers five to ten are sequentially developed through activities similar to those presented for the numbers to four. Number meanings, numeral recognition and writing, and number sequence skills are gradually extended as each new number is studied. The only new emphasis is *counting on* from sight groups to determine the number for groups having five or more objects (Figure 7.31). Sometimes children are asked to enlarge a set to show a certain number. Counting on should be encouraged in this process (Figure 7.32).

Note: If the child counts from one rather than counting on, cover the original group of three bottle caps and say: "Three here.

Figure 7.31

How many rocks?

How many in all?" Prompt, if necessary, using a numeral card as in Figure 7.33: "3—4, 5, 6." Uncover the group of three and count from one to check. Repeat with other numbers of caps.

Activities of this type play a dual role in the primary program. Their major focus is the development of basic number meanings and skills. Since they involve the *parts* which make up a *whole*, however, they also may be considered pre-addition and pre-subtraction. Counting-on skills encouraged in these activities pay off later when children count on from the greater of two addends to find a sum.

Figure 7.32

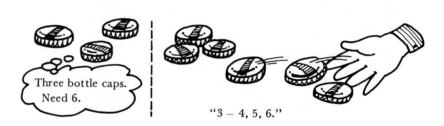

Figure 7.33

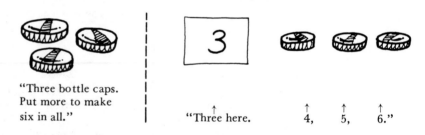

**Activity File:
Counting On 5 to 10**

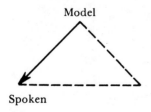

Sample Activities: Counting On in Number Development for Numbers Five to Ten

Activity 1: One More. (To introduce the next number in a sequence)

Example: (Johnny knows numbers through six) "Johnny, would you take six red chips from the box and place them on the table for me? . . . Now put just one more chip on the table." (While the child is doing this, group several objects to form a sight group that the child should recognize.) "How many chips now? Let's count." Encourage him to count on from the sight group.

Activity 2: Two of a Kind. (Reproducing the number of a given set; followup to Activity 1)

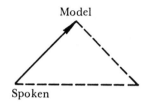

Model

Spoken

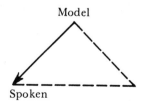

Model

Spoken

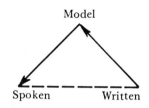

Model

Spoken   Written

"We have seven red chips here, Johnny. Can you make another group of seven using blue chips from the box?" Observe the child's counting. If he has any difficulty, be sure he places each object on the table as he counts it. It may be necessary to recount them when they are all on the table. "Are there as many red chips as blue chips?" The child should use one-to-one matching to prove that there are.

Activity 3: Tell How Many. (Identifying the number of objects in a given set; followup to Activity 2)

"Look at the rocks in this pile. How many are there?" (Seven.) Encourage the child to count on from a sight group, if possible, since this is a more efficient way of counting larger numbers of objects.

Activity 4: Clip It.

Children place paper clips, or clothespins, along the taped edges of a piece of tagboard to illustrate the numeral on the card (Figure 7.34).

Sample follow-up dialogue (Pre-addition in nature, emphasizes counting on) "Cathy, did you make 8? Two on one side, six on the other. Let's count: 2–3, 4, 5, 6, 7, 8. Yes, that's eight. Billy made his card with three on one side, five on the other. Is that eight? Yes: 3–4, 5, 6, 7, 8. Billy put eight clips on his card, too. There are many ways to make eight. Who did it a different way?"

Building on basic concepts for numbers 5 to 10. For each number introduced, the focus is first on identifying and counting out the correct number of objects. Recognizing and writing numerals comes later. Writing activities outlined earlier in this chapter, as well as suggestions for comparing and ordering numbers, can be adapted to extending children's skills for numbers five through ten.

Learning the number words. Any child who will write a check must learn to both read and write the number words *zero* through *ten*. These are usually introduced along with each number studied.

Figure 7.34   Cathy's card   Billy's card

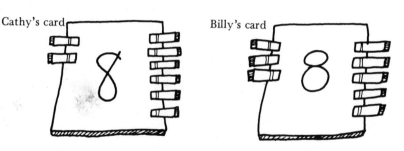

Reading techniques, including the use of phonetic cues, are generally used to help children master the words. Occasionally, for special cases, their introduction is delayed until reading skills are better developed, or they are presented as sight words.

Money and early number work. It is both natural and practical to correlate the development of basic money concepts and skills with early number work. When children understand what three means, for example, they can count out three pennies as one model for the number. After number meanings to five have been developed, a follow-up might be to teach the equivalence of five pennies and one nickel. Similarly, when children have had adequate time to develop number meanings to ten, the equivalence of ten pennies with a dime or two nickels should be introduced. Throughout, simple buying and change-making activities could be used. The idea of comparison shopping might also be introduced: "Jake's store sells the balloon for 8 cents. Al's store sells it for 6 cents. I'll buy at Al's. It costs less." Carefully structured, money activities can reinforce number meanings and skills, and help numbers to come alive in the child's world.

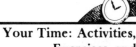

Your Time: Activities, Exercises, and Investigations

1. Be prepared to answer, as your instructor directs, the major questions on pages 153, 161, and 167 of this chapter.

2. For each of the four following situations, identify the activity, a *or* b, that normally would come first in developmental work with children.

   Situation I. Jack is given three toy cars.
   (a) "How many cars do you have?"
   (b) "Point to the numeral that shows how many cars you have."

   Situation II. The teacher points to the numeral 4.
   (a) "Jan, what number comes after this when you count?"
   (b) "Jan, draw this many balls on your paper."

   Situation III. The teacher says:
   (a) "Terry, write the numeral 2."
   (b) "Terry, would you put two pencils on my desk?"

   Situation IV. Numeral cards for 3, 5, and 8 are displayed.
   (a) Nancy is asked to match a card with three objects to the proper numeral card.
   (b) "Nancy, if each card tells the number of pennies glued to the back, which card has the most?"

3. Tom's teacher gave him a bag containing twenty-five bottle caps. Following directions, Tom made as many piles of three caps as he could: He made eight piles and had one cap left over. Tom's teacher helped him make the following record of his count:

Three's	Leftovers
8	1

Discuss the relevance of this type of activity both for reinforcing early number meanings and for subsequent numeration work (grouping by tens and ones). If possible, try this activity with a child. Observe whether the child has any difficulty perceiving a *group* of three objects as *one* entity.

For numbers 4–6, identify any special handicaps or characteristics which describe the child for which your activities are written.

4. The child's knowledge of *three* can be used to introduce the number *four*. Describe a brief activity that does this.

5. (a) Educationally handicapped children usually require instruction that is carefully sequenced, in small steps, uncluttered, with provision for extra practice and frequent review. For each of the three phases presented in this chapter, create an activity of your own which focuses on one of these requirements. The math lab activities of Appendix C may give you ideas for developing your activity.

   (b) For each activity, identify the relationship of the learning triangle you are emphasizing (see Figure 7.35).

6. Outline a sequence of related activities which helps a child recognize that when a number is one less than another it comes right before it in the counting sequence. Your sequence should use models in early activities. By the end of your sequence the child should be able to write the number that comes before any you give (to six). Assume that the child knows basic number meanings through six and can sequence numbers through five.

7. Using the learning triangle as a guide, construct a diagnostic inventory to use in assessing a child's understandings and skills related to the numbers one to ten. If possible, field test your inventory with a young child.

Figure 7.35

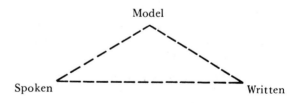

NUMERATION
EXPERIENCES

**Whole Number
Numeration:
Phase 1**

Teacher Background

Numeration means *naming numbers* using the rules of an established convention or system. Numeration systems have evolved from the need to record numbers and to communicate these numbers to others. Our own numeration system is a decimal or base-ten system and involves a "ten for one" grouping idea. This grouping enables us to express both very large and very small quantities (Figure 7.36). The *value* expressed by any numeral is determined by the *place* each digit occupies within the numeral. Hence ours is a *place value* system of numeration.

Whole number and decimal computations are based on numeration concepts. Many computational errors made by children can be traced to their poor understanding of important numeration ideas such as place value (e.g., Cox, 1975). Hence, in order to assure as many successful experiences as possible, a careful development of numeration concepts and related skills is essential. The following section focuses on this development for whole numbers. Extending numeration concepts and skills to decimals is the topic of Chapter 1

Three instructional phases characterize early work in whole number numeration: (1) Naming and writing two-digit numbers; (2) sequencing two-digit numbers; and (3) number meanings and skills beyond 100. We will discuss these phases and include activities for two special numeration topics, renaming numbers and rounding whole numbers.

There is a general pattern to instruction for each of the three phases just mentioned. *In each case instruction starts with objects as the basis for meaningfully naming and writing symbols* (Figure 7.37). When the goal is understanding and using two-digit numbers, for example, children are first asked to group objects by tens and then describe their grouping (refer to Rathmell and Payne, 1975). Children might also be given sticks and asked to lay out a certain number of tens and ones.

Learning the oral number names for multiples of tens (to 90) comes next. Then, when children learn to write the number of tens and ones, they can meaningfully describe the grouping in the standard way (Figure 7.38).

Figure 7.36

Figure 7.37

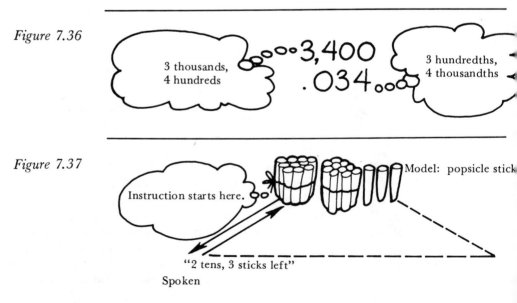

3 thousands, 4 hundreds

3,400
.034

3 hundredths, 4 thousandths

Instruction starts here.

Model: popsicle stick

"2 tens, 3 sticks left"

Spoken

Eventually the numeral itself is written in standard form: "23." Instruction continues until:

1. Hearing a number spoken, the child can use a model to show it and correctly write it (Figure 7.39).

2. Shown a model representation of a number, the child can both say and write the number in standard form (Figure 7.40).

3. Seeing a number written, the child can read the number and illustrate it with a model (Figure 7.41).

These six aspects to instruction are found in each of the major phases outlined below.

Figure 7.38

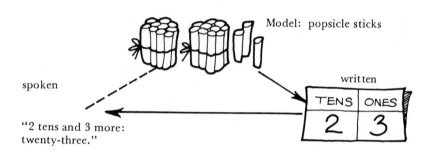

Figure 7.39

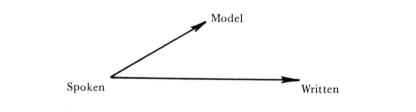

Figure 7.40

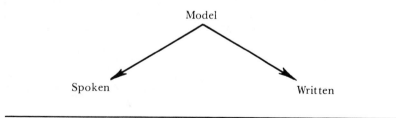

Figure 7.41

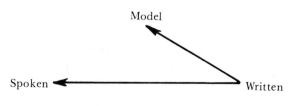

Grouping aids first

Throughout the three phases, special models are used in early stages to give real meaning to spoken or written numbers. The materials used may be called grouping aids. These are materials in which the 10 ones (10 tens, 10 hundreds) are still visible, even when grouped. Simple materials: rocks, chips, or bottle caps that can be placed in piles or in clear plastic bags; popsicle sticks that can be bundled together with a rubber band work well for illustrating two-digit numbers. Other examples of grouping materials for two-digit number activities are shown in Figure 7.42.

Unifix cubes, made of brightly colored plastic, can be interlocked easily or pulled apart by the children themselves. A child with coordination problems may find it easier to make trains of ten with these cubes than to wrap a rubber band around ten single sticks. Preglued beans or chips, counting frames, or even graph paper serve a similar purpose.

Homemade bean sticks and graph paper can also be used to represent larger numbers (see Figure 7.43). Bean stick "hundreds" can be made by gluing extra tongue depressors to the back of 10 "tens" to form a flat. Graph paper squares, 10 cm x 10 cm, can also be used to represent a hundred. Ten "hundred" squares, stapled together, form a thousand.

Certain materials, particularly place-value aids such as the spike abacus (Figure 7.44), should not be used with children until the "ten for one" grouping idea is firmly grasped. The abacus, commercially available from a number of school supply houses, is found in many primary classrooms. We feel that the abacus, if used at all,

Figure 7.42 **Grouping aids for two-digit numbers**

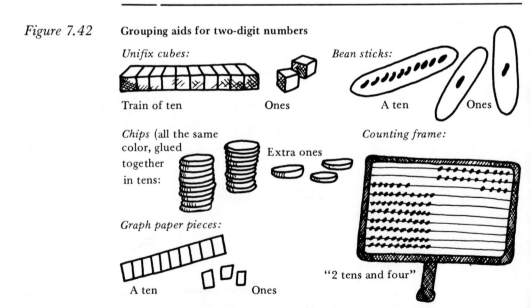

Unifix cubes:

Train of ten   Ones

Bean sticks:

A ten   Ones

Chips (all the same color, glued together in tens:

Extra ones

Counting frame:

"2 tens and four"

Graph paper pieces:

A ten   Ones

should be introduced only *after* four-digit number meanings have been developed. Some school textbooks use the abacus to illustrate larger numbers.

Younger children react strongly to perceptual differences. Their perception of size often overrides their feelings for value. The young child who chooses the nickel rather than the dime (It's bigger!) is also likely to think of the Figure 7.44 display as "seven" rather than "thirty-four." Or, children may give lip service to the fact that "3 tens 4 ones" are shown, but not really understand what they are saying. When it comes to *using* the tens-ones (place-value) idea in subtraction problems like that in Figure 7.45, they often fail because they lack numeration understandings. They have no meaningful basis for renaming 3 tens 4 ones as 2 tens 14 ones in order to correctly complete the computation. *Numeration activities and materials which emphasize grouping in early stages play an important role for success in computation and other numeration-related topics such as sequencing, estimation, and rounding.*

Figure 7.43

Grouping aids for three- and four-digit numbers

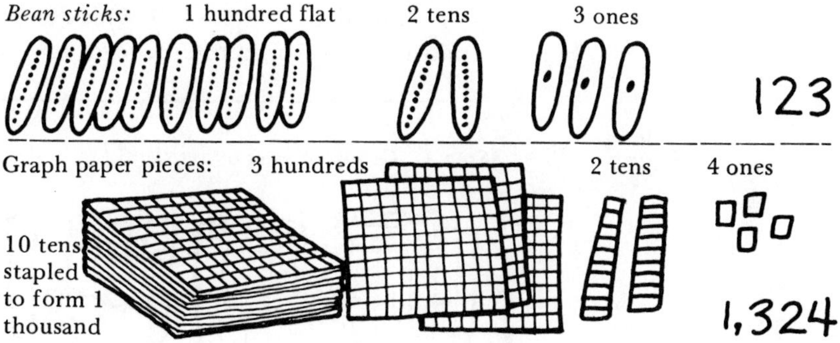

Figure 7.44

Spike abacus

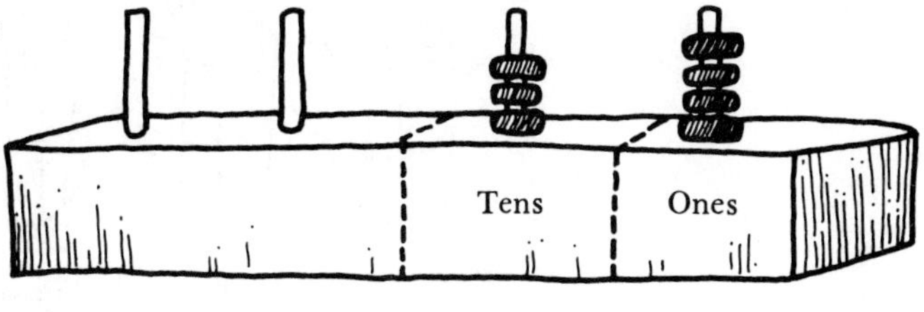

Figure 7.45

DEVELOPING
NUMERATION
CONCEPTS AND
SKILLS WITH
SPECIAL NEEDS
STUDENTS

What Do *You* Say?

1. The ideas of grouping (by tens) and place value are basic concepts of our numeration system and are focused upon in early grade instruction. In what ways are grouping by tens and place value distinct? What is the role of each in developing meaning for whole numbers greater than ten?

2. What is the critical element in the choice of numeration materials for illustrating two-, three-, and four-digit numbers?

3. What instructional sequence is basic to the development of two- and three-digit numbers?

4. How are number sequence skills developed for two- and higher-digit numbers?

5. How do regrouping or renaming experiences prepare children for computational work?

6. How does a child's ability to sequence and round numbers fit into the general goal of emphasizing mathematics for daily living with special needs students?

Naming and Writing Two-Digit Numbers

Rathmell and Payne (1975) lay out three basic steps by which children learn to represent, read, and write two-digit numbers (see

Figure 7.46

Step 1: Group objects by tens and orally describe the grouping.

Three tens, two ones left over.

Step 2: Learn oral number names for two-digit numbers.

Three tens, two ones: thirty-two

Step 3: Group objects by tens and write to describe the grouping.

TENS ONES 3 2 → 32

Figure 7.46). The three-step sequence builds on children's understanding of numbers through ten. At each step in the sequence, grouping aids are used to give meaning to the oral or written description of number. In this way an understanding of place value is gradually developed.

Because of the naming irregularities of the teens, it is suggested that these be introduced last, *not* first, as is often the case. The tens-ones naming pattern, characteristic of other two-digit numbers, does not hold for teens.

Example:

two tens, four ones   ⟶   *twen*ty-four;

four tens, four ones   ⟶   *for*ty-four;

six tens, four ones   ⟶   *six*ty-four;

but

one ten, four ones   ⟶   fourteen.

This naming irregularity causes a number of children to have reversal tendencies for teens. For instance, hearing "*four*teen," they may write 41.

The sample activities which follow reflect the three-step sequence just outlined. Grouping aids are used throughout, and teens are omitted in early work. Activities similar to each type described are necessary to help children develop adequate understandings and skills for two-digit numbers.

Activity File: 2-Digit Numbers

Activity 1: Make Ten.

Children count to find there are ten spots on the worksheet. They place a chip on each spot. When all spots are covered, they make a stack with their chips. They repeat covering and stacking until the chips given them run out. Children take turns telling the number of stacks and leftovers they have (see Figure 7.47).

Figure 7.47

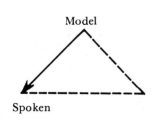

Model

Spoken

Three tens; four ones left.

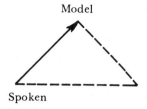

Model

Spoken

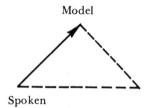

Model

Spoken

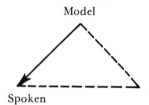

Model

Spoken

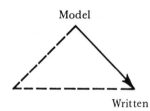

Model

Written

Activity 2: Show Me.

Using a counting frame (Figure 7.48), the teacher calls on a child to show "3 tens 2 ones," or "4 tens 6 ones," or some other tens-ones grouping.

Activity 3: You Call It. (Oral naming)

Children group by tens and use standard names to describe their grouping.

Note: At first names for multiples of ten are learned. For example, "*two* tens, *twenty*; *five* tens, *fifty*; *seven* tens, *seventy*." Any phonetic similarities to the names for numbers two through nine are highlighted. It is usually quite easy, then, to help children describe groups having extra ones: "Two tens four ones: that's twenty and four more, or twenty-four."

Activity 4: Make It Fast.

Children slide beads on a counting frame to show "thirty-four" or "fifty-seven" or some other tens-ones grouping on request. Teacher feedback along the following lines is appropriate: "Good! Thirty-four—that's three tens four ones." Children can take turns doing the activity as they leave the room during different periods of the day: for recess, for lunch, and so on.

Activity 5: Write It Down.

Repeat Activity 1, but have children write down the number of ten-stacks and extra ones they can make with the box of chips given them (Figure 7.49). Children trade boxes to repeat this activity.

Note: By the end of these activities, children should be able to write the standard two-digit numeral to describe any number of objects they count, up to 99 (see Figure 7.50).

Figure 7.48

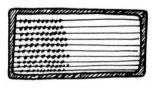

Figure 7.49

| Box 1 | Box 2 | Box 3 | Box 4 |

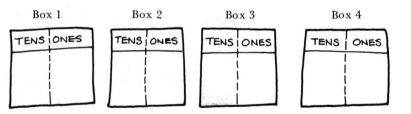

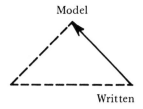
Model

Written

Activity 6: Draw 'N' Show.

Children draw a two-digit numeral card and place popsicle stick tens and ones beside it on the chalk tray to illustrate its value.

Activity 7: Name It!

Children draw a two-digit numeral card and read the numeral aloud.

Activity 8: Listen!

Children follow verbal directions such as: "Turn to page 28" or "Write 37 on the line."

Note: Two-digit number words are used in check writing. These should be introduced when children's reading and spelling skills are advanced enough to handle them.

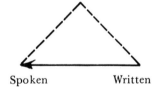
Spoken   Written

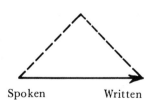
Spoken   Written

Figure 7.50

Whole Number Numeration: Phase 2

Sequencing Two-Digit Numbers

Children become involved in daily situations that require them to be able to sequence or order numbers. "The candy costs 20 cents. Joe has 23 cents. He has enough." Or "Jan walks down Cherry Street looking at the house numbers: 31, 33, 35, 37. She says 'The numbers are getting bigger. I want house 21. I'll have to turn around and go the other way.'" Sequencing skills are also a necessary prerequisite for rounding and mental estimation. Examples include estimating the cost of several items; quickly calculating the amount saved by buying during a sale or by purchasing brand X over brand Y; or even estimating to check whether a hand calculator displays a reasonable answer. Children also estimate to quick-check computational accuracy.

In sequencing numbers to ten, children often rely on counting to determine which number meant *more* or *less*. However, counting is too time-consuming and therefore impractical for comparing larger numbers. This is particularly true if the numbers are not close to each other in the counting sequence. Children can use what they know about the order of numbers zero to ten, though, to help them make comparisons involving two-digit numbers. The following examples illustrate this point.

Important goals for developing seriation understandings and skills for two-digit numbers are now presented, with grouping aids used to illustrate numbers being compared.

Goal 1: Hearing two numbers, children can use grouping aids to illustrate each and write the greater number (Figure 7.51). Right from the beginning, while working with objects, children should learn to look at the tens, the "big things" first. Only if the tens are the same do they need to look at the ones to decide which means more (or less). Sometimes children simply circle to show the greater of two numbers. Most children eventually learn to insert the correct symbol, < or >, between the numerals. It is important that children realize the meaning of the comparison, since daily situations often require this understanding.

Figure 7.51

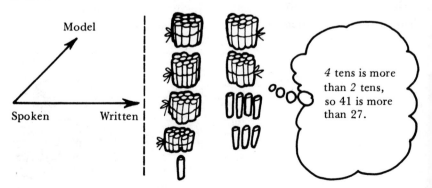

4 tens is more than 2 tens, so 41 is more than 27.

Example: A TV announcer advertises a bike. "Was $47. Now $39." Tim tells his mother: "It used to be a lot more than now."

Goal 2: Shown a model of two numbers, children can both tell and write the greater number (Figure 7.52).

Goal 3: Given two written numbers, children can tell which is greater and use grouping aids to prove their choice is correct (Figure 7.53). Sometimes aligning the numbers, tens under tens, ones under ones, helps. Then, in a manner parallel to that for placing words in ABC order, one can compare the digits, tens first (Figure 7.54). A one-to-one match between ten-bundles illustrating each number verifies the comparison: 41 > 39. Eventually children should be able to explain, without relying on grouping aids, that 41 is more than 39 because 4 tens is more than 3 tens. That they can also use materials upon request to prove the comparison supports their understanding of such statements.

When children can compare two numbers, each goal is expanded to include sequencing three or more numbers. Instruction is basically complete when, given numbers in one of the three forms (spoken, model, written), children can correctly sequence the numbers using both remaining modes. The activities that follow suggest ideas for reinforcing sequencing skills.

Figure 7.52

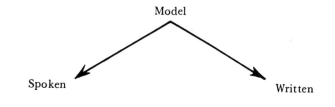

Figure 7.53

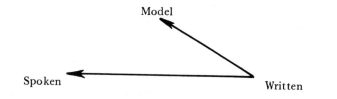

Figure 7.54

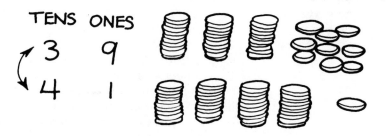

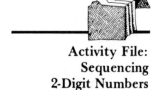

**Activity File:
Sequencing
2-Digit Numbers**

Model

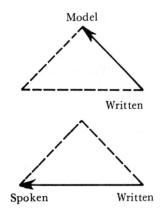

Written

Spoken   Written

Activity 1: Draw 'N' Prove. (For two or three to play)

Children draw two numeral cards, tell which is greater, and use graph paper tens and ones to prove the comparison. Children keep a tally of the times they are correct.

Activity 2: Compare! (For two or three to play)

Each child draws two one-digit numeral cards and lays them down to show the greatest number possible (Figure 7.55). The player showing the greatest number keeps all the cards. Repeat until the deck is used up.

Activity 3: 5-Draw. (For two or three to play)

Children each draw five cards from a deck of two-digit numeral cards and lay them, left to right, in the playing area. At "Go!" each child orders the five cards, low to high. First to order the cards correctly keeps the five. All other cards are placed in a discard pile for reuse as needed. Repeat until the deck is used up.

Note: If necessary, grouping aids or a meter stick can be used to check number sequence in the three activities.

Spoken   Written

Figure 7.55

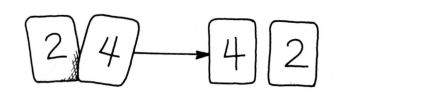

Beyond 100: Number Meanings and Skills

The instructional sequence just outlined for Phases 1 and 2, can be adapted quite easily for developing three- and four-digit number concepts and skills. Grouping aids appropriate for this purpose are illustrated in the *Teacher Background* section at the beginning of the numeration section. Since it is the physical model that will give meaning to the spoken and written symbols, continue to use these aids. Number strips might be introduced along with graph paper pieces to illustrate larger numbers (Figure 7.56). Children can then slide the strips together so just the numeral on each card shows (Figure 7.57). Have them compare the numeral they "made" with that

**Whole Number
Numeration:
Phase 3**

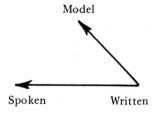

Model

Spoken   Written

first presented them, and ask them to explain, in their own words, what each of the digits means.

Reading larger numbers.

Children often need special help reading larger numbers. Sometimes, given a three-digit numeral like 234, they focus on just the first two digits and say "twenty-three, four." Children should be helped to focus on the first digit in any three-digit numeral. This digit is the only new one and tells the number of hundreds. They already know how to read the other part. Sometimes underlining the first digit is enough of a clue to reinforce this: 2̲34 ("two hun-dred thirty-four").

Similar techniques can be used to help children read four- and higher-digit numbers. A pocket chart like that in Figure 7.58 is also helpful to many students. At each bold vertical line children

Figure 7.56

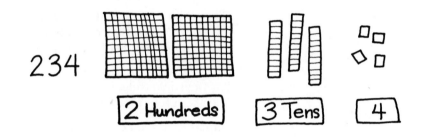

Figure 7.57

Figure 7.58

pause to give the name of the period (e.g., million, thousand) before going on. Their ability to read one-, two-, and three-digit numbers is all that is otherwise required. The number represented in Figure 7.58 would be written "forty-two million, five hundred thirty-six thousand, one hundred twenty-three." Children who can learn to read and write the full number words for larger numbers should be helped to develop these skills phonetically.

Reinforcing meanings for larger numbers.

How much is $100? How big is 1000? Children need to develop a feel for numbers beyond 100. Then, when asked to group by hundreds or thousands, when identifying the hundreds or thousands place within a numeral, when rounding to the nearest hundred, or even when reading the price tag on an expensive article, children can more realistically relate to the quantity involved. Simple activities such as the following, interwoven with other developmental work, will help children gain an appreciation for large numbers.

1. Give children old catalogs and a calculator. Let them make a "wish list" of things they'd like to have. Then let them see how many items they could actually buy with $100.

2. Let children look at a meter stick. "How many centimeters? . . . Put the stick aside. Can you mark off a line (cut a string) that is 100 cm long? . . . How close were you? . . . Now try 200 cm.")

3. Let children see how long it takes them to run 100 meters; to clap 100 times.

4. (Art project) Have children make a chain that has 100 construction paper loops. Make each "ten" in the chain a different color.

5. Let children collect 100 (1000) bottle caps. Bag them by tens (hundreds).

6. "How far is 1000 meters?" Use a trundle wheel to walk it out. This activity also gives children a feel for how long one kilometer is.

7. "How many grains of popcorn in 10 milliliters? . . . Then how many in a liter?"

Regrouping Numbers

**Whole Number
Numeration:
Special Topics**

Children sometimes need to regroup in order to complete a computation. Regrouping first occurs when adding and subtracting two-digit numbers and recurs frequently in computations with larger numbers (see Figure 7.59). The basis for children's success in meaningfully handling such situations is their firm grasp of numeration concepts. Simple trading and labeling activities can informally prepare children for regrouping.

For two-digit numbers, metric models (centimeter and decimeter strips), popsicle sticks, Unifix cubes, or other grouping materials can be used for the activities. *If* children understand the pennies-dimes money values and equivalents, these can be used. For three-digit numbers, graph paper pieces or other materials discussed above can be traded. It is important that children are convinced when a fair trade is made: one ten for ten ones or vice versa; one hundred for ten tens or vice versa.

Figure 7.59

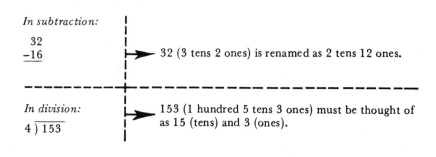

In subtraction:

$$\begin{array}{r} 32 \\ -16 \\ \hline \end{array}$$

→ 32 (3 tens 2 ones) is renamed as 2 tens 12 ones.

In division:

$$4\overline{)153}$$

→ 153 (1 hundred 5 tens 3 ones) must be thought of as 15 (tens) and 3 (ones).

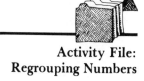

**Activity File:
Regrouping Numbers**

Model

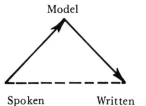

Spoken　　　Written

Activity 1: Trade!

Establish a bank with extra dimes and pennies. Give a child three dimes and two pennies. The teacher, aide, or able student plays banker. "I'm the banker and need one of your dimes. Would you trade a dime for some pennies? How many pennies must I give you to make it a fair trade?" Help children make a record of their holdings before *and* after the trade (Figure 7.60).

Note: The preceding example will be effective only for children whose basic money concepts are strong. For others, use regrouping aids. The recording is similar (Figure 7.61).

Variation A: In preparation for renaming in addition, reverse the procedure. Start with 2 tens 12 ones and trade for 3 tens 2 ones.

Figure 7.60

Variation B: Extend the activity to include three- and four-digit numbers.

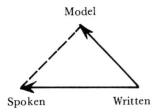

Model

Spoken   Written

Activity 2: Label It!

Sample Dialogue:

Teacher: Let's use these graph paper pieces to show 153.

Tommy: (Lays out pieces as in Figure 7.62a.)

Teacher: Tommy, can you use the label cards to tell what pieces you used?

Tommy: Sure! (See Figure 7.62b.)

Teacher: So 153 means 1 hundred, 5 tens, and 3 extras. But if I line your first two cards on top of each other (Figure 7.62c), it says we have 15 tens and 3. Is that so?

Tommy: (Gets 10 extra tens from the bank and places them on top of the hundreds square): Yes, because 1 hundred is the same as 10 tens and these 5 (pointing) make 15 tens in all. So 153 *is* 1 hundred 5 tens and 3, but it's 15 tens and 3, too.

Rounding Numbers as a Help to Mental Estimation

If children wish to estimate the cost of several small items, they normally round each cost to the nearest 10 cents and then calculate mentally. Numeration understandings are pivotal to a child's success in rounding such numbers. As with other developmental work, instruction in this area should begin with discussion and manipulation of physical objects. The activities that follow suggest ways of building skills for rounding numbers.

Figure 7.61

Figure 7.62

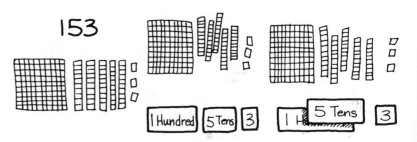

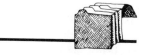

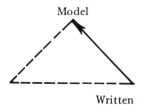

Model

Written

Activity 1: Quick Ten. (Information introduction: rounding two-digit numbers)

Need: Two decks of cards: a Bank deck and a Quick Ten deck (Figure 7.63). The Bank deck is placed in a bank box along with extra pennies and dimes. It contains one card for each multiple of 10 cents (to 90 cents). The Quick Ten deck contains cards which list amounts between 11 cents and 89 cents (no multiples of 10 cents included).

Procedure: Until children understand the activity, remove all multiples of five cents from the Quick Ten deck. Shuffle the rest of the deck and place it face down between players. In turn, children draw the top card from the deck and place it face up in the playing area. They then use dimes (as many as possible) and extra pennies to show the amount on the card. The banker selects the two cards from his deck which are closest in value to that displayed (Figure 7.64). The child in this example must decide how a "Quick Ten" can be made. Is 23 cents closer to 20 cents or to 30 cents? The child must either add extra pennies to the 23-cent pile or take some away to indicate the choice made. So either seven pennies would be added (to make 30 cents) or three pennies would be taken away (leaving 20 cents). The latter decision wins the child one point, since it involves the least number of coin moves.

When children are comfortable with the game, add the 5-cent multiples to the deck. Introduce the Banker's Rule that for these cards the "Quick Ten" is always the greater 10-cent value.

Variation: Adapt the activity for three-digit numbers (rounding to the nearest hundred) by using ten- and hundred-dollar bills. In this case children would be interested in making a "Quick Hundred."

Figure 7.63

Bank deck

Quick Ten deck (Sample cards)

Figure 7.64

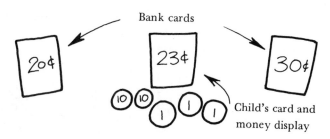

Bank cards

Child's card and
money display

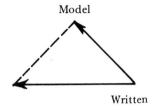

Model

Written

Activity 2: Mailman. (Informal introduction: rounding three-digit numbers)

Need: Number line segment showing house numbers to which mail is to be delivered (Figure 7.65).

Storyline: Multiples of 100 on the number line are the street names. Other numbers are those of houses in between. The mailman always parks the mail truck on the street nearest the houses to which mail is delivered. For delivery to houses 301–349, park on Street 300. For the halfway mark and beyond (houses 350–399), park on Street 400.

Procedure: Given specific house numbers, students tell where the mailtruck is parked.

Variation: Adapt the activity to rounding two- or higher-digit numbers.

Formal instruction: rounding numbers.

The number line is frequently used to help children develop rounding skills. For larger numbers, number line segments can be drawn, as in the Mailman activity. The number line dramatized *closeness to* one number rather than another. It illustrates a basic rule:

When rounding, look at the digit after. If less than 5, round down. If 5 or greater, round up.

Example: Round to the nearest ten. Examine the digit *after* that in the ten's place.

1. Round to the nearest ten (see Figure 7.66).
   The "3" in 43 is less than 5. So round down to 40.

2. Round 48 to the nearest ten (see Figure 7.67).
   The "8" in 48 is greater than 5. So round up to 50.

Follow-Up Practice Game: Rounding War. (For two to play)

Children mix and deal out all cards of a two-digit numeral deck. Each child turns over the top card and rounds it to the nearest ten. The player with the greatest "ten" captures both cards. Ties are resolved in traditional War fashion. Winner: first to capture all cards.

Variation: Adapt to three- or higher-digit numbers.

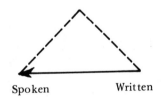

Spoken   Written

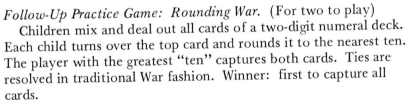

Figure 7.65

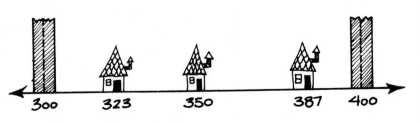

300   323   350   387   400

Figure 7.66

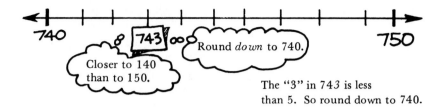

Closer to 140 than to 150.

Round *down* to 740.

The "3" in 74*3* is less than 5. So round down to 740.

Figure 7.67

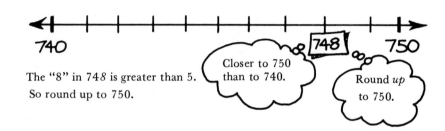

The "8" in 74*8* is greater than 5. So round up to 750.

Closer to 750 than to 740.

Round *up* to 750.

Your Time: Activities, Exercises, and Investigations

1. Be prepared, as your instructor directs, to answer the major questions on page 176 of this chapter.

2. Order the topics below as they would normally be developed with children (use 1, 2, 3, etc.).
   ____ Ordering numbers less than ten (e.g., 3 < 5)
   ____ Place value (each digit in a numeral has a value determined by its place within the numeral)
   ____ Rational counting (meaningfully counting to determine the number of elements in a set)
   ____ Giving oral names for numbers to 100
   ____ Ordering numbers greater than 1000 (e.g., 2347 < 3869)
   ____ Rote counting (e.g., singing the number song, "one, two, three" without reference to a set of objects)
   ____ Grouping objects by tens, and recording the number of tens and the number of ones left over.

3. Materials used for developing two-digit number meanings with special needs students play a major role in successful instruction. The teacher's first selection should be grouping materials—those in which ten single objects are still visible when one "ten" is formed (even though the singles be bundled or attached together). This is in contrast to place-value aids that show one thing *taking the place of* ten singles (in such a way that the ten singles are no longer visible). Do not use place-value aids with children until the place-value concept is well established.
   If possible, actually examine each of the teaching aids listed below. Which are grouping aids?
   ____ Unifix cubes
   ____ Spike abacus
   ____ Graph paper tens and ones
   ____ Counting frame
   ____ Bean sticks
   ____ Money (dimes and pennies)

_____ Bundling sticks (such as popsicle sticks or tongue depressors)
_____ Rocks, bottle caps, or chips (all the same color)

For numbers 4–6, identify any special characteristics or handicaps of children for which your activities are written.

4. On page 173 three steps were outlined for developing two-digit numbers with children.
   (a) For each of these steps, create an appropriate activity of your own.
   (b) For each activity, identify the relationship of the learning triangle you are emphasizing (Figure 7.68).

5. Create one developmental and one practice activity that can be used to help children sequence three-digit numbers.

6. (a) Briefly describe a sequence of three activities to teach a child to round two-digit numbers. You may incorporate activity ideas from the chapter in your sequence. Make at least one of your activities a *developmental* activity.
   (b) Describe two followup suggestions that reinforce the new skill by requiring children to estimate in common daily situations.

Figure 7.68

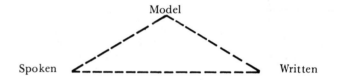

The descriptions in this section are based on classroom or clinical experiences with children having difficulty learning basic number or numeration concepts and skills. Together they represent the more common problems that daily face teachers of special needs children. If you were called upon to deal with a child who had a problem similar to any of these recounted, could you:

Number and Numeration: Common Problems

1. Diagnose the problem, giving a statement of the extent to which the child may be on the right track, as well as an accurate description of the difficulty?

2. Prescribe and then implement a sequence of related activities to redirect the child's thinking?

Case Study: Richie

The following example may help you.

The Problem: Richie "counted" the six apples on the table and told his teacher there were ten. He recounted them for the teacher in his characteristic manner, touching (most of) them as he went: "1-2, 3, 4-5-6, 7-8, 9-10. Ten apples."

Diagnosis: Richie does know the oral names and the correct counting sequence for numbers to ten. To this extent he's on the right track. He has not, however, developed any real meaning for the number names he's using; he clearly lacks the idea of associating one number name to each object counted.

Corrective procedures: Sequence of related activities (appropriate for all categories of special needs students, with obvious adaptations at times to account for certain physical handicaps).

Comment: Before Richie can be asked to focus on *numerousness*, that is, on *how many* objects are in any given set, he must be slowed down in his oral counting. Rather than teaching number meanings as such, the immediate goal of the following sequence of related activities is to help Richie relate one number name to one object, for each object counted. Tactual counting, including moving objects as they are counted, is emphasized throughout.

1. Give Richie a tin pan and four metal washers or other objects that will make a noise when dropped into the pan. Take turns playing Plunk with him. Going first, drop one washer in the pan and say "one." Drop another washer and say "two." Let Richie do it. Tell him to count the washers as you did. Prompt by counting with him if necessary. Make no special effort to have him count beyond four or five in this activity. Repeat one or two times.

2. Give Richie a heavy tagboard worm such as that of Figure 7.69. "Let's find out how many bumps are on its back." As he counts each bump have Richie push the tagboard slide down. The physical activity will help slow down his counting and reinforce the one number name to one object idea.

3. Prepare a transparency showing four balloons in a row (Figure 7.70). Place a piece of paper over three of the balloons after the

Figure 7.69

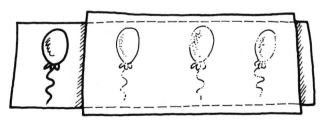

Figure 7.70

transparency has been placed on the overhead. Flash the image on the wall (Richie's height), and have him touch each of the balloons as he counts them. After Richie has said "one," slide the paper so he can touch and count the second balloon. After Richie has said "two," slide the paper again so he can touch and count the third balloon. Continue in the same manner for the fourth balloon.

Note: If Richie needs additional activities of this nature he could:

(a) Count the stairs as he goes up or down.

(b) Count rubber bands that he stretches and places around a door knob.

(c) Count blocks as he takes them from a box and places them, one by one, on a table.

(d) Bounce a ball and count the times it bounces.

(e) Clap, and *count* the claps.

(f) Set a table for four and count each chair as he pushes it in, and each knife (fork, spoon, plate, napkin, cup) he places on the table in front of each chair.

(g) Count knots tied in a rope (good for visually limited students

(h) Count pegs as he pushes them into a pegboard.

(i) Count beads as he strings them (good for visually limited students).

4. Make a series of pushbutton cards by gluing buttons to tagboard. Make at least four cards: one button, two buttons, three buttons, four buttons. As Richie counts, have him push each button.

5. Glue dots on paper, up to four or five in each cluster. Have Richie point to each dot as he counts those within each group.

6. Ask Richie to count animals or people on (simple, uncluttered) coloring book or storybook pages.

Comment: Until Richie is confident of the one-number-name-to-one-object idea, he should stay with physical and object-oriented activities. Activities requiring finer motor coordination and those dealing with pictures rather than real objects are presented late in this sequence.

Your Time

Case Descriptions

For each of the following descriptions sketch both a diagnosis and a sequence of corrective activities to help the child. Indicate any specific characteristics or handicaps that describe the child for whom your activities are especially written. If you feel your activities, with slight adaptations, are appropriate for several types of children, state this.

Use the example of Richie as a guide as you deal with each case description. Draw freely from the ideas of this chapter as you plan your activity sequence.

Case Descriptions: Common Problems in Early Number Work

1. Susan, a newcomer to your mathematics classroom, is a young primary child working with numbers to ten. Whenever Susan counts a group of objects (e.g., seven objects) she always starts from one. If you were to quickly flash the cards of Figure 7.71 before her, she would only be able to tell "how many" for card 4. She would not have time to count all the dots on the other cards. It is important for Susan to learn more efficient ways of counting now, since this will help later with addition sums. (Susan has studied the numbers to ten.)

2. When Bobby is asked to give the number that comes after six, he counts (either aloud or quietly to himself): "1, 2, 3, 4, 5, 6, <u>7</u>." *Then* he can tell you "Seven comes after six." It's this way any time Bobby is asked to tell the number that comes after another. He has developed a distaste for oral exercises which ask for the number after, because it takes him so long to answer. (Bobby has studied the numbers zero to eight.)

3. Jennifer carefully counted each button card (Figure 7.72). "Four on each card."
   Teacher (follow-up question): "Is there the same number on each card, or do some cards have more?"
   Jennifer: "These cards (pointing to cards 1 and 2) have more."
   (Jennifer has studied numbers zero to five.)

4. Janet correctly counted the balls on the sheet: six. But she wrote "9" on the line below the picture. This is a consistent tendency with Janet. She also frequently confuses the numerals 5, 3, and 8, though her counting habits are quite accurate.

Case Descriptions: Common Numeration Problems

1. Terri counted aloud: "27, 28, 29, 20-10." She commonly makes this mistake. On another occasion it may be "68, 69, 60-10."

2. Meg wrote quickly on the addition quiz. "Eight + six—that's fourteen," she thought, and wrote "41." Similarly, for seventeen she has written "71."

3. Roger has just joined your mathematics class. Examples from today's work are presented in Figure 7.73. Roger explained the first problem this way. "Seven from one, you can't do. Borrow 1 from 3—that leaves 2 with your neighbor. Now you have 10—7 from 10 is 3—1 from 2 is 1." Roger could not explain about the "10" except, "That's the way it always works."

4. You feel good about Jodi's work with two-digit numbers and with his early progress in three-digit numbers. But today you overhear him counting: "96, 97, 98, 99, 100, 200, 300, . . . 900."

5. Nancy, a newcomer to your mathematics classroom, looks carefully at the two numbers written on the paper (431, 429). She correctly answers that the first has 4 hundreds, 3 tens, and 1 one. And the second has 4 hundreds, 2 tens, 9 ones. She takes up your suggestion to show you this, using materials on the math shelf. Recognizing the

abacus, she correctly represents, in turn, each of the numbers on it. When you ask her "which means more" she thinks a bit, then answers. "The 4's are the same, and the 3 is a little bigger than the 2. But that 9 is a lot bigger than 1. So 429 is bigger."

6. Jerry looked at the directions: "Round to the nearest hundred." His completed worksheet for this section is shown in Figure 7.74.

7. When asked to find page "one hundred one" Jim, paging through his book, looks confused. Noting his expression, you tell him that you will repeat the number, and you'd like for him to write it this time *as* you say it. He writes "1001." Similarly, for "two hundred thirty-two" he writes "20032."

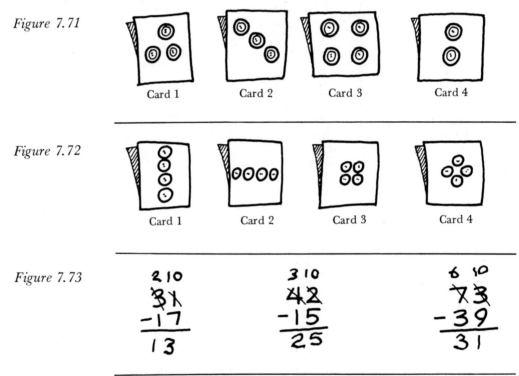

Figure 7.71

Card 1   Card 2   Card 3   Card 4

Figure 7.72

Card 1   Card 2   Card 3   Card 4

Figure 7.73

Figure 7.74

Number given	Rounded to nearest hundred
2610	2600
3468	3400
7682	7600

References

Cox, L. S. Systematic errors in the four vertical algorithms in normal and handicapped populations, *Journal for Research in Mathematics Education*, 1975, 4, 202-220.

Piaget, J. *The Child's Conception of Number.* New York: W. W. Norton, 1965.

Rathmell, E. D., and Payne, J. N. Number and numeration, in J. N. Payne (Ed.), *Mathematics Learning in Early Childhood.* Reston, Va.: National Council of Teachers of Mathematics, 1975. Rathmell and Payne include grouping by tens in the first of four teaching "units" for building numeration understandings with young children. Oral naming, writing, and sequencing two-digit numbers are topics for the other units.

8 WHOLE NUMBER ADDITION AND SUBTRACTION

Many mathematical problems can be solved by counting. When large numbers are involved, however, counting is too cumbersome. It may be more practical to estimate or, when precise answers are needed, use a calculator or do a paper-and-pencil computation. Dealing with quantitative data in daily situations requires knowing when and usually how to add, subtract, multiply, or divide.

"Knowing when" and "knowing how" are clearly two separate issues. Children must first understand an operation so they will know *when* to use it. In a given situation—which operation is appropriate? Which is most efficient? When it is impractical or not possible to use a calculator, children must estimate or personally compute to find answers. This is when knowing *how* becomes critical.

This chapter looks at concepts and skills related to addition and subtraction that prepare children to know when and how to use each in day-to-day living. Emphasis is placed on thoughtful, careful instruction that allows children to be involved in:

1. Using materials to illustrate ideas and processes

2. Verbally expressing their understanding of concepts or procedures

The importance of adequate developmental work prior to drill and practice is highlighted throughout the chapter. This development moves gradually through three phases in elementary school instruction. These phases might be labeled: (1) conceptualization, (2) fact mastery, and (3) algorithm learning.

The first phase focuses on helping children conceptualize or construct mental models for the operation. Both physical and pictorial models are used. The aim is to help children relate the arithmetic operation to a physical operation or action. For example, addition may be seen as joining objects in two or more groups; subtraction as separating objects in a group. When children are given a written number sentence, they should "see" the associated physical action. And when children encounter a problem situation involving the physical operation, they should think, "that's addition," or "that's subtraction."

A second important goal of the conceptualization phase is developing a technique or process for figuring out simple arithmetic facts. For example, given two cookies on one plate and three on another, a child might combine and *count to see how many*. Or, starting with four apples and giving two to a friend, a child might *count to see how many are left*. These processes will then be used to discover and verify easy addition and subtraction combinations.

When the first phase is successfully completed, instruction moves into a second phase which concentrates on fact mastery. At this stage children first deal with easier facts as they relate them to physical models. An important idea stressed at this time is *counting on* from the greater addend to find a sum (Figure 8.1).

Or, they might use a model to examine those facts that add up to ten (Figure 8.2).

Figure 8.1

Figure 8.2

Sometimes children can relate easy facts to familiar objects (Figure 8.3).

Instruction and learning at this state deal with organizing easy facts and stressing techniques, relationships, or cues that make memorizing them easier. Children are the prepared to move to a more advanced level of fact learning. The focus at this higher level is mastering harder facts by relating them to easier, known facts. Children might, for example, use facts that add to ten to help with harder facts. This strategy involves "making ten" first, then adding in any leftovers (Figure 8.4).

Or, children might find it easy to remember a harder fact because it is one more or one less than some known fact (Figure 8.5).

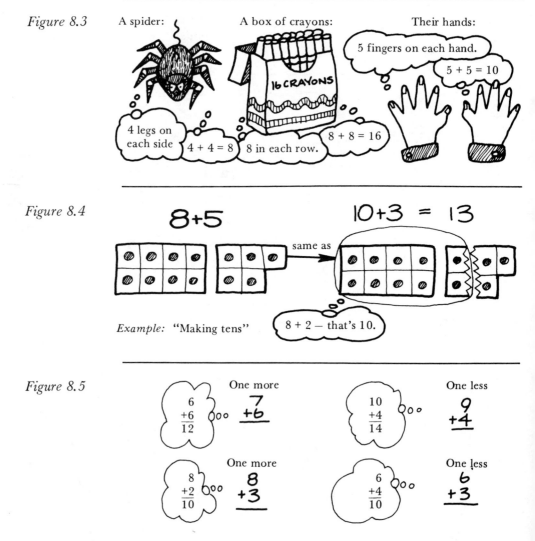

Figure 8.3   A spider: A box of crayons: Their hands:

4 legs on each side   4 + 4 = 8   8 in each row.   16 CRAYONS   8 + 8 = 16   5 fingers on each hand.   5 + 5 = 10

Figure 8.4   8+5   10+3 = 13

same as

Example: "Making tens"   8 + 2 — that's 10.

Figure 8.5   One more   One less

$$\begin{array}{r} 6 \\ +6 \\ \hline 12 \end{array}$$ ooo $$\begin{array}{r} 7 \\ +6 \\ \hline \end{array}$$ $$\begin{array}{r} 10 \\ +4 \\ \hline 14 \end{array}$$ ooo $$\begin{array}{r} 9 \\ +4 \\ \hline \end{array}$$

One more   One less

$$\begin{array}{r} 8 \\ +2 \\ \hline 10 \end{array}$$ ooo $$\begin{array}{r} 8 \\ +3 \\ \hline \end{array}$$ $$\begin{array}{r} 6 \\ +4 \\ \hline 10 \end{array}$$ ooo $$\begin{array}{r} 6 \\ +3 \\ \hline \end{array}$$

Sometimes children combine the –1, +1 relationships to figure out answers to harder facts. By doing so, they transform the harder fact into an easier one (Figure 8.6).

This stage of fact learning frequently involves additional work with physical models. *The goal is to help children concretely "see" the relationships so they will more effectively use them to master the harder facts.* Clustering facts that use the same strategy or relationship also helps in memorizing them.

When children are making final steps toward fact mastery, instruction usually moves into a third phase—that of learning addition and subtraction algorithms. *An algorithm is a computational paper-and-pencil procedure one uses to obtain an answer* (Figure 8.7). In this phase the emphasis is on the "big ideas" underlying the computational procedures. When adding, for example, there are basically just two things to remember:

1. Add like units.

2. Make a trade whenever there are 10 or more of a kind.

Rules such as lining up columns, or "ones first, then tens," should be put off until the big ideas are clearly understood. Children should see the algorithm as a combining of groups of ones, groups of tens, and so on. Like things are grouped, and regroupings are made when needed. When the initial focus is on major ideas rather than superficial rules, misconceptions are minimized. As a consequence, computational errors are fewer.

Similarly, for subtraction, two "big ideas" prevail.

1. Subtract like units.

2. If there are not enough, make a trade.

Figure 8.6

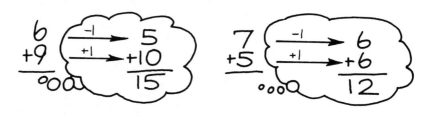

Figure 8.7

Sample addition algorithms

Sample subtraction algorithms

Early work at this level involves the use of models to illustrate the big ideas for an operation (Figure 8.8).

Figure 8.9 presents an outline, a summary of the preceding paragraphs, to guide your study of this chapter. It presents the three major phases for addition and subtraction instruction, with a capsule description of each.

We now move to the three-phase development of basic sequences and procedures leading from conceptualization to fact mastery to algorithm learning for addition and subtraction. *Understanding this development and ways of implementing it is a prerequisite to adapting to special needs of students.* The case studies and trouble shooting sections of this chapter give specific suggestions for meeting special needs. At times, for example, it may be necessary to follow the standard sequence but include a great deal of visual reinforcement, or limit writing requirements. Ideas for anticipating and avoiding common difficulties are also highlighted in the sections that follow.

Figure 8.8

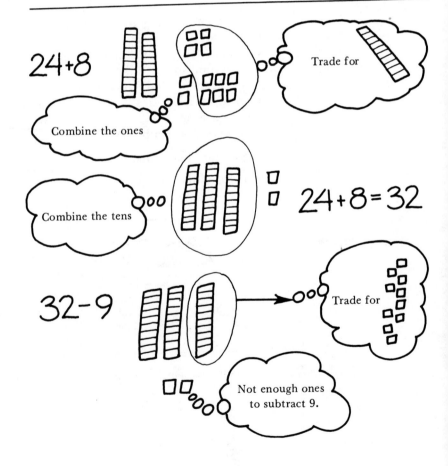

Figure 8.9   **Instruction for Whole-Number Addition and Subtraction**

Phase 1: Conceptualization
- Relate arithmetic operation to physical operation.
- Develop a process for finding answers.

Phase 2: Fact Mastery
- Easy facts
    (1) Figure them out using process already developed.
    (2) Organize them in many ways to see relationships.
    (3) Memorize them.
- Develop more efficient techniques for remembering facts that
    (1) Are *mental* processes.
    (2) Use easier facts already memorized.
- Harder facts
    (1) Figure them out using the new efficient processes.
    (2) Organize them.
    (3) Memorize them.

Phase 3: Algorithm Learning
- Deemphasize superficial rules.
- Stress big ideas and relate them to concrete models.
- Then develop computational expertise apart from models.

What Do *You* Say?

1. What prerequisite concepts and skills ready a child for success with early addition and subtraction?

2. How does active involvement with physical materials help a child during early work with addition and subtraction?

3. What role do "hiding activities" play in reinforcing the addition-subtraction relationship and in preparing children to memorize subtraction facts?

4. What is the distinction between developmental and practice activities, and what is the role of each in developing basic meanings for addition and subtraction? In helping children learn the facts?

5. When, in a normal teaching sequence for addition (subtraction) does focus on concept development stop and work toward fact mastery start?

6. What is the teacher's role in helping children master the basic facts for addition and subtraction?

7. How can teachers keep children from adding or subtracting on their fingers? What clues or thinking strategies might help?

8. How does one assess whether children have *mastered* the basic facts?

Even when children are learning meanings and skills related to the numbers zero to ten, they are building prerequisites for success in both addition and subtraction. Chapter 7 discussed ways in which children can be encouraged to search for patterns as they group objects.

Besides learning basic meanings for numbers that will be used in simple addition and subtraction combinations, children also discover different ways they can combine smaller groups to make a larger group. Informally, as part of the early number program, they also experiment with the reverse process. They partition a larger set into smaller subgroups (Figure 8.10).

They also learn to complete a set to "make" the number needed (Figure 8.11).

Activities of this type are pre-addition or pre-subtraction in nature. Their primary goal is to reinforce basic number concepts and skills for numbers through ten:

1. Basic number meanings

2. Visual (and meaningful) recognition of numerals

3. "Counting on" skills

These concepts and skills are also critical in early addition and subtraction. The activities further provide an experiential base that readies children for the writing of number sentences to describe the combining or separating actions. At first, combining four beans and two beans is a very different action from partitioning a set of six

Figure 8.10

Figure 8.11

beans into a group of four and a group of two. By repeatedly engaging in such activities, children gradually internalize, or bring these two actions together in their minds. Later they will express this relationship more formally (Figure 8.12).

Addition: Phase 1

Basic Concepts and Models: Addition

Problems that involve finding the total number of objects in two or more sets can be solved by adding. Since young children's thinking is typically concrete, learning to add entails, in the beginning, a great deal of active involvement with physical objects. Blocks, beads, rocks, buttons, chips, beans, coins, and wooden spools—these and other familiar objects can be used to help children learn the meaning, the new vocabulary, and the symbolism for addition. Instruction starts by having children join sets of objects. Children tell the number in each set and then combine them (Figure 8.13).

Children learn that "+" means *put together*. New vocabulary words *plus* and *add* are introduced in the context of physically combining groups of objects. First experiences are framed to help children relate the arithmetic operation *add* to the physical action of combining. Given a situation in which objects are combined, children should think *addition* and they should be able to write the expression to match. Further, given a written expression such as 2 + 3, children should be able to use objects to dramatize the addition.

When children are comfortable with these ideas, a second task emerges. It now becomes necessary to help them develop a process for finding the number in the total set. At first children will count the number of objects in each set, put the two sets together,

---

Figure 8.12

A Number "Family"

2 + 4 = 6   6 – 4 = 2
4 + 2 = 6   6 – 2 = 4

Figure 8.13

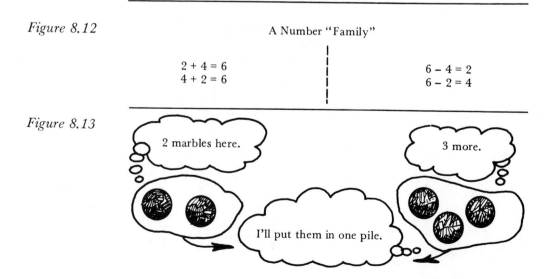

2 marbles here.

3 more.

I'll put them in one pile.

---

then recount all the objects again to find the total number (Figure 8.14). The new term *equals* is introduced, and children learn to write the complete number sentence using the equal sign. At first the horizontal form, modeled in Figure 8.15, is used. It reinforces the left-right reading movement and contains the new sign. Eventually children must comfortably handle the vertical form as well (Figure 8.16a), since it is a common form for computational problems. To help children who have abstract reasoning difficulties, some teachers use a double bar, simulating the equals sign, to separate addends from the sum (Figure 8.16b).

As children combine more and more groups of objects, they learn to hold in mind the number of the first set while they lay out or count the second. Then, when they join the two sets, they only need to *count on* from the number in the first set to arrive at the total. In the example with the bars above, a child might think "2–3, 4, 5 in all."

Figure 8.14

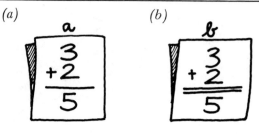

Figure 8.15

Child pushes bars together:

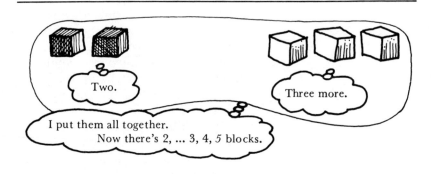

$$2+3=5$$

Figure 8.16

(a)   (b)

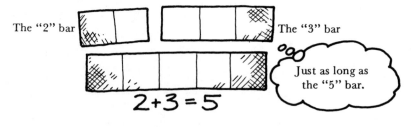

Any addition sentence such as 2 + 3 = 5 occurs in a variety of situations, and can be represented in many ways. Not only will children combine groups of objects, using grooved bars they will place a "2" bar and "3" bar end to end. They will dramatize the addition by combining a group of two students and a group of three students. They will also discuss pictures: "Two red birds in the photo, and three bluebirds. Five in all."

Interaction with the teacher or peers is necessary at this stage. As children manipulate and count objects, as they work with pictures for addition situations, they should be encouraged to tell what they find. *Verbalizing what they are doing while they are performing the activity is important. It helps translate the motor act into a mental process which becomes internalized as a mathematical concept.* Providing familiar objects stimulates children to use everyday language to create and tell addition stories (Figure 8.17).

Experiences like those just described help children conceptualize addition. In the beginning, models are used (Figure 8.18). Emphasis on the spoken ⟷ model relationship gives meaning to the *process* of addition. By relating the physical action of joining to the arithmetic operation, the new vocabulary for addition is given a meaningful base. Instruction at this level also links the physical combining of groups of objects to the written expression (Figure 8.19).

Figure 8.17   **Johnny's "story"**

'My little brother took two cookies.
Then he wanted one more.
In all he had three cookies.''

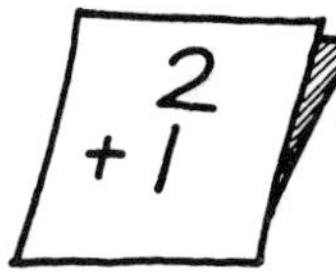

Figure 8.18

Figure 8.19

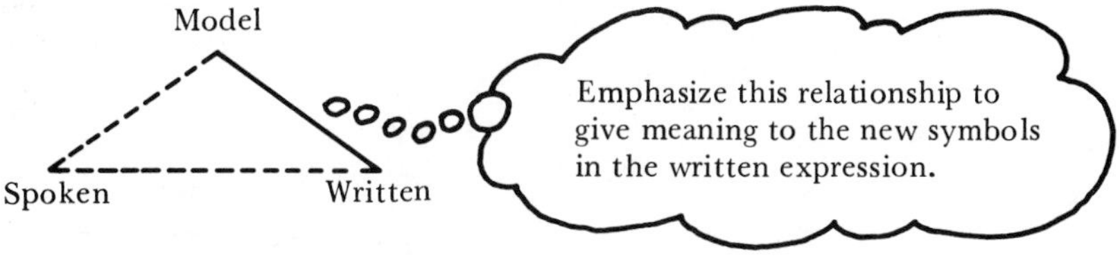

Instruction focusing on developing the concept of addition is basically complete when:

1. Hearing a description of an addition situation, the child is able to use objects to illustrate the joining and writing of the matching number sentence to tell how many in all (Figure 8.20).

2. Shown groups of objects being joined, the child is able to identify the situation as *addition* and write the correct addition sentence to describe the situation (Figure 8.21).

3. Given a verbal written addition expression such as 2 + 3, the child is able to create a verbal situation to dramatize the addition and use objects or draw a picture to tell the total number (Figure 8.22).

Note: Some children have difficulty verbalizing ideas and may not be able to describe common situations which use or illustrate addition combinations. While encouraging children to relate addition to familiar experiences is important, be careful not to overemphasize expectations for their doing so.

As a general guide, successful performance of each of the above tasks by children demonstrates their understanding of the addition concept and their readiness to move on toward mastering the addition facts.

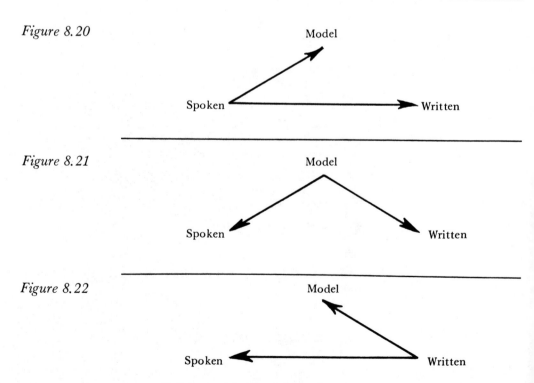

Figure 8.20

Figure 8.21

Figure 8.22

The number line.

Number lines are often introduced prematurely in the primary grades. Certainly they *can* be used to illustrate addition. But sometimes children are required to use them when simple counters would better serve the purpose. Or they are used when little or no preparation is provided. Some of the more common problems that occur in number line use are illustrated in Figure 8.23.

Some teachers minimize these difficulties because they anticipate trouble spots. They avoid using number lines in very early sessions for addition, and use blocks, chips, or other counters instead. When they do use them, their introduction is careful and deliberate. They introduce a large, walk-on number line to dramatize its use. They ask children to take steps to dramatize number sentences on the floor model. They use a storyline and frog-like character to illustrate the number-line movements on small models (Figure 8.24).

Figure 8.23

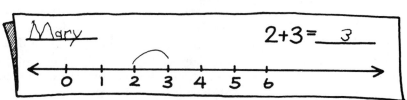

Mary simply connected the "2" and the "3".

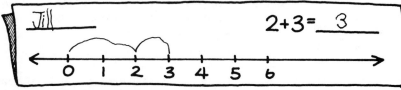

Jill went to the "2", then on the "3".

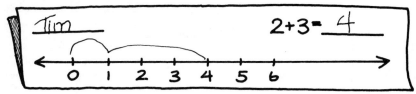

Tim counted slashes, not unit segments.

Figure 8.24

Example: Frog takes 4 hops, then 3 more—   or 7 hops in all.

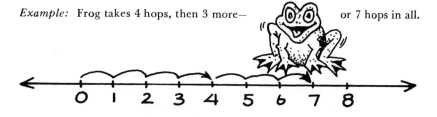

Chapter 8   Whole Number Addition and Subtraction   207

Perhaps the most natural use of the number line in the primary grades is as a reference for number sequence. It is far more valuable in middle and upper grades for illustrating integers, rounding, and certain fractional relationships. For some children, particularly those with size, spatial, or perceptual difficulties, number lines may be entirely *in*appropriate.

Addition: Phase 2

Toward Mastery of Basic Addition Facts

There are an infinite number of addition facts (1 + 1 = 2, 7 + 2 = 9, 13 + 5 = 18, 276 + 30 = 306, 198 + 17 = 215, and so on). Since it is not possible to memorize them all, we memorize a small body of facts and develop a procedure which uses those facts to figure out the rest. The computational procedure is called the algorithm. *The facts required by the algorithm are called the basic facts. They are basic because they constitute one of the bases for finding all other facts* (Figure 8.25). (It is interesting to note that if we were to change the algorithm we might also change the set of required, or basic, facts. It is possible to teach an algorithm, for example, that would only require that a student memorize facts with sums through ten.) A preceding section summarized tasks children should be able to perform to demonstrate their understanding of the addition *concept*. These tasks involved the model, spoken, and written forms for addition (Figure 8.26). When given an addition situation in any of these three forms, the child should be able to illustrate it using both other forms. Children who can do so have an adequate *conceptualization* of addition and are ready to work toward fact mastery.

A teacher's role in this process is important. At the beginning level, for simple sums through ten, teachers can encourage children

---

Figure 8.25   Basic addition facts   Algorithm using these facts
(memorized during the primary grades)

$$\begin{array}{c} 7 \\ +2 \\ \hline \end{array} \qquad \begin{array}{c} 3 \\ +6 \\ \hline \end{array} \qquad \begin{array}{c} 73 \\ +26 \\ \hline \end{array}$$

Figure 8.26

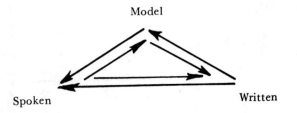

Model

Spoken   Written

to *count on* from the greater addend for facts they can't immediately recall (Figure 8.27). This approach is much faster than counting from one each time. As a *mental process*, it is a first big breakthrough in helping children find fact sums quickly. It is a process children can rely on until, through study and repeated use in exercises and activities, the simpler facts are memorized.

Some children at this beginning level have the maturity to use even more advanced mental strategies for figuring out sums that they cannot immediately recall (Figure 8.28). Focusing on sums for addition doubles or for facts that "make ten" is a strong base for this kind of thinking.

A *few* children will adopt strategies like these independently. Some will not need strategies or cues at all because they memorize quickly and easily. *Most will need more than drill alone to master the basic facts.* First they may need to develop confidence in their ability to memorize facts. Many will need teacher support in discovering and using relationships, patterns, and techniques that make memorizing easier. Then drill, in the form of motivating games and worksheets, can be used to build speed of recall and insure retention of known facts. The goal is to get children to give sums *quickly* and *accurately*, and to be *consistent* in performance over time.

Organizing the facts into clusters for easier learning is a first major step toward achieving this goal. Several approaches are charted in Figure 8.29. Throughout, the commutative form of each fact would also be studied.

Each of the examples illustrates a way of organizing the one hundred basic addition facts to make it possible for children to memorize them faster and retain them longer. Clearly, other variations are possible.

The first two examples are basic sequences. For young beginners, the emphasis is first on learning the easier sums to ten. The

Figure 8.27

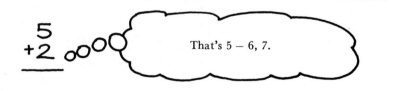

Figure 8.28

Examples: *One more* than a known fact.   *One less* than a known fact.

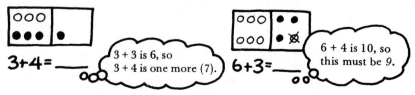

strategy of *counting on* from the larger addend is formally taught and later extended to other facts having 1, 2, or 3 as an addend. Doubles and ten-sums receive special attention, since these characteristically are easy for children and can serve as a springboard to learning other harder facts. The sequence of Example 1 then looks separately at two sets of facts related to doubles: near doubles and doubles plus two. The second example takes a more general approach. It focuses on identifying and using facts children know to learn a whole new set of related facts. Specifically, children are helped to study facts which are just "one more" than an easier, known fact. This process is applied over and over until most facts are memorized. Both examples then present addition 9's as a separate cluster.

Example 3 may be regarded as a sample remedial sequence for older students. Rather than starting with a review of easy sums to ten, doubles and related facts are studied. When the students have mastered doubles and near doubles, a third set of facts is presented: $6 + 4, 4 + 6; 5 + 7, 7 + 5; 6 + 8, 8 + 6; 9 + 7, 7 + 9;$ —facts whose

---

Figure 8.29   **Addition Fact Cluster**

Example 1:
- Sums to ten with emphasis: counting on from greater addend
  (a) Doubles and 10-sums also focused upon
  (b) Random discovery of other strategies for remembering facts
- Other count-ons
- Doubles (with sums greater than 10, such as $8 + 8, 6 + 6$)
- Near doubles (facts 1 more than a double, such as $6 + 5, 7 + 8$)
- 9's (e.g., $9 + 3, 3 + 9; 6 + 9, 9 + 6$)
- Three last facts: $4 + 7, 4 + 8, 5 + 8$

Example 2:
- Sums to ten with emphasis: counting on from greater addend.
- Other count-ons
- Doubles
- One more than facts:
  (a) 1 more than a ten-sum
  (b) 1 more than a double
  (c) 1 more than some other known fact (this includes "building" on ten-sums and near doubles once these facts are learned)

- 9's

Example 3 (Focus on remedial needs):
- Doubles
- Near doubles
- Facts whose addends differ by 2 (e.g., $6 + 8, 5 + 7$—3 strategies to choose from)
- 9's (3 strategies to choose from)
- Review of count-ons and +0 facts
- Review of ten-sums, then three last facts: $4 + 7, 4 + 8, 5 + 8$

addends differ by two. Several strategies for learning the facts in this set are illustrated in Figure 8.30. The approach easiest for the student would be used.

The 9's cluster is presented next. Again, some children find one strategy more helpful than others in the "getting ready to memorize" stage (Figure 8.31). Use the strategy a child finds easiest.

Now, in the remedial sequence, review the easier facts: the plus-zero facts, count-on's, and ten-sums. At this point it is highly effective to share with students how far they've come in mastering the one hundred basic addition facts. An addition table like that shown in Figure 8.32 might be used to color-highlight known facts. Mastery of doubles and related facts, 9's, count-on's, and the plus-zero facts leaves just three facts to be studied: $4 + 7$, $4 + 8$, and $5 + 8$. Many children have found it helpful to relate these three facts to known ten-sums.

Figure 8.30

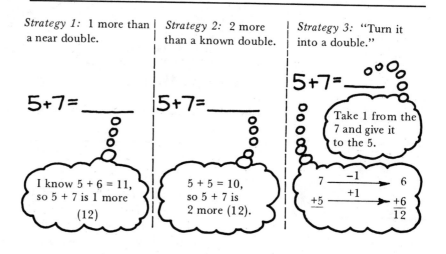

Strategy 1: 1 more than a near double.

Strategy 2: 2 more than a known double.

Strategy 3: "Turn it into a double."

Figure 8.31

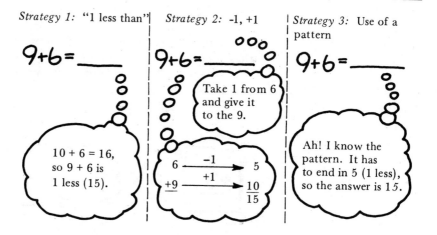

Strategy 1: "1 less than"

Strategy 2: -1, +1

Strategy 3: Use of a pattern

Organizing the addition facts into clusters such as those just described is a critical first step in helping children memorize them. Most children, such as Timmy and Pete in the following case study, also profit from carefully structured activities that teach or reinforce specific techniques or relationships to help them learn the facts.

Figure 8.32

+	0	1	2	3	4	5	6	7	8	9
					Addition Table					
0					Add 0					
1										
2					Count ons					
3										
4									Doubles + 2	3 last facts
5							Near Doubles			
6	Turn arounds						Doubles			9's
7	(commutatives)									
8	of other facts.									
9										

Case Studies

Counting On

Background Information. Timmy had just transferred into the fourth-grade class. He was ten years old and functioning at a low second-grade level in mathematics. Records, which were fairly complete, indicated that Timmy's work in addition had focused on sums to ten. Pete, already in the class, was near Timmy's age and level in basic facts. Pete, however, always counted on his fingers to find addition sums.

Result of teacher check on Timmy. A quick oral check showed that Timmy *did* count on, mentally, from the "big one" (as he called it) to find simple addition sums (see Figure 8.33).

Figure 8.33

Flash Card

$$\begin{array}{r} 2 \\ +8 \\ \hline \end{array}$$

Timmy's Approach

That's 8 . . . , 9, 10.

Teacher's plan for Timmy and Pete. To allow for larger blocks of instructional time, Timmy's teacher decided to subgroup within the mathematics class rather than work strictly one on one. Timmy would team with Pete for basic facts. The teacher grouped the facts into clusters to make for easier memorizing. She created several developmental and practice activities for each cluster, and planned to supplement as needed. She intended to assign adequate written exercises to reinforce the activity sessions, and to give frequent quizzes. She also planned to provide regular cumulative reviews in the form of activities, written exercises, and quizzes. Here is her basic outline for helping Timmy and Pete with the Count On cluster.

Focus: To teach Pete the "count on" idea; to help Timmy extend the idea to teen sums having 1, 2, or 3 as addends.

Note: Strictly speaking, count-on's are facts having 1, 2, or 3 as addends. The idea of counting on from the greater addend is one which is better, any time, than counting from 1 to find addition sums. Until facts are memorized, or better strategies learned, the counting-on technique can be used.

Sample Developmental Activity: Say 'N' Peek! (To help Pete count on from *any* number)

Need: Ruler and paper arrow card (see Figure 8.34).

Procedure: One player places a paper arrow card to the right of a ruler number. The other counts to twelve from the number pointed out. This player says the number, then peeks to check. Take turns.

Sample Developmental Activity: Oh Big One! (To help Pete recognize the greater addend)

Need: Deck of fact cards which have 1, 2, or 3 as addends.

Procedure: In turn, draw a card and give the sum *right away* for two points. Or, point to the big one (greater addend) and count on from there for one point. (No need to start counting from 1 each time.)

Note: At this point help the boys to see that, for example, 2+9 and 9+2 have the same answer. Turn-arounds always give the same answer —when you know one, you know the other. Using the commutative property cuts by nearly 50 percent the number of facts to be studied.

Figure 8.34

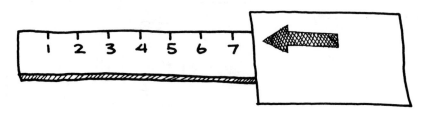

Sample Practice Activity: Spino! (To help Timmy and Pete count o[n]
for the facts having 1, 2, or 3 as an addend whenever they don't kno[w]
the sum right away)

Need: Two spinners (pictured in Figure 8.35).

Procedure: Both spinners are spun. First to say the sum wins
the point.

Winner: The one with most points at the end of ten rounds.

Variation: Take turns. Spin as often as possible within a time
limit. At each spin give the sum of the numbers on the two spinners. Tally one point if correct. Use a key to check if needed.
Keep track of "rights," and try to improve this record.

Discussion. The case study serves to illustrate the need for
both developmental and practice activities during the study of
basic facts. Sample activities for reinforcing other strategies, and
for providing extra practice with facts, are described below and in
the math lab activities of Appendix C. Oral and written exercises,
drills, and quizzes would also be used regularly in a program of
fact study.

Figure 8.35

Activity 1: Handstand. (Developmental activity for 10 sums)

Immediate goal: For the children to discover or review the many different combinations that add to 10: 7 + 3, 6 + 4, and so on.

Long-range goal: For the children to use effectively strategies like *one more than, two more than,* or *one less than* for memorizing fact sums near ten. For example, 7 + 3 = 10, so 7 + 4 is one more (11).

Need: Handstand board (Figure 8.36), pipe cleaner, and key listing all possible addition combinations for ten using two of the digits, 1-9 (e.g., 1 + 9 = 10, 9 + 1 = 10; 2 + 8 = 10, 8 + 2 = 10).

Procedure: Write all addition facts which have ten as a sum. Place the pipe cleaner between fingers on the handstand board, if necessary, to find any missing combinations. Figure 8.36, for example, shows 2 + 8 = 10. There are nine different facts that "make ten." Try to find them all.

Activity 2: Make 10. (Practice activity for 10 sums)

Need: Set of 81 special dominoes, like those shown in Figure 8.37, numerals on one side, dots on other; bag to store dominoes.

Figure 8.36

Figure 8.37

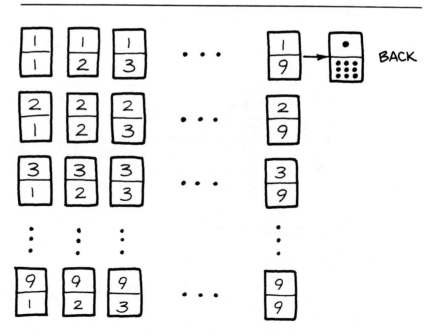

Procedure: Mix the dominoes in the bag, then draw one and place it, numeral side up, between players. Now each player draws seven dominoes. In turn, try to make a domino match by pairing two numerals that "make ten." When a match cannot be made, the player must draw from the bag until a playable domino is found. If need arises to check an answer, players can turn the dominoes over and count dots. Game play ends when a player runs out of dominoes or when the game is blocked because no player can make a match.

Winner: Player holding least number of dominoes when game play ends.

Activity 3: Find It. (Developmental Activity for Doubles)

Need: Addition flash cards for doubles; "touch-me" cards described in Figure 8.38.

Note: Textured cards like these are helpful to students with perceptual and memory disabilities. They can close their eyes and feel to reinforce the visual association being made.

Procedure: Match doubles cards to the touch-me cards.

Figure 8.38

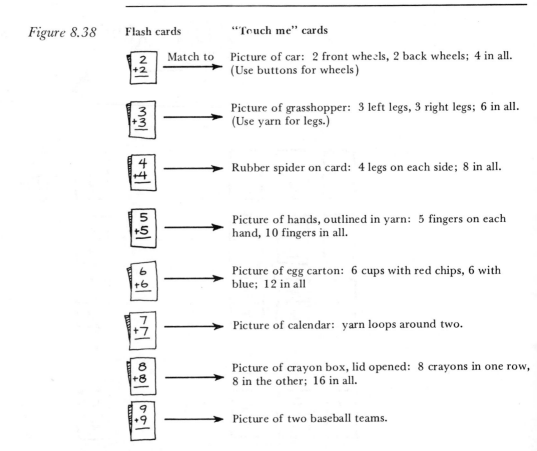

Flash cards   "Touch me" cards

Match to — Picture of car: 2 front wheels, 2 back wheels; 4 in all. (Use buttons for wheels)

Picture of grasshopper: 3 left legs, 3 right legs; 6 in all. (Use yarn for legs.)

Rubber spider on card: 4 legs on each side; 8 in all.

Picture of hands, outlined in yarn: 5 fingers on each hand, 10 fingers in all.

Picture of egg carton: 6 cups with red chips, 6 with blue; 12 in all

Picture of calendar: yarn loops around two.

Picture of crayon box, lid opened: 8 crayons in one row, 8 in the other; 16 in all.

Picture of two baseball teams.

Activity 4: On The Double. (Practice activity for doubles)

Need: File folder with game track drawn inside as shown in Figure 8.39; die marked 1, 2, 3; game markers for each player; key.

Procedure: To start, place markers on *in* and roll the die. High point goes first. Take turns. Roll and move as many spaces as the die says. Tell the double of the digit in each space as you go. If a miss is made, place marker beside that space until your next turn.

Winner: First to reach finish.

Note: An answer key for doubles, included with materials for the activity, makes it possible for children to check on each other. This or some alternate manner of providing answers frees the teacher to work with others in the class while On The Double is being played.

Activity 5: Spot The Double. (Developmental activity for near doubles)

Note: This activity can be adapted for stressing the *one more than* relationship of other facts as well.

Prerequisites: Mastery of doubles; ability to tell the number after another (to 18). If necessary, review Say 'N' Peek (p. 213). Require the child to say just the number after, then peek to check.

Need: Cards for doubles and near doubles—fact on front side, dots on the back as in Figure 8.40. (Textured dots are suggested;

Figure 8.39   Pocket for game materials.

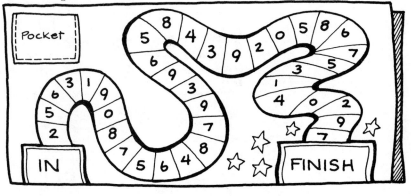

Figure 8.40

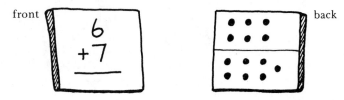

they make it possible for students with perceptual or memory disabilities to close their eyes and feel the relation of the double and near double. Go over the dots with glue and let it dry. Use sandpaper, cotton, or cloth dots.)

Procedure: Cards are mixed and placed inside a used mailing envelope. Take turns. Draw a card and tell the sum. If you don't know an answer right away, turn the card over and *spot* the double—cover one spot so top and bottom look alike (Figure 8.41). Count from this double to tell the sum. Write the double and near double.

Variation: Give the sum right away: two points. Give the sum by counting on, aloud, from a double: one point.

Note: From this point on, for every fact studied, the matching fact with addends in reverse order would be included in a cluster (e.g., 6 + 7 and 7 + 6). *An important idea, to be reviewed often, is that turn-arounds give the same answer.* When you know one, you know the other.

Activity 6: In And Out. (Practice activity for near doubles)

Need: Cardboard potato chip or English muffin can, cut as shown in Figure 8.42. Deck of near-doubles cards: problem on top, answer on bottom.

Procedure: Mix deck and place it in can holder. Look at the problem. Say *only the answer out loud.* See how long it takes to get through the whole deck. Log the time and try to improve it.

Figure 8.41

Figure 8.42

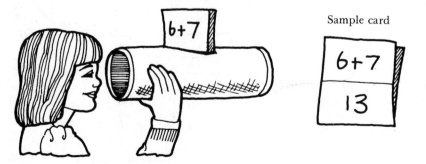

Sample card

Activity 7: Spot The Double. (Developmental activity for doubles + 2)

Note: Doubles + 2 are addition facts two more than a known double: 6 + 8, 5 + 7, 7 + 9, 7 + 5, 6 + 4, and so on. In each case, the addends differ by two.

Need: Cards for doubles and doubles + 2—number on front side, dots on the back.

Procedure: As in the previous Spot The Double activity, but adapt to facts *2* more than an easier, known double (see Figure 8.43).

Activity 8: Sort. (Developmental activity for doubles and related facts)

Need: Flash cards for doubles, near doubles, doubles + 2 facts.

Procedure: Sort cards into three piles: doubles, near doubles, doubles + 2 facts.

Variation: Include other facts as distractors. If Timmy and Pete are successful with this sort, they will be able to help themselves on written review exercises which contain a mixture of problem types.

Activity 9: Try It. (Practice activity for doubles plus two)

Need: Laminated cards like that shown in Figure 8.44; thirteen markers for each player (each player has a different color).

Figure 8.43

Examples:

$$6 + 6 = 12 \xrightarrow{\text{so}} 6 + 8 = 14; \ 8 + 6 = 14;$$

$$5 + 5 = 10 \xrightarrow{\text{so}} 5 + 7 = 12; \ 7 + 5 = 12.$$

Figure 8.44

Try it !!				
5+7	6+8	6+4	5+7	9+9
9+7	5+5	8+8	8+6	7+9
4+6	7+9	7+5	6+6	6+8
7+7	8+6	7+5	9+7	6+6
9+9	7+9	8+8	6+4	7+7

Procedure: Pick a game card. Take turns. Point to a problem and see if your partner can tell the sum *right away*. If so, the player claims it with a marker. If not, you have the chance.

Winner: Player claiming most squares when card is covered.

Activity 10: Make It Easy. (Developmental activity for addition 9's)

Prerequisite: Mastery of addition 10's (10 + 4 = 14; 10 + 6 = 16; and so on).

Need: Counters (two colors); flash cards for addition 9's.

Procedure: In turn, draw a 9's card. Give the answer *right away* for two points, *or* use counters to turn the problem into an easy 10 for one point (Figure 8.45).

Note: There are several other patterns for remembering addition 9's. Some children, for example, use the idea that 10 + 4 = 14, so 9 + 4 = 13 (1 less); 10 + 6 = 16, so 9 + 6 = 15 (1 less); and so on. Others notice the pattern that the one's digit of the sum is always one less than the number being added to nine (Figure 8.46). If the children find either of these ideas easier than that presented in the activity above, a change would be made to incorporate the easier strategy.

Discussion. This section has focused on ways of helping children memorize the basic addition facts. Children good at memorizing or devising their own strategies for recalling answers to harder facts don't need much teacher intervention. Most children, however, are helped when teachers organize facts into clusters which stress natura

---

Figure 8.45   *Example:*

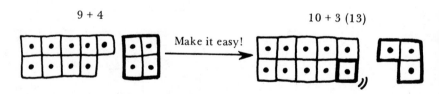

Figure 8.46   *Examples:*

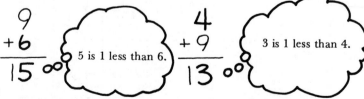

relationships or patterns. As they get ready to memorize, many children also need teacher support and encouragement to discover and use strategies or techniques which make memorizing easier.

Just telling these students to memorize the facts without more help on *how* to do so, is not enough help. The teacher's role in suggesting cues, hints, or other mental strategies to make memorizing faster and to insure retention is an important one. Equally important is the need to be open to ideas children themselves present for remembering certain facts. Most of these ideas will center on using "easier," known facts to help memorize "harder" combinations.

Both developmental and practice activities belong in a program of fact study. The case study of Timmy and Pete and the activities that followed presented ideas for this purpose. Most activities described can be adapted for use with several fact clusters. Additional work at the practice level would be extended to include oral and written exercises, flash card and other rapid-drill activities, and timed quizzes over each new set of facts. Frequent reviews and cumulative quizzes are also needed as students progress toward fact mastery.

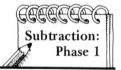

Subtraction: Phase 1

Basic Concepts and Models: Subtraction

Subtraction as "take away." Most elementary-school mathematics textbooks present subtraction after addition sums to give have been studied. The easiest and most common approach is to introduce subtraction through *take-away* situations. These examples start with an intact group, such as the five birds pictured in Figure 8.47a. To illustrate, or model the problem 5 – 3 = _, three birds are removed from the group (Figure 8.47b). This movement shows, in a dynamic way, the meaning of "5 subtract 3." One can now ask how many birds remain. Rapid sight recognition of the remaining group would be emphasized.

$$5 - 3 = 2$$

In early introductory work, children can use objects to model many different take-away situations for minuends five or less. This

Figure 8.47   *(a)*   *(b)*

is the only way they will *see* what is meant by *subtract*. This is the only way the new sign "–" will take on meaning. Given a physical situation in which objects are removed from an intact group, the child should be able to write the matching subtraction sentence (Figure 8.48). And, given a problem card such as 4 – 3 = _, the child should be able to use objects to show that only one now remains. Creating oral "story" situations to dramatize number sentences is also appropriate at this stage.

Relating addition and subtraction. Conceptually, subtraction is closely related to addition (Figure 8.49). The four number combinations illustrated may be regarded as a fact "family." When the whole and one part are known, the other part can be found. Teachers can help make children aware of the natural relation between addition and subtraction. Having children use counters or draw pictures and then write the fact family for given situations is one way of doing so. Activities like the following may be used.

Figure 8.48

Before | After

4 Jacks

4 – 3 = 1 Jacks

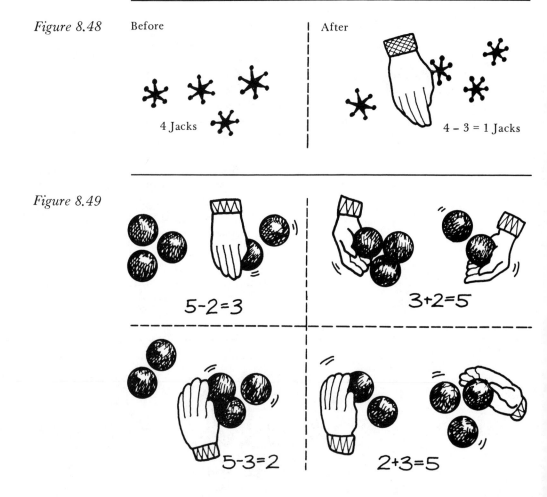

Figure 8.49

5-2=3

3+2=5

5-3=2

2+3=5

Sample Developmental Activity: Outlaw. (To focus on families of related addition and subtraction facts)

Need: Card deck of addition and related subtraction facts; one Outlaw card (Figure 8.50); ten counters for each player.

Procedure: Follow traditional Old Maid rules, but lay down all cards that belong to the same addition-subtraction family each time.

Examples: 2 + 2; 4 – 2.
3 + 2; 2 + 3, 5 – 2, 5 – 3.

Each set of related addition-subtraction cards could be color coded alike. As cards are laid down, answers to each fact problem should be given. Children could also be required to use counters to demonstrate how the chosen cards belong to the same fact family.

Activities with counters, verbal feedback from the teacher, and written work such as that just suggested will help children recognize families of addition-subtraction facts. Such experiences will pay off when children are required to memorize the subtraction facts. It usually is easier for children to give a subtraction answer when they can think of the related addition fact (Figure 8.51). This addition-subtraction interface is emphasized by most commercial mathematics textbooks as children are introduced now to:

1. Sums for 6

2. Related subtraction facts

3. Sums for 7

4. Related subtraction facts . . . and so on

Figure 8.50

Figure 8.51

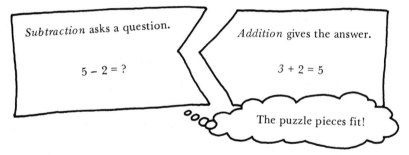

Subtraction as comparison. When children have developed confidence in the basic subtraction idea that is modeled by take-away situations, other subtraction problem types can be posed. These problem situations involve *comparisons* in which answers can be found by subtracting. An example follows.

Comparison Type 1: How much more does one have than another? Bill has 8 nickels. Sue has 5 nickels. How many more coins does Bill have?

The problem might be visualized as in Figure 8.52. The problem-solver immediately matches Sue's coins with Bill's and finds that Bill has three more. Intuitively, comparison situations of this type can be modeled and solved by one-to-one matching of the smaller set with a subset of the larger, and recognizing how many are left over. Children might also count on to find the answer. In this example ". . . 6, 7, 8," gives a count of 3 for the difference. Eventually, children learn that they can simply subtract to find this difference.

Another type of comparison situation is suggested by the following problem.

Figure 8.52

Bill's nickels:

Sue's nickels:

Figure 8.53

Ted needs:

Ted has:

Here the comparison is between what Ted has and what he needs
(Figure 8.53). Tactics similar to those presented for the previous
problem type might be used to solve this situation.

Another approach is one which directly focuses on *addition*. It
presupposes adequate mastery of addition facts used.

1. The teacher would write an addition number sentence on the
   board (Figure 8.54a), then

2. Cover one addend with a card (Figure 8.54b), and

3. Ask students to tell what's missing.

After repeating this with several addition facts, teachers might
divide the class into two groups. While half the class keep heads
down, the other checks that the fact written by the teacher is correct. Then one addend is covered with a card, as before, and the
first group tries to identify it.

Figure 8.55 illustrates how children with this background might
solve the example.

Whatever the approach, it is critical to prepare children to handle
comparison situations like those described. Too often a child's first
experience is reading about them in second-grade books. A typical
response, then, is to be misled by the word *more* into adding the
two numbers involved and reporting the sum. Introducing comparison situations earlier gives children time to learn ways of handling
them.

Figure 8.54

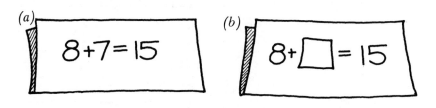

Figure 8.55

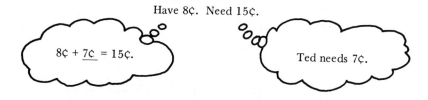

Assessing concept mastery. Several critical questions guide a teacher in evaluating a child's grasp of subtraction concepts. Being able to give correct answers to simple written combinations is not enough.

$$\begin{array}{cc} 5 & 6 \\ -3 & -2 \\ \hline \end{array}$$

"Right" answers to problems like these may indicate rote, rather than meaningful, understanding. One needs to probe deeper. Perhaps the most important question is the following.

1. Can the child model spoken or written problems with objects (Figure 8.56)?

   A child's ability to model subtraction situations with objects dramatizes a strong conceptual grasp of the operation. Children might be asked, for example, to use counters to prove a written solution is correct. Or, they might use objects to illustrate given problem situations.

   Teachers can also seek answers to the following questions.

2. Can the child write appropriate subtraction sentences to illustrate and solve simple situations that are presented orally or by concrete models (Figure 8.57)?

Figure 8.56

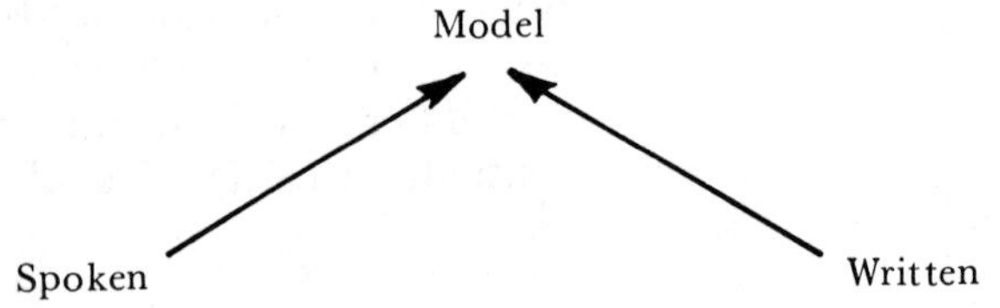

Figure 8.57

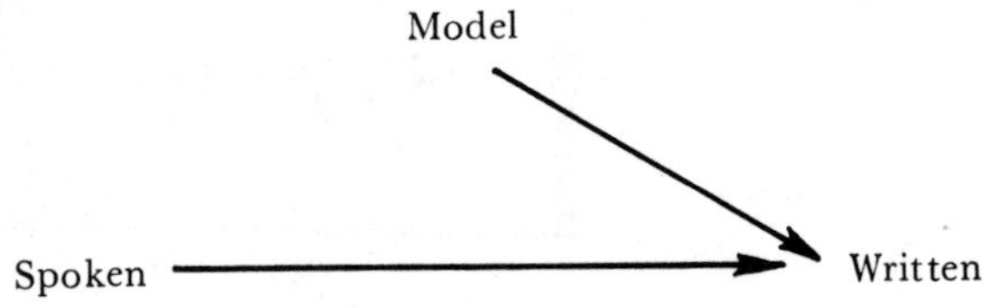

Figure 8.58

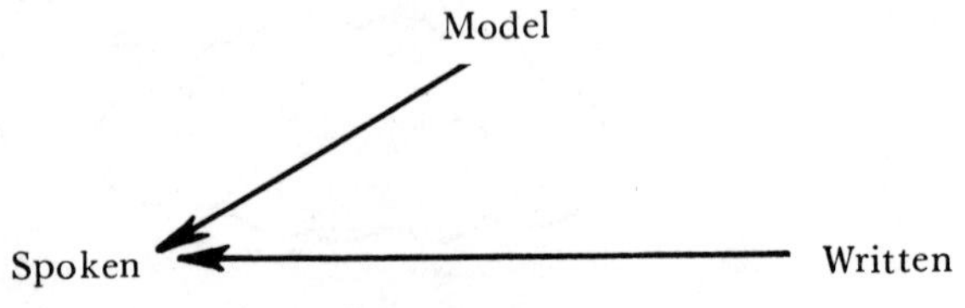

Teaching Mathematics to Children with Special Needs

A child's ability to do so demonstrates a recognition of subtraction in real-life applications.

3. Can the child create simple oral problems to interpret a physically modeled or written subtraction situation (Figure 8.58)?

Children who have mastered the concept and can use it may still have some trouble with this task. Creating or verbalizing examples may be difficult. When oral problem-solving has been interwoven during instruction, it will be more natural for a child to describe subtraction situations that relate basic facts to familiar experiences.

Yes answers to at least the first two critical questions indicate that children are ready to begin memorizing the basic facts.

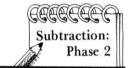

**Subtraction:
Phase 2**

Toward Mastery of Basic Subtraction Facts

For many children, memorizing subtraction facts is harder than mastering addition facts. Not only are they unable to remember answers to given facts; they also frequently rely on inefficient or faulty strategies for deriving fact answers. Being sensitive to the problem may prevent it from being inevitable.

The outline near the beginning of this chapter summarized a basic approach to the problem of fact mastery. *The natural starting point is memorization of easy facts, beginning with minuends through 5.* Concrete models are helpful at this stage. Repeated visualization of the simpler facts using the take-away model leads to random learning of many of these facts.

Subtraction facts related to ten-sums or to addition doubles are often included among the "easy" facts by children. These, too, can be modeled. The case study of page 231 presents a way of using fingers to visualize subtraction facts having ten in the minuend. The Touch Me cards of Figure 8.42 suggest ideas for modeling those facts related to addition doubles (see Figure 8.59). Eventually most children are able to close their eyes and picture the

---

Figure 8.59

Full box of crayons:

Take 8 crayons (1 row) out:

16
16 CRAYONS

16 CRAYONS

16
− 8

8 crayons
(1 row) left.

model as an aid to recalling these easier facts. Repeated use of the facts in oral and written exercises and in activities leads to their memorization.

This background readies children to move on to the development of efficient *mental* strategies for learning harder subtraction facts. As in addition, a first step in this process is *organizing* the facts in a manner which emphasizes patterns and natural relationships. Children generally note patterns involving zero and one rather quickly (Figure 8.60). Many children at this stage begin using the addition-subtraction relationship to retrieve subtraction answers (Figure 8.61). Other children are helped by an organization which relates a "hard" fact to an easier, known fact. The *one more than* strategy is one basis for this organization (Figure 8.62). Though not as common, some children use the *one less than* idea for figuring out facts

Figure 8.60

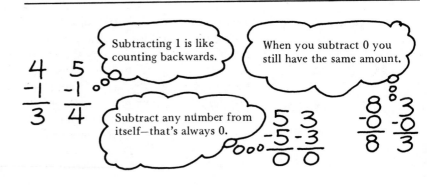

Figure 8.61

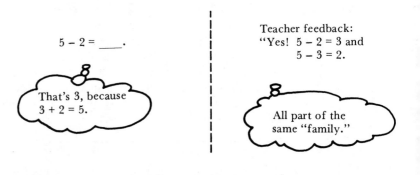

Figure 8.62   **Examples using the "one more than" idea:**

Easy fact		Unknown fact	Easy fact		Unknown fact	Easy fact		Unknown fact
5 −2 3	so →	6 −2 4	10 − 3 7	so →	11 − 3 8	16 − 8 8	so →	17 − 8 9

(Figure 8.63). Teachers can help children focus on relationships such as these by:

1. Illustrating them with counters

2. Placing examples side by side on worksheets

3. Discussing the relationships when they naturally occur during class sessions

The final goal in this process is mastery of all one hundred basic subtraction facts. Children meet this goal when they can give *accurate* and *quick* responses to the facts, and can do so *consistently* over time.

The following paragraphs elaborate on these and other suggestions for helping children memorize the basic subtraction facts. Hiding activities are presented first. They serve as a bridge between:

1. Early work with objects in which children can count the *number left* in subtraction situations, and

2. Later work without objects that requires answering written, or orally presented, subtraction facts.

A look ahead: Hiding activities. In the familiar take-away situation, children can see and count, if necessary, the number of objects in the remaining set (Figure 8.64). After repeated opportunities to remove objects from intact groups, children good at memorizing will eventually remember this number for solving written or orally presented facts. Other children, however, can benefit from a class of activities called *hiding activities*. These activities show objects in

---

Figure 8.63　　**Examples using the "one less than" idea:**

Easy fact		*Unknown fact*		*Easy fact*		*Unknown fact*
12		11		10		9
− 6	so	− 6		− 4	so	−4
6	→	5		6	→	5

Figure 8.64

3 are left.

Chapter 8　Whole Number Addition and Subtraction　　**229**

the intact group, but "hide" the number in the remaining subgroup. Since they challenge children to think about rather than see or count the number remaining each time, hiding activities provide a stepping stone to work without objects and memorization of subtraction facts. The following example illustrates this point.

Sample Hiding Activity

Prerequisite: Basic concept of subtraction, as modeled in take-away situations, well established.

Need: Small box or other container; pennies or other counters.

Procedure: Teacher calls on students to carry out the steps illustrated in Figure 8.65. Children should quickly tell, even guess, the number left hidden in the container *before* emptying it or peeking to see. After several rounds of the activity, children should be required to *write* the subtraction sentence that describes the situation.

Oral problems like the following might also be used with hiding activities.

Put 10 pennies in a purse and go to the store. Take out six cents to buy gum. How many pennies in the purse now?

Discussion. Hiding activities force children to *think* about answers to subtraction facts rather than rely on seeing the number in the residual group. An important step of the activity is having students tell the number left *before* looking. When possible, emphasize strategies such as those illustrated earlier for recalling fact answers. The feedback of emptying or peeking into the container will reinforce children's attempts to obtain subtraction answers through mental strategies. If children continue to guess wildly, or if they consistently blank out, the approach illustrated in the following case study may be necessary.

Figure 8.65   (a) Put 5 pennies in;   (b) Take 2 pennies out;   (c) Tell how many are left.

5

5 − 2

How many left?

5 − 2 = ☐

**Case Study:
Bob**

Counting Up

Background information. Bob, a first grader, was slower than others in learning the basic facts. He *had* demonstrated his understanding of the subtraction concept, but found it very hard to memorize even the easier subtraction facts. He was easily frustrated in both oral and written exercises and often went blank, unable to respond, when given a simple subtraction problem to answer.

Teacher's approach to the problem. The teacher decided to show Bob how to *count up*, as illustrated in Figure 8.66, to help him figure out subtraction answers. Given the total set and one part, one *can* count to find the number in the remaining part. The teacher planned to provide feedback, when possible, that would help Bob focus on more mature *mental* strategies for figuring out subtraction answers (Figure 8.67).

Discussion. Counting up in this manner to find subtraction answers involves some finger counting, hence is less desirable (also slower) than using other strategies discussed in this section. Still, it *is* a start for those who go blank or resort to even less efficient strategies. In this problem, to keep the child from counting up "6, 7, 8" and giving three as the answer, instruct that a pause be made after saying six and then to continue with "7, 8."

Teacher feedback, as illustrated in Figure 8.67, is crucial when the counting-up technique is used. Reminding children of other strategies or relationships as often as possible gradually leads many of them to independent recall of fact answers. In the process, of course, finger counting would be dropped altogether.

Figure 8.66   *Sample problem:*   "6, — 7, 8. The answer is 2, since 2 more make 8."

Figure 8.67   *Example 1:* Relate to addition.
"Yes (pointing),

2 and 6 more
make 8."

Example 2: 1 more than a known fact.
"Yes, you know that:

$$\begin{array}{c} 7 \\ \underline{-6}, \\ 1 \end{array} \quad \text{so} \quad \begin{array}{c} 8 \\ \underline{-6} \\ 2 \end{array} \text{."}$$

Subtraction facts: a drawback to success in computation. Sometimes teachers are faced with situations in which inadequate mastery of subtraction facts severely impedes progress in computation. This situation is the subject of the following case study.

Case Study: Jennifer

Using a Subtraction Table

Background information. Jennifer, a fourth-grader, is regarded as a slow learner in mathematics. At the time of this case study, her slowness in figuring out answers to subtraction facts was interfering with progress in computation.

Teacher's approach to the problem. The teacher drew up a subtraction table showing all one hundred basic facts. Jennifer was then tested to determine which facts she has mastered. Together with the teacher, Jennifer blacked out all known facts on the table of Figure 8.68.

Each week, the teacher identified three or four facts still on the table, each of which was *one more than* some fact Jennifer already knew. The teacher made sure Jennifer was aware of the relationship between new and known facts, and then suggested that Jennifer use the idea to help memorize the facts by the end of the week. "Then we'll black these facts out, too."

At the week's end the teacher blacked out the three or four facts, as promised. Only a bite-sized task was given each week and this made learning the facts a reasonable task. If for some reason Jennifer fell behind and failed to learn the facts for the week, they were still blackened. Jennifer was then left to her own resources for figuring them out when needed. Since the teacher always chose facts with answers one more than an easier, known fact, this was not too difficult. In the long run, however, Jennifer decided not to get behind. She found it easier to do a bit each week. In the meantime, she was able to make greater progress with computation by referring to the table for facts not yet memorized.

Figure 8.68

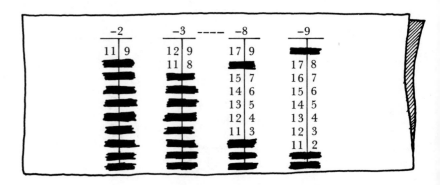

Discussion. The approach used by Jennifer's teacher is one that has helped many students in similar situations. Two goals are met:

1. Children can make progress with more difficult computational tasks by using the chart when necessary. Since all known facts are blacked out, the chance of becoming over-dependent on the chart is eliminated.

2. Children do, in the meantime, master unknown facts. And they do so rather quickly.

In the case study the teacher helped Jennifer use the *one more than* idea to learn facts she didn't know. Other students may find different strategies easier. A technique like that illustrated in the following case study, which uses 10 to help with teen minuends, has proved useful to many children.

Case Study: Beth

Special Help With Teen Minuends

Background information. Beth's major learning difficulties are auditory sequencing and short-term memory. Though in the fifth grade, she is functioning at a beginning to middle second-grade level in mathematics. At the time of this study she was having difficulty with teen minuends in subtraction. Her learning strengths are reasoning and the ability to profit from instruction that is highly visual and tightly sequenced in small increments, with adequate provision for overlearning.

The following four-step sequence was used to help Beth:

1. Review of subtraction facts having ten in the minuend.

2. Review of relationship between teen numbers and ten. (1$\underline{4}$ is $\underline{4}$ more than ten; 1$\underline{6}$ is $\underline{6}$ more, and so on.)

3. Use of ten in the minuend to help with teen minuends.

4. Use of games, activities, written exercises, and timed quizzes to promote speed and provide necessary "overlearning."

Beth had mastered most subtraction facts having minuends ten or less when the teacher started the four-step sequence.

Step 1

Review of subtraction facts having ten in the minuend. Since Beth's teacher wanted to use the ten fingers as a visual model, Beth was first checked for *instant sight recognition* of any number of fingers the teacher might display (Figure 8.69). The goal was to avoid any counting of fingers during the activity.

Assured that Beth's sight recognition for this task was strong, the teacher tied the activity to subtraction facts having ten in the minuend. She first showed all ten fingers to Beth, then asked the child to close her eyes. "You saw my ten fingers. Now I'm putting two fingers down. How many are still up?" Beth was encouraged to quickly tell—even guess, before looking to check (Figure 8.70). With Beth watching, the teacher dramatized the finger action again. Beth then told what subtraction fact could be written to describe the situation (Figure 8.71). The activity was repeated for other facts having ten in the minuend. The teacher and Beth traded roles from time to time in the activity.

The teacher used follow-up flash card drills, written fun sheets, timed quizzes, and self-check activities, and Beth was encouraged to think of her fingers if she didn't know a fact right away.

Sample Practice Activity: Flip And Check.
Need: Flip pad like that shown in Figure 8.72 on one side, answer on the back of each sheet. The pad is bound with a large ring binder.
Procedure: Say the answer, then flip to check. Time yourself. Try to improve your record on your next try.

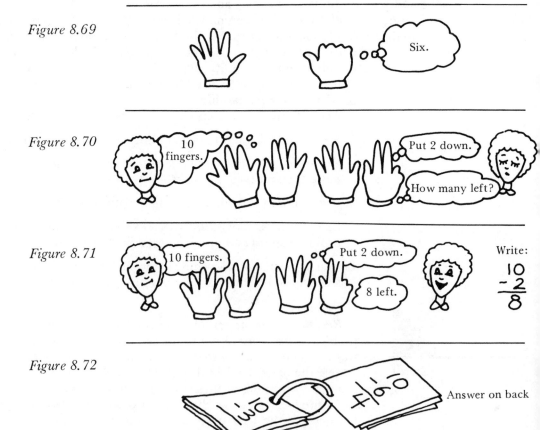

Figure 8.69

Figure 8.70

Figure 8.71

Figure 8.72

Answer on back

Step 2

Review of relationship between teen numbers and ten. The teacher next made sure that Beth could compare ten with other teen numbers. In particular, the teacher wanted Beth to look at any teen number she wrote and tell how much greater than ten it was. Beth's teacher wrote teen numbers on cards, and showed them to Beth one by one. A centimeter ruler was used to check. Beth moved a button from ten to the teen numeral on the ruler, counting the number of moves made (Figure 8.73).

Step 3

Use of ten in the minuend to help with teen minuends. Beth's teacher helped her analyze a subtraction fact with a teen minuend (Figure 8.74). Beth and the teacher repeated this analysis with other teen minuend facts. Then the teacher placed a "0" card over

Figure 8.73

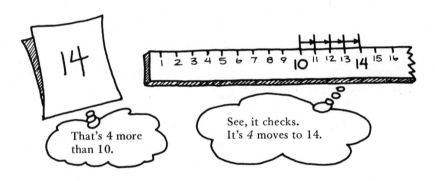

Figure 8.74

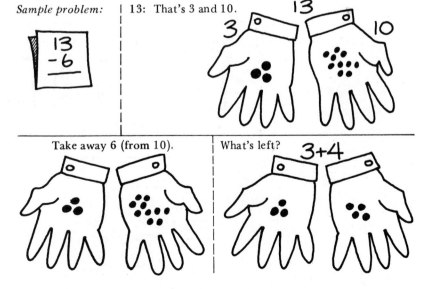

the one's digit, and asked *two easy questions* (Figure 8.75). Other examples reinforced the two-question approach to teen minuends (Figure 8.76). Beth's work in steps 1 and 2 of the sequence readied her with prerequisites needed for success at this step. Beth discovered that her index finger could be the "0" card (Figure 8.77).

Figure 8.75

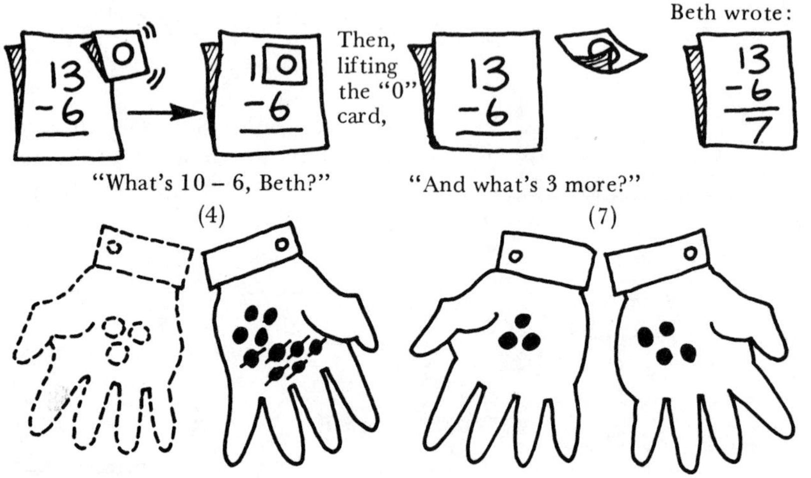

Beth wrote:
13
-6
10
-6
Then, lifting the "0" card,
13
-6
13
-6
7
"What's 10 – 6, Beth?"
(4)
"And what's 3 more?"
(7)

Figure 8.76

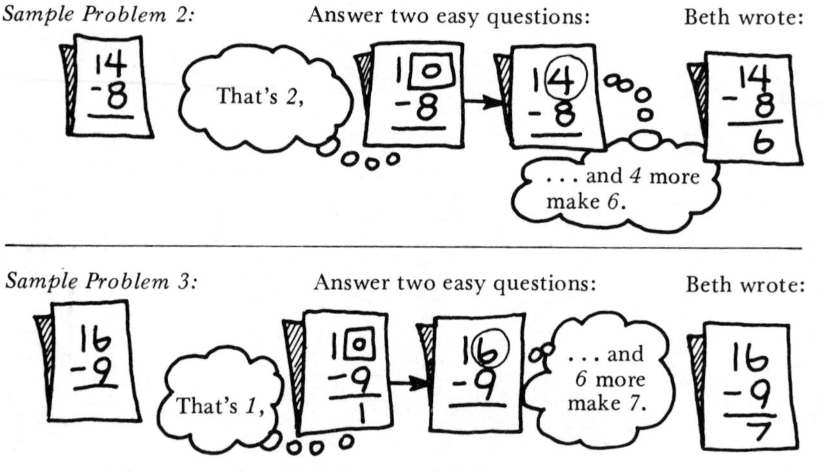

Sample Problem 2:
Answer two easy questions:
Beth wrote:
14
-8
That's 2,
10
-8
14
-8
... and 4 more make 6.
14
-8
6
Sample Problem 3:
Answer two easy questions:
Beth wrote:
16
-9
That's 1,
10
-9
1
16
-9
... and 6 more make 7.
16
-9
7

Figure 8.77

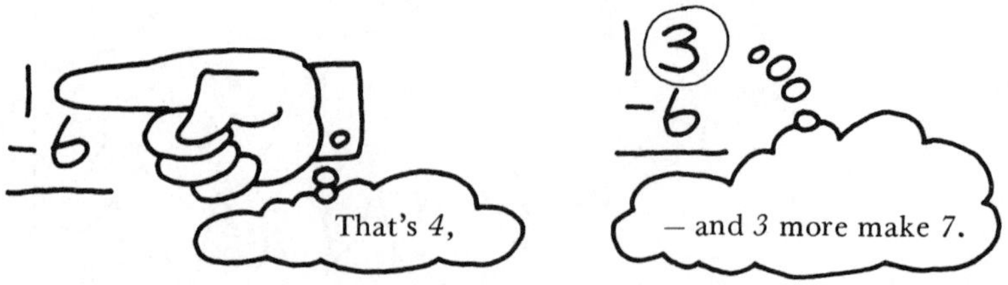

1
-6
That's 4,
13
-6
– and 3 more make 7.

Whenever possible during this stage Beth's teacher confirmed each answer by referring to the related addition or an easy subtraction fact. Beth was using a strategy to help her out of difficulty with teen minuends. The strategy involved skills she had previously mastered: subtracting from ten and then adding in extras. If Beth could learn to use the cue of a related addition fact, or relate a harder teen-minuend fact to an easier known fact, this two-step strategy would be unnecessary. She would be able to give the subtraction answer right away.

Step 4

Use of games, activities, written exercises, and timed quizzes to promote speed and provide necessary overlearning. When Beth was confident with the strategy, the teacher promoted speed with "ten-minuend" facts in many ways. Besides written and oral exercises, timed quizzes and games, activities like the following were useful.

Sample Practice Activity: Draw And Take
    Need: Subtraction flash cards (see Figure 8.78).
    Procedure: Draw a card and tell the answer. Turn card over to check. Make two piles: those you know *right away*; and others. Make a bar graph of the number of cards in the "I know it" pile. Try another time to better your record.

Discussion. This section has presented ideas for helping students master the basic subtraction facts. The case studies further illustrated special tactics that may be necessary to meet special needs. For children who have difficulty, *the need to cluster facts for easier memorizing is critical.* Doing so will help them discover and use relationships, patterns, or techniques which enable them to figure out, as quickly as possible, answers to unknown subtraction facts.

At times it is helpful to use the addition sequence as a guide for clustering subtraction facts. For example, if the *doubles, near doubles* sequence is followed, subtraction facts related to doubles might be studied before those related to near doubles (Figure 8.79).

Some teachers, particularly in remedial cases, make sure that students have mastered *all one hundred* basic addition facts before

---

Figure 8.78   Sample Cards

Front

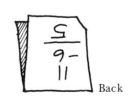
Back

turning to teen minuends in subtraction. This generally makes it easier for them to use addition to help find subtraction answers.

Another approach is to help children use easier, known subtraction facts as a basis for learning harder facts (Figure 8.80).

Sometimes children discover patterns which help them remember certain difficult facts (Figure 8.81). Such approaches should be encouraged or even suggested whenever they seem to help.

Many ideas of this section are included to help children as they prepare to memorize the facts. *Final mastery is not achieved until all facts are memorized and tested for retention over time.* When facts are mastered, responses are immediate rather than delayed by intermediate strategies. Repeated use of subtraction facts in motivating games and activities, oral and written exercises, speed drills and timed quizzes is necessary in the final states to secure fact mastery.

Figure 8.79

Subtraction facts related to addition doubles —"easier than"→ those related to near doubles.

$$\begin{array}{r}12\\-\ 6\\\hline\end{array}\quad\begin{array}{r}16\\-\ 8\\\hline\end{array}\quad\begin{array}{r}14\\-\ 7\\\hline\end{array}\qquad\qquad\begin{array}{r}13\\-\ 7\\\hline\end{array}\quad\begin{array}{r}13\\-\ 6\\\hline\end{array}\quad\begin{array}{r}11\\-\ 5\\\hline\end{array}\quad\begin{array}{r}11\\-\ 6\\\hline\end{array}$$

Figure 8.80

Example 1: "One more than"

$$\begin{array}{r}16\\-\ 8\\\hline 8\end{array}\ \xrightarrow{\text{so}}\ \begin{array}{r}17\\-\ 8\\\hline 9\end{array}\qquad\qquad\begin{array}{r}10\\-\ 7\\\hline 3\end{array}\ \xrightarrow{\text{so}}\ \begin{array}{r}11\\-\ 7\\\hline 4\end{array}$$

Example 2: "One less than"

$$\begin{array}{r}12\\-\ 6\\\hline 6\end{array}\ \xrightarrow{\text{so}}\ \begin{array}{r}11\\-\ 6\\\hline 5\end{array}\qquad\qquad\begin{array}{r}14\\-\ 6\\\hline 8\end{array}\ \xrightarrow{\text{so}}\ \begin{array}{r}13\\-\ 6\\\hline 7\end{array}$$

Figure 8.81

Example: Subtraction 9's

$$\begin{array}{r}14\\-\ 9\\\hline 5\end{array}\qquad\begin{array}{r}17\\-\ 9\\\hline 8\end{array}\qquad\begin{array}{r}16\\-\ 9\\\hline 7\end{array}$$

The answer is always 1 more than the one's digit in the minuend.

As children work to master basic facts, teachers can use informal problem situations to alert them to the many kinds of problems they have learned to solve. Children can be encouraged to finish short stories based on number sentences of their own choosing. Gradually, teachers can use pictograms or simple vocabulary to create problem situations or stories for which children just tell whether to add or subtract. Creative teachers can also find ways to use dramatizations to involve students in problem-solving situations.

Later, children will be introduced to written problem situations in their mathematics books. Then drawing pictures to illustrate given facts can be encouraged. Sometimes a picture will suggest a solution (Figure 8.82). An extension of this activity is to have children draw pictures to help solve problems they personally create.

Figure 8.82

Jack has 6 boats.

Terry has 4 boats.

Who has more boats? _____

How many more? _____

Dramatizing storylines can also be an enjoyable experience. Sometimes children enjoy the challenge of filling in their own numbers, and then showing that their answers make sense (Figure 8.83).

Many problems can have several correct solutions. One purpose of problem solving is to promote creativity and resourceful thinking. It is a good idea in early work to assign only a few problems at any one time. It is also good to reserve class time regularly to discuss different ways of solving given problems. Thoughtful, careful approaches like these can go a long way toward creating happy and fruitful problem-solving experiences for children.

Figure 8.83

_____ girls went to the park
_____ boys went.
How many children went?
Were there more boys or girls? _____
How many more? _____

Cost: _____
Have: _____
How much more money
is needed? _____

**Basic Facts
and Assessment**

Two levels of assessment.

Basic fact mastery involves giving *correct* answers *quickly*, and being *consistent* in performance over time. Any assessment in the area of basic facts is based on these criteria.

Level 1 assessment: initial mastery of a cluster of facts. The first level of fact assessment relates to a child's attempts to master each new cluster of facts studied: count-ons, doubles, near doubles, and so on. When children have had the chance to investigate cues and hints for remembering difficult facts in a cluster, both oral and written quizzes could be used. Jiffy quizzes are good for this purpose. These quizzes are:

1. Short—only 6 to 8 facts; if timed, only 15 to 30 seconds

2. Frequent—even daily over a period of several days

3. Limited at first only to the facts in the new cluster

If children have severe learning difficulties, avoid unannounced quizzes. These children feel more secure, do better, and tend to develop more positive attitudes toward fact study when they can practice a fact set before being tested. Telling the students they will be quizzed on a selected subset of facts so that they can study in advance helps accomplish two goals:

1. It minimizes frustration over being ill-prepared

2. It builds in success experiences by maximizing opportunities for them to do well

These goals are particularly important for special students.

Teachers will need to establish some criterion for children to meet before moving on to a new cluster of facts. Some teachers require that children have perfect or near-perfect scores on a written and an oral quiz within the same school week. Others require perfect or near-perfect scores on three consecutive written quizzes. Some give regular timed quizzes and require a certain minimum number correct in the given time over a period of several days. Whatever criterion is adopted should be discussed with and understood by the children. They tend to work harder and do better when they know what is expected of them.

Level 2 assessment: cumulative reviews. The Level 1 assessment just described does not evaluate long-term retention of facts. *Real mastery implies that, over time, children consistently exhibit both accuracy and speed with facts.* Further, they must maintain this performance when facts are mixed in reviews with others previously learned. Can a child, for example, still give ready answers to addition 9's when they are mixed with count-ons, doubles, and other facts already studied? Cumulative review quizzes serve to test mastery at this level. We recommend that jiffy quizzes again be used for this purpose. Only occasionally would longer two- or five-minute quizzes be given. Note that pencil-and-paper quizzes such as these can only tell whether children can find answers quickly. They don't pinpoint which facts are more rapidly answered than others. Nor do they reveal whether the answers are memorized or just quickly figured out. Timed or rapidly paced quizzes may be entirely inappropriate for students with poor motor coordination, integrative processing, or expressive language disabilities.

During oral quizzes some children with auditory handicaps may need to repeat the fact problem before giving an answer. Ordinarily this would not be allowed; only *immediate responses* would be counted.

For written quizzes, a mastery level in terms of number correct per minute could be set. This level is really an individual matter and should be placed just below a child's writing speed. These speeds vary due to differences in age and motor coordination. An individual's writing speed can be determined by counting the number of digits a child can write in one minute. A child who can write fifty digits per minute should be able to correctly write answers to thirty basic facts per minute. Experience has shown this 5 to 3 ratio to be an appropriate one. It accounts for the two-digit responses of many addition facts and the greater difficulty children characteristically experience with subtraction.

Individual goal charts.

It is important to keep children attuned to their progress and goals for improvement. One effective technique for doing this is to have children keep a personal bar graph of the number of *rights* on basic fact quizzes (see Figure 8.84).

Experience has shown these graphs to be highly motivating. They are personal records, so they encourage children to meet individual goals and monitor personal progress. There is no pressure to compare progress to that of others in the class. A record is good not because it betters what others have done, but because it shows the child has met or even exceeded personal goals.

Teachers may find it useful to keep a master list, such as that illustrated in Figure 8.85, to record children's progress through the fact clusters. Note how the chart allows the recording of individual student's performances on particular facts. Knowing exactly which facts a student knows will help the teacher in grouping children and selecting appropriate instructional activities for them. Both initial mastery and retention can be charted through cumulative review quizzes. Teachers can either checkmark or date entries to track a child's progress. The master list information can assist teachers in parent conferences or IEP evaluation and planning for handicapped students.

Other teachers prefer to use a separate addition table for each child (Figure 8.86). Initial mastery for each fact is dated or checked. Demonstrated retention is recorded using a different-color pen.

Your Time: Activities, Exercises, and Investigations

1. Prepare, as your instructor directs, answers to the major questions on page 201 of this chapter.

2. Describe *one* set of materials you might use for early work in addition. Using these materials, construct a sequence of developmental activities to introduce addition. Be brief but concise in your writeup. Make clear how you would go about introducing the "+" and "=" signs.

3. Review the case study of Timmy and Pete (p. 212).
   (a) What ideas from this sequence can be used with beginners for learning addition sums to 10?
   (b) Briefly describe a practice activity that might be used to supplement that given in the case study.

4. Consult the sample curriculum given in Appendix A. From the list of objectives,
   (a) Identify the grade level by which students are normally expected to have memorized basic addition facts with sums ten or less.
   (b) Identify the grade level by which students are normally expected to have memorized basic addition facts with sums through eighteen.
   (c) Tell what expectations are laid out, by grade, for mastery of basic subtraction facts.

Figure 8.84

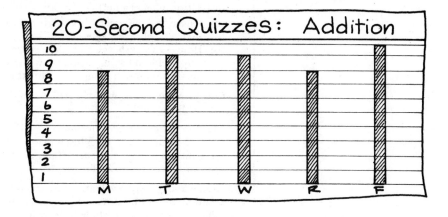

Figure 8.85   **Sample Master List for Addition—Basic Facts**

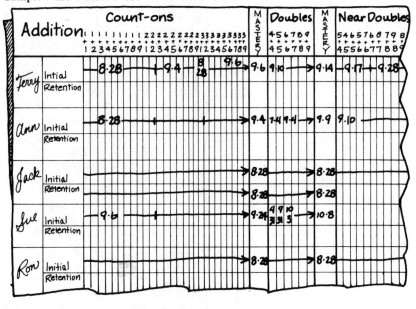

Figure 8.86

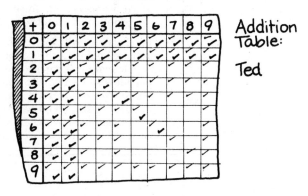

5. A worksheet shown in Figure 8.87 emphasizes patterns of thinking or relationships as a means of helping children learn addition facts.
   (a) Where in a sequence for studying the basic addition facts might such a worksheet be used?
   (b) Design a worksheet which emphasizes relations or patterns to help children having difficulty with addition facts one more or one less than ten. Sample fact for this page: $8 + 3 = 11$.

6. Compare the role of worksheets like that illustrated in Figure 8.88 to that of Figure 8.87.

7. The one hundred basic addition facts are laid out in the addition table of Figure 8.32. If children make use of the commutative property, and no others, how many facts are left to learn? What if adding ones and zeros is also included?

8. Design a diagnostic/evaluative test for assessing a child's grasp of addition and subtraction *concepts*. Use the comments of pp. 206 and 226–227 as a guide. Include both written and interview questions. If a young primary child is available, try out your test.

9. Write a word problem to represent each of the following subtraction situation types:
   (a) Take away   (b) Comparative

10. Refer to *one* of the problems you just wrote. Suggest two different techniques or strategies that might help children solve it.

11. Review the case study of Beth on p. 233. Suppose the teacher saw the need for additional practice activities. Briefly describe two activities that might be used:
    (a) A self-check activity Beth could do alone.
    (b) A second activity Beth could do with one or two friends.

12. Consult the teacher's edition of a current mathematics text for grades 1, 2, or 3 (recent copyright date).
    (a) Use the text as a guide for creating one developmental *and* one practice activity for basic facts.
    (b) If either of your activities can be used for *both* addition and subtraction, describe any adaptations necessary.
    (c) Share the activities with others in your class.
    (d) If possible, try out the activities with a young child.

13. Refer to the master list of Figure 8.85.
    (a) Design a retention quiz for Terry which includes count-ons and zeros; doubles and near doubles.
    (b) Briefly describe the criterion you would use to decide whether, in fact, retention through this cumulative review is demonstrated.
    (c) How much time do you think should elapse between initial mastery and a retention check for basic fact clusters? Discuss this problem with others in your class and with your instructor.

14. Teachers normally expect they will need to supplement the exercises in the school mathematics book for children to master the basic facts. Supplementing is particularly necessary for children with special needs. The ideas and activities of this chapter provide suggestions in this regard. Select a current second-grade mathematics textbook and analyze the section on basic facts. Identify *three* places where ideas, techniques, or activities from this chapter might be used to supplement the activities or exercises of the text.

Figure 8.87

Figure 8.88

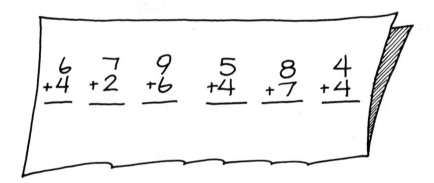

ADDITION AND SUBTRACTION COMPUTATION

What Do *You* Say?

1. What big ideas pervade developmental work for the addition and subtraction algorithms?

2. How can teachers prepare children to regroup in addition?

3. How do fair trade activities prepare children for regrouping in subtraction?

4. What types of motivational worksheets can be provided to make computation practice more interesting?

5. What basic types of self-checking activities can be provided for this purpose?

6. How does one assess computational mastery?

7. What basic guidelines can one follow in remediating computational difficulties?

Addition Algorithms: Phase 3

Big Ideas First

A basic understanding of place value and an adequate mastery of basic facts ready a child for computing larger sums. While school mathematics textbooks vary slightly in their instructional approaches to addition computation, a fairly standard sequence can be noted. Throughout this sequence, two big ideas prevail. First, children must understand that like units are added (ones to ones, tens to tens, hundreds to hundreds, and so on). Second, they must learn that when there are too many of some unit to write it down in one position, a trade must be made. Popsicle sticks or other grouping aids can be used to illustrate these ideas (Figure 8.89). If the addition gives ten or more of any unit, a trade must be made (Figure (8.90).

Figure 8.89

$4+43$ *Two easy questions:* *How many ones?   *How many tens?

*To find the sum, you add like units.

$4+43$

I'll rewrite so like units are in columns. This will make it easier to add them.

TENS	ONES
	4
+ 4	3
4	7

Figure 8.90

TENS	ONES
3	6
+ 1	7
	13

Too many to write in one place, so a trade must be made.

Trade   for

TENS	ONES
3	6
+ 1	7
5	3

Regrouping in Addition: A Close Look

Teachers can prepare children for regrouping in addition before presenting problems which require this. The idea is to use materials to help children realize that a number can be represented in different ways (Figure 8.91). Activities such as the following develop this idea.

Figures 8.91

**Activity File:
Preparing to Regroup**

Sample Activities: Preparing to Regroup in Addition

Activity 1: 10 Wins. (Developmental activity)
  Need: Regular die; popsicle stick tens and ones in a bank.
  Procedure: Each player rolls the die. High point goes first. In turn, roll and take that many ones from the bank. Whenever you get 10 ones, trade for a ten.
  Winner: After ten rounds, the player with the most tens wins.

Activity 2: Spin 'N' Take. (Developmental activity for 3 to play)
  Need: Bank of popsicle stick tens and ones; two spinners (children supply the pencils); recording sheet for each player (see Figure 8.92).

Figure 8.92

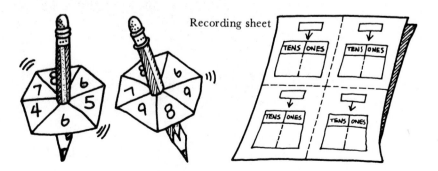

Recording sheet

Procedure: Each player spins. High point goes first. Spin both spinners and add the numbers you get. Take this number of sticks, in tens and ones, from the bank. Keep a record of each "take." Next player has a turn. After four rounds, look at your record sheet. Count the number of tens. Check that this number matches the number of ten bundles you have.

Winner: Player who has the greatest number of tens.

Note: Both activities above can be adapted for use with three-digit numbers. Use graph paper hundreds, tens, and ones instead of popsicle sticks.

Activities of this type, together with textbook exercises of supplementary worksheets, prepare children for regrouping in addition (Figure 8.93).

Many school mathematics textbooks present transitional forms for computation (Figure 8.94). Experience suggests that teachers

Figure 8.93

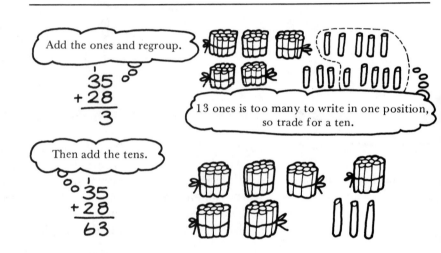

Figure 8.94

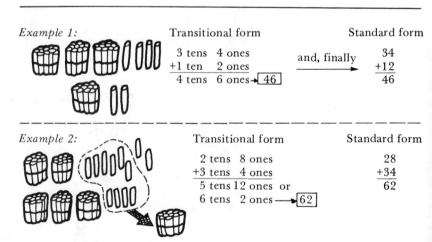

should *avoid* such forms with children having learning difficulties in mathematics. It *is* helpful, instead, to:

1. Use popsicle sticks or graph paper pieces to illustrate the addition in early developmental sessions (Figure 8.95)

2. Use *column labels* (hundreds, tens, ones) to make a record of this activity.

Activities like the following have also proved useful. Though illustrated for two-digit addends, each can easily be adapted for use with three-digit addends.

Activity File: Computation— Addition

Activity 1: Make It Big. (Developmental activity for two or three to play)
   Need: Deck of numeral cards for 0–9, five of each numeral; graph paper pieces (tens and ones) in a bank.
   Procedure: Draw two cards and use graph paper pieces to picture the greatest number possible, as in Figure 8.96. Then draw two more cards and follow the same procedure. Use the graph paper pieces to compute the sum of the two numbers, regrouping if necessary (trading 10 ones for 1 ten with the bank). Player with the greatest sum wins all the cards.
   Winner: Player who has most cards when deck is used up.

Figure 8.95

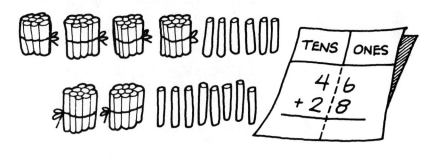

Figure 8.96   *Example:*

Activity 2: Put Together. (Developmental activity for one or two to play)

Need: Graph paper tens and ones in a bank; problem cards like that shown in Figure 8.97 (upper left corner notched so it is easy to tell when a card deck is stacked, problem side up).

Procedure: Draw a card and use graph paper pieces to compute the problem on the card. Write each problem and its answer. Regroup if necessary, trading 10 ones for 1 ten. When finished, turn the card over to check. Keep a tally of the times you are right.

Figure 8.97   Front  Back

Your Time

Analyzing an Addition Sequence

1. Study the mini-tasks presented on the cards of Figure 8.98. Clearly, the tasks are now out of order.

2. Work with another member of the class. Use the identifying letter in the upper left corner of each card to tell how you would resequence the cards into an appropriate learning sequence for children.

3. Check your sequence with that of another class team, then write the letter to describe your sequence on the chalkboard. Compare your sequence with that given by others in the class.

Several points possibly emerged in your discussion while sequencing the cards:

1. Developmental activities with manipulatives or picture cues should precede practice sets. So cards G and H come before A, and card B precedes J.

2. Problems with renaming follow those with no renaming in the sequence. So card B comes after G, H, I, and A.

3. Three-digit problems with no regrouping (card F) are "easier than" regrouping in two-digit addition, so it can come before exercises that require this.

4. There are several possibilities for placing card E in the sequence. Since the problems on the card involve single-digit addends and

teen sums, the card could come first--before computation of two-digit numbers. On the other hand, if a teacher wished to develop a careful sequence for column addition, it may better serve the purpose to place it at the beginning of that sub-sequence, after card J.

Figure 8.98

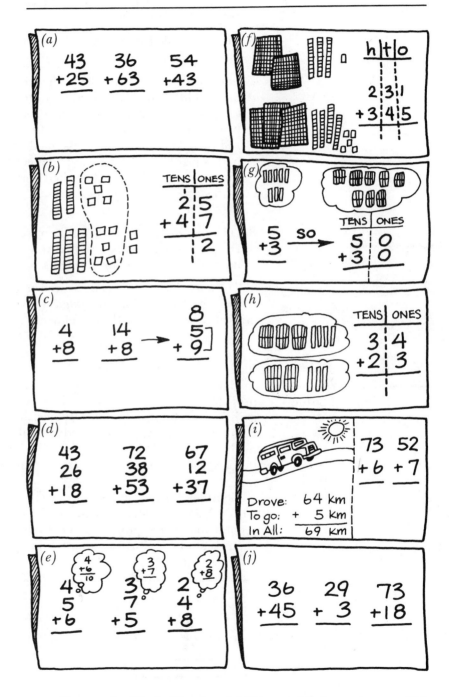

5. There are two possibilities for placing card I. Logically, one may tend to place the card before A in the sequence. Many children, however, find problems on this card harder than dealing with the more symmetric, "filled" column arrangement of the problems on card A. Being sensitive to the children helps determine the sequence in this case.

6. There is evidence that problem solving is interwoven with computational practice (card I).

7. Some tasks are omitted from the sequence. For example, column addition involving both one- *and* two-digit numbers in the same problem are missing. The problems presented are clearly pencil-and-paper. The activity of introductory teaching and followup sessions must be assumed. Work with three-digit numbers is just begun. One would expect this sub-sequence to extend the approach taken for two-digit computation and gradually deal with single, then multiple regroupings. No mixed or cumulative reviews are shown.

8. One possible sequence for the cards: G, H, A, I, F, B, J, E, C, D. Consult an elementary textbook series to see how the authors have sequenced similar problems.

The analysis above was presented to emphasize the importance of breaking instruction into *small steps*. Mastery of each step is critical for children with special needs in mathematics.

Trouble Shooting

Ideas for Column Addition

Many children have difficulties with column addition. These problems arise for many reasons, some associated with learning disabilities. To help children overcome these difficulties, the following column addition approach, made popular by Hutchings (1975), might be tried. This algorithm is called a "low-stress" algorithm because the information-processing requirements are kept to a minimum.

Consider the problem shown in Figure 8.99. To work it, one has to add 5 and 6 (11), and then add 4 to get the final sum of 15. This second addition requires that a child remember the first sum and then mentally perform an addition beyond that of basic facts.

Figure 8.99

The low-stress algorithm for column addition builds on the regular addition algorithm and employs the big idea of making tens. The example of Figure 8.100 uses this algorithm. Note that one never has to go beyond basic facts to complete the problem.

1. First add 5 and 6 (sum is 11);

2. Write a "1" to the right of the column to indicate the number of ones and a "1" to the left of the column to indicate a ten.

3. Then add 1 to the 4 (sum is 5).

4. Write the "5" to the right of the column and below the line as well.

5. Now go back and count the tens written to the left of the column during the addition (in this case, 1). Write this number in the tens place to complete the sum.

Note that this approach does not use any ideas which are new or difficult for children.

The low-stress algorithm can be extended to finding sums in larger-digit problems, such as those shown in Figure 8.101. When this is done, the columns must be written a little further apart and the addends need to be spaced out a little more. Note that the

Figure 8.100

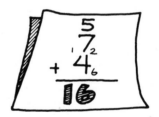

Figure 8.101

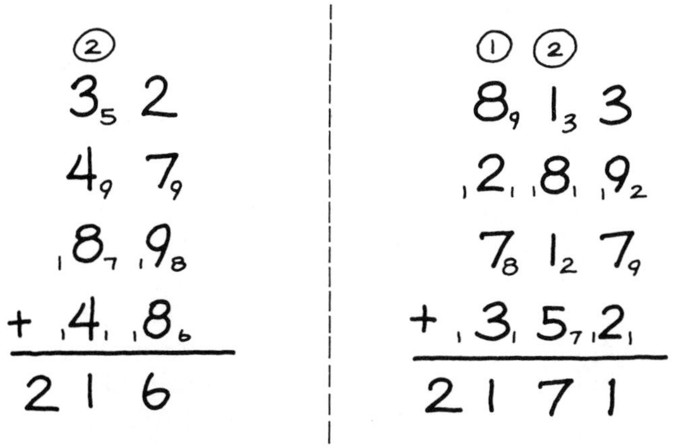

number of tens carried into the next columns is circled when placed at the top of the column. This helps the teacher check over the student's work at a later time.

One of the advantages of this method is that it allows teachers to spot fact errors easily. These errors are not always apparent in other approaches to column addition. While it may take slightly longer to write each intermediate sum, difficult additions and mental arithmetic are virtually eliminated. These factors compensate to allow for an increase in speed as one masters this approach. Try a few yourself for practice.

Subtraction Algorithms: Phase 3

Just Two Big Ideas

The basic instructional approach to subtraction of two- and higher-digit numbers parallels that of addition. Like addition, there are just two big ideas children need to learn:

1. Always subtract like units

2. If there are not enough of some unit to do the subtraction, a trade must be made (see Figure 8.102).

Informal introductory sessions in which children use manipulatives to illustrate subtraction computation are helpful. Popsicle sticks in bundles of ten with extra singles, graph paper tens and ones, and money (*when* concepts and skills are strong), are commonly used for this purpose. It is suggested that activities such as the following be included in early lessons.

Figure 8.102

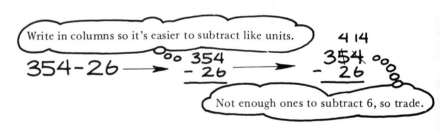

Activity 1: Your Pile, My Pile. (Developmental activity for two to play)

Activity File: Computation— Subtraction

Need: Subtraction problem cards—no renaming involved; tongue depressors in bundles of ten with extra singles; nailboard like that shown (see Figures 8.103 and 8.104). The nailboard device is really three boards, with string stapled along the ends that hold them together. The board is hung by this string.

Procedure: In turn, draw a card and picture the problem on the nailboard. The example of Figure 8.104 shows 3 tens, 5 ones. To play Your Pile, My Pile, the sticks are separated into two piles: twelve in one pile, and the leftovers in a second pile. Children take turns claiming the pile they want.

The child claiming the twelve will subtract like units, taking a group of ten and 2 ones from the sticks on the top row of the nailboard. When the 2 ones are taken, they are placed on the middle row of the nailboard. The remaining three are hung on the bottom row (Figure 8.105); "3" is written on the problem card to describe the number in the leftover pile.

The process is repeated for tens. In this case the child takes 1 ten, leaving two for the bottom row leftovers. The number of tens in the

Figure 8.103

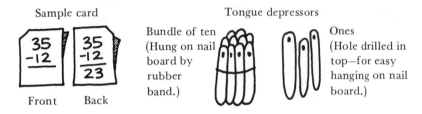

Sample card

<table>
<tr><td>35
-12</td><td>35
-12
23</td></tr>
<tr><td>Front</td><td>Back</td></tr>
</table>

Tongue depressors

Bundle of ten
(Hung on nail
board by
rubber
band.)

Ones
(Hole drilled in
top—for easy
hanging on nail
board.)

Figure 8.104

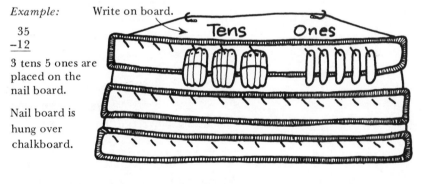

Example:

35
−12

3 tens 5 ones are placed on the nail board.

Nail board is hung over chalkboard.

Write on board.

Tens   Ones

Figure 8.105

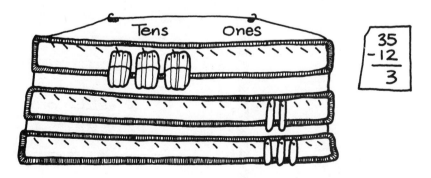

Tens   Ones

| 35
-12
3 |

last row is written on the problem card (see Figure 8.106). The child can then turn the card over to check. Repeat with other problem cards.

Note: Getting children to note that the two piles together make what they started with informally prepares them for adding to check on subtraction answers. When children show they understand the process for subtraction computation with no regrouping, practice exercises from the school mathematics textbook, supplementary worksheets, and activities like the following can be used to build computational skills.

Activity 2: In And Out. (Practice activity for two or three to play)

Need: Maze board like that shown in Figure 8.107 and a deck of problem cards for subtraction; die marked 1, 2, 3; a game marker for each player, answer key. Store all game materials in a vinyl pocket attached to back of gameboard. (Vinyl wallpaper from sample books is suitable.)

Procedure: In turn, players draw a card and compute the difference. If correct, they roll the die to see how many spaces they can move.

Winner: First to get in and out of the maze.

Figure 8.106

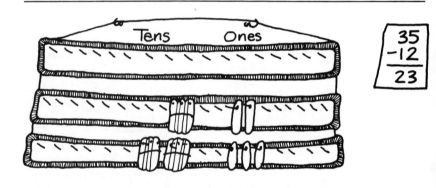

Figure 8.107

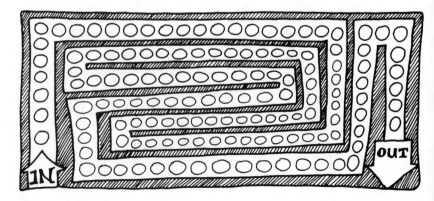

Regrouping in Subtraction

In subtraction, as in addition, it is necessary to prepare children in advance for the regrouping idea. Even as children are working with subtraction requiring no regrouping, they can review numeration concepts and skills which later will help in the regrouping process. The critical issue is realizing that numbers can be represented in different ways. Since children are concrete rather than abstract thinkers, it is important to dramatize this point with materials (Figure 8.108).

Fair-trade activities like the following can be used to help build this realization. They serve as the middle step between subtraction with no regrouping and that which requires it (Figure 8.109). This

Figure 8.108

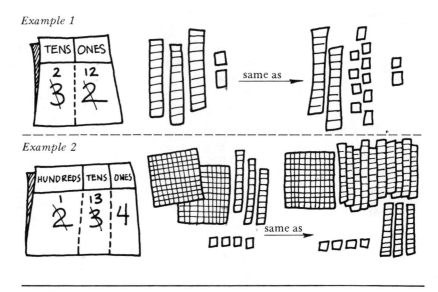

Figure 8.109

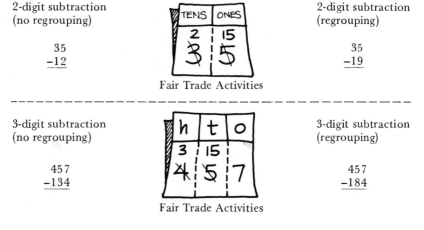

2-digit subtraction
(no regrouping)

35
−12

Fair Trade Activities

2-digit subtraction
(regrouping)

35
−19

3-digit subtraction
(no regrouping)

457
−134

Fair Trade Activities

3-digit subtraction
(regrouping)

457
−184

important middle step is often neglected. Experience suggests that fair-trade activities are critical for children with learning difficulties in mathematics. For these children it is helpful to use *column labels* and the format illustrated to record the renaming, since this transfers most directly to regrouping in the standard algorithm.

**Activity File:
Subtraction with
Regrouping**

Sample Activities and Worksheets: Fair Trades; Subtraction with Regrouping

The following fair-trade activity is illustrated for two-digit numbers. It can be adapted easily for use with three-digit numbers. Activities like this during an instructional period are normally reinforced by textbook exercises or supplementary worksheets.

Activity 1: Trade! (Developmental activity)

Purpose: To provide a natural, meaningful transition into two-digit subtraction with renaming.

Need: A bank of graph-paper pieces (tens and ones); deck of laminated two-digit numeral cards like that shown in Figure 8.110.

Procedure: In turn, children draw a card and use graph-paper pieces to picture it. The teacher leads the following discussion. "The bank needs another ten. Will you trade one of your tens for ones?" ... "How many ones make a fair trade?" (10 ones for 1 ten) ... "Write what you have now." Repeat with other cards, making a record of each trade.

Your Pile, My Pile activities can now be played for problems requiring renaming. The storyline, as a followup to trade activities, is a good one for cueing children *against* subtracting the smaller digit in a column from the larger digit in that column (Figure 8.111

Figure 8.110

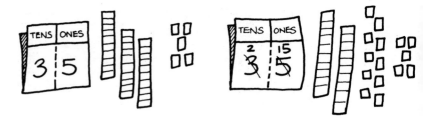

Figure 8.111 Common error

$$
\begin{array}{r}
35 \\
-17 \\
\hline
22
\end{array}
$$

The storyline can also be adapted to three-digit subtraction problems, as the following activity illustrates.

Activity 2: Your Pile, My Pile. (Developmental activity in three-digit subtraction for two or 3 to play)

Need: A bank of graph-paper pieces: hundreds, tens, and ones; deck of laminated numeral cards similar to that shown in Figure 8.112; grease pencil and wipe-off cloth.

Procedure: (illustrated for three-digit numbers, one regrouping, hundreds to tens) In turn, children draw a card and use graph-paper pieces to picture it.

"The bank needs another hundred. Will you trade one of yours for tens?" . . . "How many tens make a fair trade?" (Ten) . . . "Write what you have now."

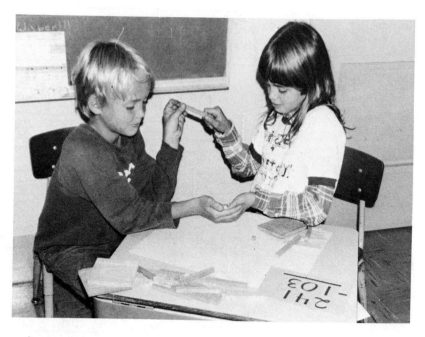

Figure 8.112   Front   Back

Sample Follow-up Worksheets:

1. Sample for Codes, based on riddles from library books (Figure 8.113)

2. Sample for Odd Ball (Figure 8.114)

3. Sample for "Work Till You Bingo" (Figure 8.115)

Figure 8.113

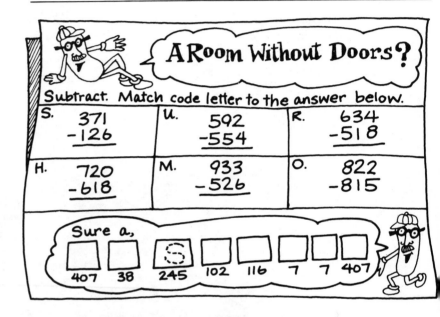

Figure 8.114

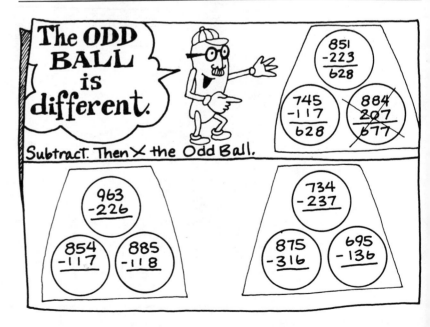

Figure 8.115

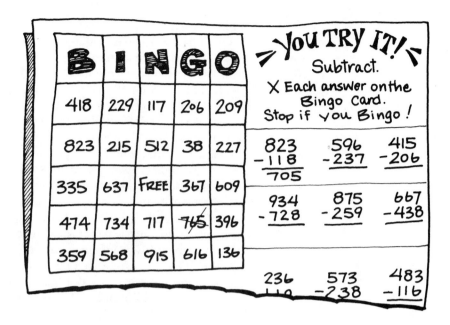

Help with Troublesome Zeros

Case Study: Nancy

Background information. Nancy is in grade 6, but is presently functioning at a low to middle fourth-grade level in mathematics. She knows the basic addition and subtraction facts and can handle most three- and four-digit subtractions. She can even illustrate the trades with materials, except when zeros are in the minuend. Then she tends to lose track in the process and make mistakes.

Teacher's solution: Simplify the regrouping (Figure 8.116).

Figure 8.116

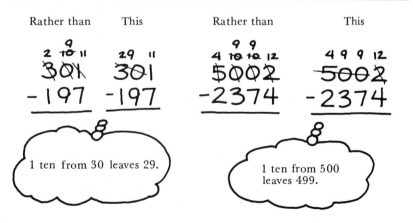

Addition and Subtraction Computation: Assessing Mastery

Backing Down

One approach to assessing mastery for addition and subtraction is to start with more difficult problems and then to "back down" to simpler types if children hesitate or have difficulty (see Figure 8.117). For children at a third-grade level, present two- and three-digit problems. If children are likely to make basic fact errors, use easier facts where possible. This approach allows a teacher to assess more accurately a child's grasp of the computational procedure itself. Ideally, when time permits, at least two problems of each type would be given. This makes for greater ease in sorting out careless errors from others, and provides the chance to note consistency in performance.

If children can handle more difficult problems, one may choose not to give additional test items. However, if they have difficulty or or if you suspect low conceptual understanding, further testing is necessary. One can establish a baseline by giving easier problems with no regrouping or those that use manipulatives to illustrate the computation. If children have a low self-concept or are easily frustrated, use the following approach.

Forward, Then Back

A second approach often used to assess computational mastery is to lay out a sequence of problems that very gradually become more difficult. (Small step-size is the word!) Then every third or fourth problem in the sequence is given until an error is made. If one knows a child well enough to suspect carelessness or trouble with facts, a second similar problem can be given. Otherwise one would immediately give problems that were skipped, to establish a baseline of what *can* be done, before proceeding further.

Sometimes children show a checkered achievement pattern. They can do problems at various points in the sequence, but have many loopholes in learning. Studying these loopholes often reveals one basic cause. The problem may be poor numeration understandings, inadequate mastery of facts, simple carelessness or, more often, rote rule instruction. Incongruous achievement patterns may also be caused by illness or emotional or attitudinal problems.

Figure 8.117   Sample problems — grade 4 level

Multiple regroupings	Staggered columns	Multiple regroupings	Zeros in the minuend
3646 +1987	9336 + 79 + 489	5312 −2176	3010 −1237

Trouble Shooting

Common Difficulties

Every teacher, as a troubleshooter, is on the watch for problems children may have with computation. The sooner they are noted and corrected, the less damage they cause. Fact errors are common (Figure 8.118). Sometimes the errors are just careless ones. Figure 8.119 shows an example where addition is incomplete. Zeros often cause difficulty, especially after children have studied multiplication facts (Figure 8.120). Sometimes error patterns for problems with zeros emerge, as in Figure 8.121, in which the child subtracted the smaller from the larger digit in a column. Other typical problems recur. An example is regrouping when unnecessary (Figure 8.122).

Figure 8.118

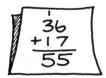

Figure 8.119

Figure 8.120

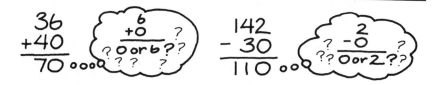

Figure 8.121

Figure 8.122

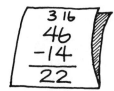

This problem commonly occurs after children complete a unit which focuses only on regrouping. Anticipating the problem, some teachers give children mixed problem types and ask them to just circle those that require regrouping (Figure 8.123). Answers can be computed later. Many common errors, like those illustrated in Figure 8.124, can be traced to inadequate numeration understandings.

Figure 8.123

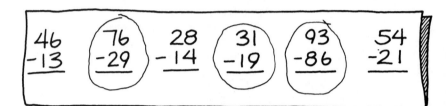

Figure 8.124

a. 36
   +18

   44

b. 73 (with small 2 above)
   +54

   19

c. 46 (with small 4 above)
   +18

   91

d. 26
   +18

   314

e. 43 (with small 1 above)
   + 9

   142

f. 43
   - 6

   43

g. 82 (with 7 10 above)
   -17

   63

h. 536 (with 3 13 16 above)
   -199

   247

i. 4390 (with 8 above)
   - 2173

   2212

Your Time

Identify the Errors

In which addition problem of Figure 8.124 does the child:

1. Add ones to ones *and* to tens?

2. Reverse tens and ones when writing the answer?

3. Handle each column as a separate problem instead of regrouping?

4. Add left-right?

5. Fail to regroup during the addition?

In which subtraction problem does the child:

1. Think: *2 less hundreds = 10* more tens, *10* more ones?

2. Just subtract the smaller from the larger digit in a column?

3. Forget to add in extra ones while regrouping?

4. Regroup, but subtract the smaller from the larger digit anyway, treating the "10" as "1"?

Remedying Problems

As a general rule, when conceptual misunderstandings are a cause, the remedial sequence would be based on developmental work prior to drill and practice. The procedures and sequences laid out in this chapter can serve as a guide to needed remediation. Many of the following ideas were included in the body of the chapter. They are summarized here because of their relevance to remediation.

1. Make sure children clearly understand learning goals. Then give them a means to monitor progress made (e.g., use of a personal bar graph).

2. Involve children *actively* in the remediation:
   (a) Illustrating concepts or procedures with materials
   (b) Verbalizing personal understanding
   (c) Creating oral situations that apply the mathematics to common real-life uses.

3. Structure instruction in *small* steps—developmental *then* practice.

4. In early developmental work, emphasize understanding and accuracy over speed.

5. When possible, provide activities as well as paper-and-pencil exercises for drill and practice.

6. Regularly spread practice time out over short periods each day.

7. When possible, provide practice activities that give immediate feedback.

8. Provide sufficient cumulative reviews to maintain concepts and skills.

9. Give frequent quizzes, but keep them *short*. Include interview as well as written questions.

10. Help the child *organize* important ideas or procedures; approach learning in a way that is sensitively individual.

Sometimes careful instruction of this nature was never carried out in the first place. At other times illness, emotional or family problems, or other special learning difficulties have created loopholes in learning.

Special Help for Special Needs

Some special learning problems require special treatment. Problems and solutions for sample cases are outlined below.

Problem 1: Forgetting the sequence for steps in a computational procedure.

Sample solutions:
a. Use materials to illustrate the procedure (Figure 8.125).
b. Use flip charts which sequence sample problems step by step (Figure 8.126).
c. Use verbal cues (Figure 8.127).
d. Color cue one's digits green. Relate to the traffic signal, where green means "Go," or "Start here" (Figure 8.128).
e. Use visual directional cues (Figure 8.129).

Figure 8.125

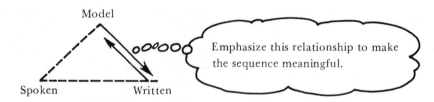

Model

Spoken   Written

Emphasize this relationship to make the sequence meaningful.

Figure 8.126

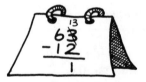

Figure 8.127

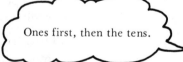

Ones first, then the tens.

Figure 8.128

Problem 2: Subtracting smaller from larger digit in a column, regardless of placement.

Sample solutions:
a. Review Your Pile, My Pile activities.
b. Provide a sample problem correctly completed at the top of the worksheet.
c. Use visual cueing (Figure 8.130).

Problem 3: Losing the place in column addition.

Sample solutions:
a. Train child to color highlight a column before adding (Figure 8.131).
b. Provide square centimeter paper for the child to use (Figure 8.132).

Figure 8.129

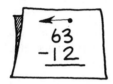

Figure 8.130

Figure 8.131

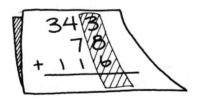

Figure 8.132

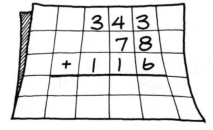

Problem 4: Blanking out on known facts within larger problems.

Sample solutions:
a. Have child finger trace and say fact to self, *then* write answer.
b. Have child write fact to the side to trigger recognition (Figure 8.133).

Problem 5: Not completing worksheets due to high distractibility, hyperactive tendencies, or frustration.

Sample solutions:
a. Provide fewer problems per page.
b. Create several standard formats for worksheets and provide black construction-paper masks that blot out all but 1/4 or 1/3 page at a time.
c. Actually cut the worksheet into fourths (or thirds) and assign only one small section at a time.

Problem 6: Taking too long or inaccurately copying the problems from the board or text (visual-spatial-motor difficulties).

Sample solutions:
a. *Limit written problems!*
b. Provide special lined paper or masks to mark problems which are copied.
c. Require *no* board copying.

Problem 7: Adding instead of subtracting on a page of mixed addition and subtraction problems.

Sample solutions:
a. Use visual cueing on the page itself (Figure 8.134).

Figure 8.133

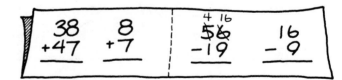

Figure 8.134

b. Have child circle all addition problems green and do these first, before turning to the subtraction problems on the page.

c. Have child finger trace the sign before computing.

One generally builds on learning *strengths* when designing remedial strategies to meet special needs. Approaches for children with strong visual association abilities differ from those with high auditory, low visual strengths. Children must then gradually be taught techniques which help them compensate for learning disabilities. Two examples from the preceding suggestions: finger tracing and color highlighting to compensate for visual dissociations or imperceptions. Children with these learning difficulties can also trace, then *say* answers to basic facts they don't recognize visually.

Your Time: Activities, Exercises, and Investigations

1. Prepare, as your instructor directs, answers to the major questions on page 245.

2. Refer to the card sequencing task of Figure 8.98.
   (a) Suggest a word problem that might be used to introduce card D. (Hint: Use card I as an example.)
   (b) Briefly describe a developmental activity that might precede card F.
   (c) Briefly describe one practice activity that might be used to reinforce addition skills with three-digit numbers, no regrouping.

3. Refer to the Trouble Shooting section for column addition on p. 252. Briefly describe a motivational practice activity a teacher might use to followup instruction in the low-stress algorithm.

4. Adapt the trade activity of p. 258 for use with three-digit numbers (one regrouping, hundreds to tens). Provide sketches to illustrate your description.

5. Sequence the following problems, placing easiest problems first. Briefly explain your sequencing.

   (a) 5010   (b) 3774   (c) 4713   (d) 6287
       −1976   −2136   −1986   −3428

6. Nonverbal problems can be used to suggest problem contexts for computation. One example is shown in Figure 8.135. Create three additional examples, one for each of the following categories:
   (a) Primary level (grades 1–3)
   (b) Intermediate level (grades 4–6)
   (c) Remedial middle school (grades 7–9)

7. The best practice activities for computation are those which:
   (a) Make it possible for each child in the group to be *actively* participating throughout.
   (b) Are self-checking.
   (c) Provide *many* encounters to practice computation.
   Refer to the case study for Nancy (p. 261). Using the criteria of exercise 7, design a practice activity to build Nancy's skills in subtracting when zeros appear in minuends.

8. Select a current third-grade mathematics textbook and find a chapter on addition or subtraction computation.
   (a) Choose any *one* lesson in the chapter and design a pre-book activity for that lesson—an activity that can be done with children *before* they open the book.
   (b) Design a followup activity to reinforce the written exercises of the textbook lesson.
   (c) If there is a test or quiz at the end of the chapter you select, decide:
      (1) How many questions relate to the lesson you have chosen
      (2) Whether, in your judgment, additional problems or questions are necessary. If you feel the need for extra items, construct them. Indicate whether they are interview or paper-and-pencil items; indicate how you might sequence them along with the other given problems. Turn in the name of the textbook, publisher, and year of publication so the book can be referred to when this exercise is graded.

9. Discuss how an emphasis on estimating to check reasonableness of computational answers might help a child with problems like those on the bottom of Figure 8.124.

10. Refer to the problems and sample solutions of pp. 266–267 to:
    (a) Design a half-page worksheet which incorporates *one* of the solution ideas.
    (b) Describe the learning difficulties *and* strengths of a child for whom your worksheet would be appropriate.

Figure 8.135

Reference

Hutchings, B. Low stress subtraction. *The Arithmetic Teacher*, 1975, *22* (3), 226–32.

9 WHOLE NUMBER MULTIPLICATION AND DIVISION

Teacher Background

We have just examined addition and its inverse operation, subtraction. This chapter focuses on the concepts and skills of multiplication and its inverse operation, division. Early work in developing the concept of multiplication is made meaningful by relating it to addition. We then relate division to the operations of multiplication and subtraction. Emphasis is placed on developmental laboratory activities involving concrete materials that assist children to acquire a basic understanding of multiplication and division. Following this carefully sequenced developmental work, we present practice activities and application problems, and a section of ideas for helping children master multiplication and division facts.

We next turn to the computational algorithms for multiplication and division. A meaningful understanding of the place value system underlies learning and using these algorithms. Teachers need to spend time working with multiples of tens and hundreds to prepare children for multiplication or division computations. A basic understanding of tens and multiples of ten is important to avoid guesswork.

In both multiplication and division the digit zero causes problems. Children make strange mistakes and get wrong answers in problems that involve zeros. The middle zero or terminal zero in the quotient of a division problem tends to be difficult for children. Another common problem involves aligning the columns properly for computation. Suggestions for dealing with these and other common difficulties children experience are included in a Trouble Shooting section which concludes the chapter.

A careful development of the basic concepts and skills related to multiplication and division moves through the three stages of instruction focused on in this chapter: (1) conceptualization, (2) fact learning, and (3) algorithms. Conceptualization involves the instructional aspect where a child constructs a mental image of an operation. The teacher relates physical operations with concrete materials to the arithmetic operation. For example, a child can associate combining sets of equal size with the operation of multiplication. Division being the inverse operation of multiplication, a child can see the process of separating a set into equal-sized parts. When a physical model is presented and used, a child should think of a particular arithmetic operation and recognize the association between a physical situation and the operation.

The other important facet of the conceptualization stage is that after performing the physical operation related to the arithmetic operation, the child must have developed a process for finding the answer. This process might be some counting procedure. However, by this time, children will have normally learned to add and subtract. In order to allow the child to use the most sophisticated process available, repeated addition or subtraction are usually taught as the process for finding answers for multiplication and division in this stage.

After the conceptualization stage, instruction proceeds to relate basic fact learning to physical models. Understanding the multiplication concept must precede work on fact memorization. Organization of relationships and patterns is emphasized with fact learning so that a child can use the easy facts to build upon enroute to the more difficult basic facts. For example, a child may reason what 8×7 is by thinking, $8 \times 5 = 40$ and $8 \times 2 = 16$ so $8 \times 7 = 40 + 16$ or 56. In this process children can use the basic facts that they have already memorized. The difficult facts can be found by extending the organizational pattern. Children need to achieve immediate recall of all basic multiplication facts. Activities where children use physical models provide a basis for memorization of the basic facts of an operation.

When a child is in the final steps of mastering the basic facts, instruction may proceed to the stage of learning the associated algorithm. The major important ideas need to be emphasized and related to physical models. For example, in multiplication basic grouping, place value, and renaming concepts are the fundamental ideas related to the algorithm. As the algorithm for multiplication is developed, step by step, multiplication by 10 and the use of partial products constantly recur. Hence these ideas should be emphasized.

Within instruction one focuses on grouping by tens with the physical models. In particular, a child can think of 3×60 as already knowing $3 \times 6 = 18$, so 3×6 tens = 18 tens. 18 tens = 180, so 3×6 tens = 180. Ones times tens, ones times hundreds, and tens times tens can be directly related to what the child has already learned, the basic facts of multiplication. Computation difficulties can be alleviated if important ideas like these are emphasized in the instructional development of an algorithm.

Whenever multipliers are larger than 10, children should see that partial products are used to break the problem into easier problems that allow a solution using basic facts and multiplication by 10. For example, 42 x 4 can be thought of as 40 x 4 plus 2 x 4. The partial products are then combined to find the final product.

Teachers who understand the developmental sequence from conceptualization to fact learning to algorithms can more easily adapt their teaching to meet the special needs of children. This chapter includes specific suggestions for relating to these needs throughout multiplication and division instruction. Figure 9.1 summarizes the development of multiplication.

What Do *You* Say?

1. Which manipulative aids, concrete physical materials, or activities could be used to build a child's readiness for multiplication and division?

2. How does one distinguish between developmental and practice activities:
   (a) In developing basic concepts for multiplication or division?
   (b) In helping children learn the basic facts?

3. What types of everyday situations can serve as examples for the repeated addition model or the array model for multiplication?

4. What thinking strategies might help children learn the basic multiplication facts?

5. How does a teacher assess whether a child has mastered the basic facts of multiplication?

Figure 9.1   **Instruction for Whole Number Multiplication and Division**

Phase 1: Conceptualization
1. Relate arithmetic operation to physical operation.
2. Develop a process for finding answers.

Phase 2: Fact Mastery
1. Easy facts
   (a) Figure them out using process already developed;
   (b) Organize them in many ways to see relationships;
   (c) Memorize them.
2. Develop more efficient techniques for figuring out facts that:
   (a) Are *mental* processes;
   (b) Use easier facts already memorized.
3. Harder facts
   (a) Figure them out using the new efficient processes;
   (b) Organize them;
   (c) Memorize them.

Phase 3: Algorithm Learning
1. Deemphasize superficial rules;
2. Stress big ideas and relate them to concrete models;
3. Then develop computation expertise apart from models.

Multiplication Models:
Phase 1

Basic Concepts and Models: Multiplication

The concept of multiplication is closely related to the concept of addition, as two or more sets are combined to form a new set. In multiplication, however, each of the groups joined must have the same number of objects.

Practical problems from the child's world can be used to introduce the concept of multiplication. Some examples follow.

Example 1: There are six plates on the table for treats in the afternoon (Figure 9.2). Each plate has four cookies on it. How many cookies in all?

Six plates, 4 cookies on each plate. $6 \times 4 = $ _____.

Example 2: Beth had nine friends at her birthday party. Each received two candy bars (Figure 9.3). How many candy bars were there altogether for her friends?

Nine sets of 2 candy bars. $9 \times 2 = $ _____.

In each example described above, same-size groups of everyday items were described in a problem-solving situation.

There are several models which can be used to represent multiplication. The following models will be considered: (1) repeated addition, (2) the set model, (3) the array model, and (4) the number line. In teaching special children, it is important to choose *one* model and develop the concept of multiplication through a series of closely related activities and practical story-problem situations based on that one model. These developmental experiences for multiplication are then extended gradually from introductory work to memorizing basic facts to making estimates to doing paper-and-pencil computations and, finally, to solving problems. Children have a new sign (x) and new vocabulary words (times, multiply, factor, product) to learn. In written work, children have to deal with both horizontal and vertical formats (Figure 9.4). Each new term, symbol, or form should be introduced one step at a time.

Figure 9.2

Figure 9.3

This gives children with learning difficulties a chance to experience success. This procedure also helps them focus on small parts of problems, which they can handle.

Repeated Addition

Various examples with the repeated addition concept can be used. Children should realize that 5 x 3 represents the combining of five groups of three objects. It must be emphasized that each group contains the same number of objects. Since children will already have associated the combining of groups with addition, the combining of several groups of the same size should easily be related to successive addition. Children see the importance of multiplication in situations involving repeated addition as addition becomes too cumbersome a task, especially when they are asked to find answers to problems like 13 x 24. Multiplication, when mastered, is far more efficient than counting or repeated addition.

The Set Model

The joining of equivalent sets can be modeled by using various items such as bottle caps, beans, buttons, beads, blocks, or counters. Physical situations can be used to give the children a variety of experiences in grouping equivalent sets of objects (Figures 9.5 and 9.6).

Figure 9.4

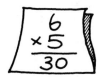

Figure 9.5

2 teams of children in a game.
5 children on each team.
How many children in this game?

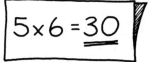

Figure 9.6

4 soda drinks.
2 straws in each glass.
How many straws in all?

The Array Model

Children investigating various arrangements of counters, shells, beans, pegs on a hundreds board, or nails on a geoboard arranged in the form of rectangular arrays soon discover that the number of objects in the total array is equal to (number of rows) x (number in each row). Some everyday examples of items packaged in arrays include cartons of pop, packages of buttons, spools of thread, party favors, cupcakes, candy, and cookies (Figures 9.7 and 9.8).

Multiplication problems can easily be displayed as rectangular arrays on a hundreds board. For example, 5 x 7. Five rows of seven pegs can be looped or the pegs can be turned over or removed to indicate 5 x 7. The child can then draw this representation on dot paper or graph paper (Figure 9.9).

Figure 9.7

4 rows of children
6 children in each row
How many children?

Figure 9.8

An egg carton is filled with eggs.
2 rows
6 eggs in each row
How many eggs?

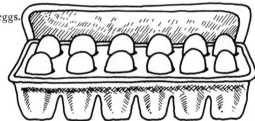

Figure 9.9

5 rows of 7 pegs
5 x 7 = _____

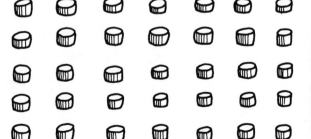

Number Line

The number line can be used as a model for children to use in solving multiplication situations. Use either a floor number line, with the children taking a certain number of jumps, or a desk number line. Figure 9.10 gives an example of the use of a number line to find the solution to a multiplication situation.

Summary

The conceptualization of multiplication can be considered complete when the child can represent multiplication using a physical model, verbalize multiplication and indicate multiplication with written symbols. When a multiplication example is given to the child in one mode, the child should be able to give it back in the other two modes. When the child sees a number of equivalent sets or groups combined, the child should be able to verbalize the corresponding multiplication situation and write the corresponding multiplication example (Figure 9.11a). When given a written multiplication example, the child should be able to read the example and model it with physical materials (Figure 9.11b). After hearing an example read, the child should be able to write it and model it (Figure 9.11c).

Figure 9.10   5 x 3 = _____   5 x 3 would mean 5 jumps of 3 units

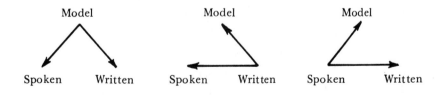

Figure 9.11

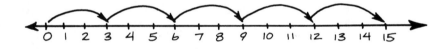

Model	Model	Model
Spoken Written	Spoken Written	Spoken Written

Toward Mastery of Basic Multiplication Facts

Various strategies can be used to facilitate a child's learning of the basic multiplication facts. Children can master the facts more quickly if (1) fact clusters are carefully sequenced, (2) cue prompts and thinking strategies to aid memorization are suggested, and (3) motivational activities and worksheets for drill are used after adequate time has been provided for mastery of the basic facts. Emphasize to children that they need first to master the easy facts and then use them to study the more difficult ones.

The Twos

Give the children many opportunities to discover some basic multiplication facts and then construct various multiplication tables by using groupings of objects. In beginning multiplication the children need opportunities to work with common objects and group them into equal-sized sets. When children are ready to work toward mastery of the facts, the twos are among the easiest to learn. They are closely related to doubles in addition. Some examples include: 2 bicycle wheels; 2 shoes; pop bottles in an eight-pack, 2 x 4; and egg in an egg carton, 2 x 6. Many different visual examples like these can help the child learn the twos facts.

At this point, a discussion of the equality of 2 x 3 and 3 x 2 is in order. When children realize that the commutative property of multiplication holds, it eases their mastery of the fact clusters. Many children need to be SHOWN rather than merely *told* about commutativity. Figure 9.12 illustrates how rotating an array 90° conveys this idea. The commutative property of multiplication is used freely in the development of the following fact clusters.

Cluster 1

Figure 9.12

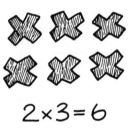

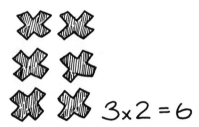

Cluster 2

The Fives

Since children are familiar with fives in terms of money (5 pennies = 1 nickel, 5 nickels = 1 quarter) the "five" facts can be introduced next. For example,

2 nickels = _____ cents   4 nickels = _____ cents
5 nickels = _____ cents   7 nickels = _____ cents

Other examples follow:

5 x 5 . . . "Think of 5 nickels, which make a quarter (25 cents)."
4 x 5 . . . "Think of 4 nickels, which makes 20 cents."

When children are learning to tell time, they count minutes around the clock in fives. Relate multiplication fives to clock times. For example,

6 x 5 ... "Think of the half hour. 6 x 5 = 30."
8 x 5 ... "Think of quarter hour times after two o'clock.
When the big hand is on the 9, it's 9 x 5 or 45, or *2:45* (9 x 5 = *45*).
When the big hand is on the 8, it's 5 minutes less: *2:40* (8 x 5 = *40*).

Check the examples of Figure 9.13. Then examine the problems of Figure 9.14. Some children have discovered that five times an even number equals half of the even number with an added zero (Figure 9.15). Try this rule for five times some two-digit even number, too.

Figure 9.13

9:15   11:30   3:20

Figure 9.14

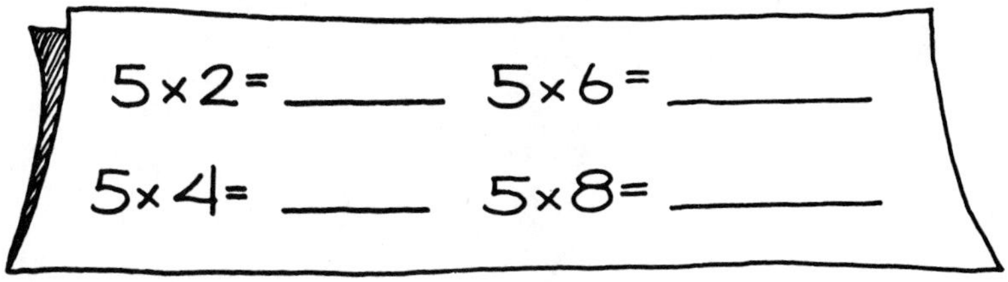

5×2= _____   5×6= _____

5×4= _____   5×8= _____

Figure 9.15

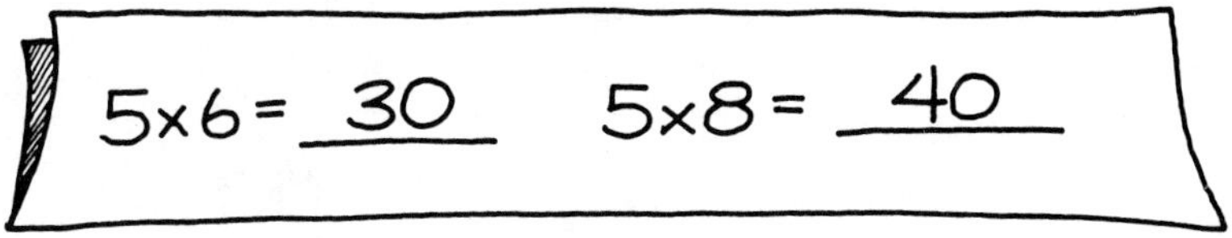

5×6= __30__   5×8= __40__

Cluster 3

The Nines
    Children can enjoy finding patterns among the digits in the facts related to nine. Encourage them to look for a pattern. The product for one-digit multipliers greater than one starts with a ten's digit one

less than the multiplier (see Figure 9.16). Also, the sum of the digits in the product of nine times any one-digit number is equal to nine (Figure 9.17). Some children may discover that these two patterns, put together, will help them with the basic facts of nines.

Finger multiplication by nines is another approach which can be used to figure out the basic facts of nine. Label the fingers on your two hands consecutively 1, 2, 3, 4, . . . 10. Some examples are shown in Figure 9.18. To multiply 3 times 9, bend the third finger. Two fingers represent the tens and 7 fingers represent the ones.

$$3 \times 9 = 27$$

To multiply 7 times 9, bend the seventh finger. Six fingers represent the tens and 3 fingers represent the ones.

$$7 \times 9 = 63.$$

Figure 9.16   9 x *3* = *27* (2 is one less than 3)
9 x *5* = *45* (4 is one less than 5)
9 x *8* = *72* (7 is one less than 8)

Figure 9.17   9 x 4 = *36* (3 + 6 = 9)
9 x 7 = *63* (6 + 3 = 9)
9 x 6 = *54* (5 + 4 = 9)
9 x 2 = *18* (1 + 8 = 9)

Figure 9.18

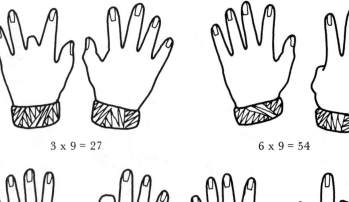

3 x 9 = 27   6 x 9 = 54

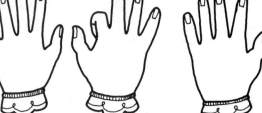

7 x 9 = 63   8 x 9 = 72

 **Cluster 4**

Skip Counting

Skip counting works quite well when practicing the basic multiplication facts of twos, threes, fours, and fives. Multiples of a number can be stated aloud by the children in a matter of a minute or less. These sequences can be practiced at various times throughout a day. Skip counting by threes can be recited as 3, 6, 9, 12, 15, 18, 21, 24, 27. Skip counting is not rapid recall but does help children to obtain an answer. Children can number their fingers 1, 2, 3, . . . , 10 and manipulate them while skip counting. The child would bend the "1" finger when saying 3, the "2" finger when saying 6, and so on. This illustrates 1 x 3 = 3, 2 x 3 = 6, and so on.

 **Cluster 5**

Easy Parts

Easy parts can be a super strategy for children who already know the easier basic facts. The distributive law of multiplication over addition can be very useful to children learning the basic facts. The more difficult facts to learn can be broken down into easier basic facts already known. For example, if a child knows that 5 fours is 20, then 6 fours can be reasoned as 5 fours plus one more set of fours (Figure 9.19). If children know a multiple of a number, they can find the next succeeding multiple by adding a single-digit number.

Figure 9.19

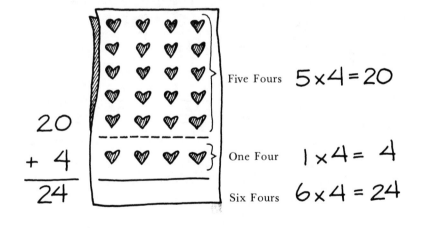

 **Cluster 6**

Twice As Much

A special form of the use of the distributive law appears in finding facts which are twice others. These facts can be learned by thinking of doubling or taking twice as much as an already learned fact. For example, 4 x 3 equals twice 2 x 3 or 12. Figure 9.20 shows how a child might use this idea to derive 4 x 6 = __?__ .

Chapter 9 Whole Number Multiplication and Division   281

Properties of Multiplication

Some of the properties of multiplication were referred to informally earlier in this chapter. Now we'll take a more careful look at each of these properties. Work at this stage should be a review of the discussion and manipulation which occurred during early concept development of the operation. Use of the properties now is important since this can facilitate children's learning of the basic multiplication facts. When children learn to use the properties, they'll find how to make shortcuts in their work.

Commutative Property of Multiplication. As mentioned earlier, the order in which two whole numbers are multiplied has no effect on the product: $a \times b = b \times a$. This order property considerably lessens the number of facts the child needs to learn (see Figure 9.21). The related basic multiplication facts (e.g., $7 \times 3 = 21$ and $3 \times 7 = 21$) can be seen as turn/arounds, mirror images, or reflections of each other about the line of symmetry shown in Figure 9.21.

Figure 9.20

2 x 6 = 12, so just double 12 to get

2 rows of 6 = 12

2 rows of 6 = 12

$4 \times 6 =$

4 rows of 6 = 24

Figure 9.21

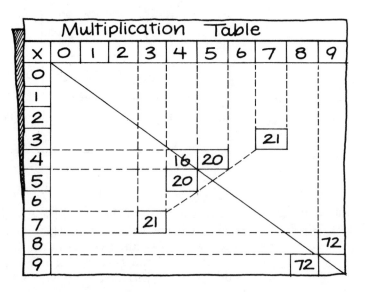

Teaching Mathematics to Children with Special Needs

Multiplicative Identity. When 1 is multiplied by a number, the product remains the same as the original number: a x 1 = a. There are eighteen basic multiplication facts involving 1 as a factor. Recognizing this relation, a x 1 = a = 1 x a, that any number multiplied by 1 is itself, children have learned nineteen basic multiplication facts. The number 1 is called the multiplication identity because it identifies the other factor as the product. The related cluster of basic facts is illustrated in the basic multiplication fact table of Figure 9.22.

Associative Property. When multiplying three numbers, the product of two of the numbers is multiplied by the third number. In general, (a x b) x c = a x (b x c), where changing the grouping does not affect the product. This fact about the manner of grouping factors in computing an indicated product is called the associative property of multiplication (Figure 9.23).

Figure 9.22

X	0	1	2	3	4	5	6	7	8	9
0		0								
1	0	1	2	3	4	5	6	7	8	9
2		2								
3		3								
4		4								
5		5								
6		6								
7		7								
8		8								
9		9								

Figure 9.23

$$(2 \times 3) \times 4 = 2 \times (3 \times 4)$$
$$6 \times 4 = 2 \times 12$$
$$24 = 24$$

(2 x 3) x 4   2 x (3 x 4)

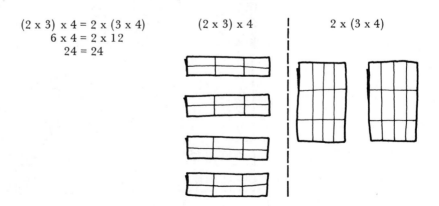

Distributive Property. The distributive property of multiplication over addition is a powerful property underlying the multiplication algorithm. In general, it is as follows: a x (b + c) = (a x b) + (a x c). For example, 8 x 7 can be thought of as in Figure 9.24.

The Basic Facts of Multiplication: Summary

The clusters of basic multiplication facts—(1) the twos, (2) the fives, (3) the nines, (4) skip counting, (5) easy parts, and (6) twice as much—can be used as an organizational pattern for teaching the basic facts in beginning multiplication. Children can use the easy facts to build to the more difficult ones. Thinking strategies and patterns described in the clusters will help children learn these basic facts. Encourage the children to construct or complete their

Figure 9.24

8 x 7 = 8 x (5 + 2)
  8 x 5 + 8 x 2
  40 + 16
  56

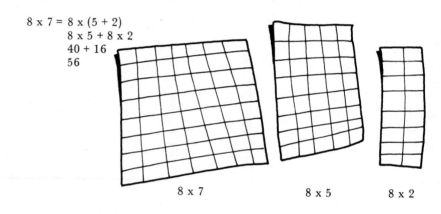

8 x 7   8 x 5   8 x 2

Figure 9.25

X	0	1	2	3	4	5	6	7	8	9
0	0	0	0	0	0	0	0	0	0	0
1	0	1	2	3	4	5	6	7	8	9
2	0	2	4	6	8	10	12	14	16	18
3	0	3	6	9		15				27
4	0	4	8		16	20				36
5	0	5	10	15	20	25	30	35	40	45
6	0	6	12			30	36			54
7	0	7	14			35		49		63
8	0	8	16			40			64	72
9	0	9	18	27	36	45	54	63	72	81

- Squares: along diagonal line of symmetry
- Easy Twos
- Fives are easy (money, time)
- Nines are easy (many patterns)

own multiplication table and to share with others the relationships or strategies they use to remember the harder facts. The commutative property will enable the children to notice symmetry in the multiplication table (Figure 9.25). A colored strip of transparent material can be used for the diagonal line of symmetry on the chart. The square number facts can be highlighted by drawing boxes around their answers (Figure 9.25).

Children who know the multiplication facts related to twos, fives, nines, and squares, and who can multiply by zero and one, have only ten of the one hundred basic multiplication facts left to study. These last ten facts are: 3 x 4, 3 x 6, 3 x 7, 3 x 8, 4 x 6, 4 x 7, 4 x 8, 6 x 7, 6 x 8, and 7 x 8. Not all of the strategies presented in this chapter can be used efficiently with all of the basic facts. But many of the ideas suggested using the distributive property and related methods will help children master these last ten facts.

Activity File:
Basic Multiplication
Facts

Activity 1: Spin And Discard
   Need: Spinner with digits 2, 3, 4, 5, 6, 7, 8, 9 on it, deck of eight cards per player, with two each of 1, 2, 3, and 4.
   Procedure: In turn, children spin the spinner, say the product of 5 times the number that shows on the spinner. Match the ten's digit of the said product with one of your eight cards. Example: Spin a 6. 5 x 6 = 30, say 30. Put the 3 card in the center of the table.
   Winner: First player to turn in each of the eight cards.
   Note: Focusing the children's attention on the ten's digit could help them memorize the basic multiplication facts for five. Also point out that each of the products for the fives ends in a 0 or 5.

Activity 2: Spin To 200
   Need: Spinner with digits 2, 3, 4, 5, 6, 7, 8, and 9 on it.
   Procedure: Multiply by nine the number that appears on the spinner. Keep a running sum of your products of nine.
   Winner: First to get a sum of 200 or more.
   Note: Provide an answer card with the products of nine to use as a reference check if necessary.

Activity 3: Order 'Em: 3 At A Time (For two to five players)
   Need: Deck of cards labeled for all the basic facts (e.g., 3 x 6), using twos, threes, fours, and fives as factors; answer key including a segment of the basic-facts multiplication table.
   Procedure: The leader shuffles the cards in the deck, face down, and stacks them in the center of the table. The first player then draws three cards and orders them by their products. If correct, the player keeps the cards. If incorrect, they are randomly placed back

in the original stack of cards. Players take turns and the game ends when all the cards are used.

Winner: The player with the most cards.

Sample of play: Player draws the cards 3 x 7, 4 x 8, 9 x 2, says the product for each card and then orders the cards by size of product (i.e., 9 x 2, 3 x 7, 4 x 8). Player keeps those three cards.

Activity 4: Using A Number Line

Using a number line, have a child illustrate 2 x 3 and then 3 x 2 (Figure 9.26).

2 x 3 = _____ (2 jumps of 3) = _____
3 x 2 = _____ (3 jumps of 2) = _____

Children should try other basic facts and then be asked to describe the pattern they noticed in the products for a x b and b x a.

Figure 9.26

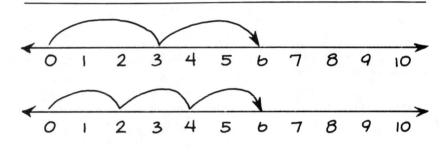

Figure 9.27

Try 3 x 4 = _____   Then, try 4 x 3 = _____
3 rows, 4 columns   4 rows, 3 columns

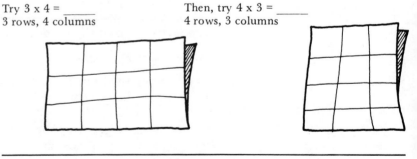

Figure 9.28

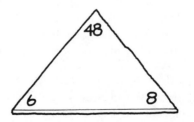

Cover card

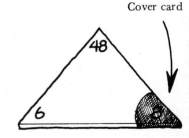

Activity 5: Using A Geoboard

Have the children use a geoboard and make rectangular arrays (Figure 9.27).

Try 3 x 4 = _____   Then, try 4 x 3 = _____

Then have the child transfer the geoboard result to squared-centimeter graph paper by coloring in the appropriate squares. Each rectangular array or model has the same number of squares (3 x 4 = 12, 4 x 3 = 12).

Variation: A hundred pegboard can be used in a similar fashion.

Activity 6: Flash Cards

Triangular flash cards can be made and used by the children. Two children can practice the basic multiplication facts by having one child cover the product, the other child find, for example, 8 x 6. These cards can also be used for the basic division facts. One of the factors can be covered. In this example, the child could cover the 8 and the other child would only see the 6 and the 48 (Figure 9.28).

Basic Concepts and Models of Division: Phase 3

Like multiplication, division is a process which involves equal-sized groups. In multiplication the equal-sized groups were combined. Since division is the inverse operation of multiplication, division can be related to the physical operation of separating a group into equal-sized parts. Children experience division problems in everyday situations. One example is afternoon "snack time." Mother gave Alice nine carrot sticks to share with her three sisters. Alice takes out two carrot sticks for herself and gives two carrot sticks to each of her sisters. She finds that from the nine carrot sticks she gave two to each of four people with one carrot stick left over, which she gives to her mother. A child develops division concepts by experiencing division in common situations like these. Many of the physical materials that were used in beginning multiplication work can also be used to help children in early work with division.

Encourage children to talk through division problems, since experiencing problem-solving situations is an important developmental step. Verbalizing a division problem-solving situation will enhance a child's understanding of this new concept. Interaction with peers and with a teacher should be emphasized.

Use practical everyday examples to introduce division examples. An example from the child's world is that of a child placing blocks in a wagon. "Jane has a small wagon that will hold exactly one layer of twenty-four blocks. If the wagon will hold four rows of blocks, how many will fit into each row?" By placing the blocks into the wagon, Jane found that when twenty-four blocks were divided into four rows, six blocks fit into each row (Figure 9.29). So $24 \div 4 = 6$. Be sure to provide a variety of physical models for children to use in beginning division.

Two Types of Division

Children will encounter two types of division situations: sharing (partitive division) and grouping in equal quantities (measurement division). *In partitive division we know the number of groups. In measurement division we know the size of the groups.*

Teachers need to build the concept of division upon the sharing (partitive) idea. Children use this type of division earlier than the measurement type. To dramatize sharing division, common objects such as fruit, candy bars, marbles, cards, or toys can be shared or distributed. When the objects are passed around to a group of children, each child takes one item until there are not enough left for another round. This is a mathematical experience of sharing. In sharing, a fixed amount is divided among a certain number of people. Children should be given many opportunities to share things with other children and then record the results.

Measurement division—grouping fixed amounts—needs to be experienced by children at a later time. Children can experience grouping of objects by counting out subgroups of a known quantity or fixed number of items. Each of these groups is part of a larger group where the size of each is known and equal but the number of groups is not known (Figures 9.30 and 9.31).

The relationship between sharing and grouping problems should be understood. Both partitive and measurement division can be related to multiplication. The product is known and one of the factors is unknown. However, at the concrete level these two division types are different and a variety of examples needs to be experienced by children.

To solve either type of division problem, children can recall related multiplication facts. Rather than being treated as an isolated operation, division can be introduced informally in realistic problem situations during the development of multiplication. For example, consider $5 \times 2 = 10$. Think of the number of different ways you could discuss this basic multiplication fact in relating division to multiplication. How many twos are there in ten? When ten objects are shared among five children, how many objects does each child get?

Let's examine more closely the relation between multiplication and division. For multiplication, factor $\times$ factor = product. For division, product $\div$ factor = factor.

When the concept of division as the partitioning of a group into equal-sized parts is established, help children focus on the multiplication-division relationship. Allow children to use objects at this stage.

Figure 9.29

Figure 9.30

Practical Examples in Division

Sharing	*Grouping*

Sharing

8 candy bars
4 children

How many candy bars will each get?
Give 1 candy bar to each child until
none are left.

8 ÷ 4 = 2

16 crayons
3 children

How many crayons will each get?
Give 1 to each in turn until none
are left.

16 ÷ 3 = 5 with 1 left

Grouping

12 oranges are to be given away.
2 oranges to each child

How many children will get oranges?
Give 2 oranges to each, until
no more.

2 ÷ 2 = 6

15 candy bars
2 candy bars to each child.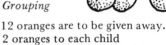

How many children will get
candy bars?
Give 2 candy bars to each, until
no more.

15 ÷ 2 = 7 with 1 left

Figure 9.31

Sharing

Sometimes called
Partitive Division
We know how many groups.

$$\frac{\text{Total Quantity}}{\text{No. of groups}} = \text{No. in each group}$$

Grouping

Sometimes called
Measurement Division
We know the size of the groups.

$$\frac{\text{Total Quantity}}{\text{No. in each group}} = \text{No. of Groups}$$

Example: Three rows of four chips is twelve in all (3 x 4 = 12). Twelve chips in three rows means four in each row (12 ÷ 3 = 4).

Children should be able to orally describe the arrangement and write the appropriate multiplication or division sentence being modeled. Such experiences will help children recognize families of related multiplication and division facts. For example,

$$3 \times 4 = 12 \qquad\qquad 12 \div 4 = 3$$
$$4 \times 3 = 12 \qquad\qquad 12 \div 3 = 4$$

Children who can "see" the relatedness of facts like these will find it easier in later work to memorize the division facts. They can simply use multiplication facts they know to help.

The basic multiplication chart of Figure 9.32 can also be used to emphasize the relatedness of division to multiplication facts children know. In using this chart for division, the product inside the chart is called the dividend. The top row contains the divisors. The left column gives the answer or quotient of each fact. Of course, the first column inside the chart must be ignored since division by zero is undefined. For example, there is no number that, when multiplied by zero, will give six as a result. Hence, we can never divide six by zero. Dividing by zero makes no sense, for if $6 \div 0 =$ _____ then _____ x 0 would have to equal six. This follows because division is defined in terms of multiplication. To emphasize the fact that division by zero is undefined, shade or remove the first column inside the chart when using it for division.

Figure 9.32

X	0	1	2	3	4	5	6	7	8	9
0	0	0	0	0	0	0	0	0	0	0
1	0	1	2	3	4	5	6	7	8	9
2	0	2	4	6	8	10	12	14	16	18
3	0	3	6	9	12	15	18	21	24	27
4	0	4	8	12	16	20	24	28	32	36
5	0	5	10	15	20	25	30	35	40	45
6	0	6	12	18	24	30	36	42	48	54

Figure 9.33

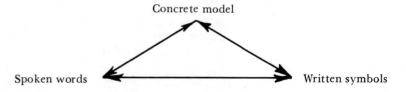

Concrete model

Spoken words ⟷ Written symbols

Summary. The triangular learning-teaching model is an important guide for developing the division concept with children (Figure 9.33). Experience with division at this early stage needs to occur in three modes, with the children using concrete models, talking about them and then writing in symbols what they have just done. In the beginning, division interpreted as "sharing" is the basis for developing meaning for the ÷ symbol and the relative positions of the dividend, divisor, and quotient in the sentence a ÷ b = c. These terms, as well as remainder, are gradually phased into the child's vocabulary. Later, grouping situations for division are introduced, and the relating of division to multiplication is emphasized as a general procedure for finding answers to simple division problems.

Mastery of Basic Division Facts: Phase 4

Toward Mastery of Basic Division Facts

Once the basic division concept is grasped and the new symbols, a ÷ b and $\frac{a}{b}$, are introduced, then the children need to memorize the basic division facts. Although there are one hundred basic multiplication facts, there are only ninety basic division facts, since division by zero is undefined. Of course the organizational problems outlined in Figure 9.1 help reduce the number of actual memorization tasks for the child.

As was pointed out in the preceding section, relating division to corresponding multiplication facts will make memorizing them easier. For example:

> 15 ÷ 5 = _____
> How many fives make fifteen?
> Think, _____ x 5 = 15.
> *3* x 5 = 15, so 15 ÷ 5 = 3.

Activities during early work with division in which children wrote multiplication and division sentences for models like that of Figure 9.34 "pay off" now. Such experiences form the basis for children being able to *independently* recall the multiplication fact needed to solve a given multiplication problem. Practice activities like the following will also help.

Figure 9.34

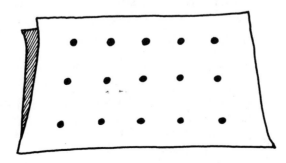

Activity: Outlaw

Purpose: To focus on families of related multiplication and division facts.

Need: Card deck of multiplication and related basic division fact one Outlaw card (Figure 9.35).

Procedure: Follow traditional Old Maid rules, but lay down all cards that belong to the same multiplication-division family each time. Examples:

$$2 \times 3, 3 \times 2, 6 \div 3, 6 \div 2$$
$$8 \times 6, 6 \times 8, 48 \div 6, 48 \div 8$$

As cards are laid down, answers to each fact problem must be given.

Variation: Each set of related multiplication-division cards could be color-coded alike.

Figure 9.35

Assessment of Basic Multiplication and Division Facts: Phase 5

Assessment of Mastery of Basic Multiplication and Division Facts

Basic fact mastery requires accuracy, speed, and consistency. Both initial mastery and retention through cumulative reviews need to be assessed.

Level 1 Assessment: Initial Mastery of a Cluster of Facts

The first level of fact mastery includes a child's attempts to master each new cluster of facts. After a child has had opportunities to explore, learn, and study the facts, both oral and written quizzes should be given. These quizzes should be:

- Short—only six to eight facts; or, if timed, only 15-30 seconds.

- Frequent—daily over the period of fact mastery.

The quizzes should be limited at first only to the basic facts in a given new cluster. Establish a criterion that children must meet before proceeding to a new cluster of facts. Whatever the planned criterion, discuss it with the children so that what is expected is clearly understood.

Level 2 Assessment: Cumulative Reviews

Cumulative review quizzes can be used to test retention mastery at this level. The children must consistently maintain knowledge of basic facts with both speed and accuracy even when fact clusters are mixed with earlier learned facts. These quizzes should be kept short.

For oral quizzes, some children with auditory handicaps will need to repeat the fact problem. Immediate recall of fact answers would be required of all other children. For written quizzes, children's speed may vary due to differences in age and eye-hand motor coordination.

Both levels of assessment should be recorded on a chart. It may be helpful to keep a master record of children's progress on specific facts through both the fact clusters and the review quizzes.

Your Time: Activities, Exercises, and Investigations

Questions and Activities

1. Use blocks or chips for the following and illustrate the process in solving each example:
   (a) four groups of six
   (b) 12 ÷ 3

2. Why is multiplication by zero and one often omitted in beginning multiplication work? What questions could you ask children to help them develop the concept of multiplication by zero?

3. Describe an activity or a physical model that children could use to learn about the concept of multiplication.

4. How would you help a child use the distributive property of multiplication over addition? For example, illustrate 5 x 8 = 40 or some other basic fact.

5. Write an example illustrating the commutative property of multiplication. Describe how to make this property meaningful to children.

6. Create a developmental activity and a practice activity for basic multiplication facts. Focus on a single fact cluster.

7. Examine three elementary mathematics textbooks. Select multiplication activities and adapt them to meet the needs of children with visual handicaps.

8. (a) Write an example of a story problem using division as a sharing situation. Illustrate this example.
   (b) Write an example of a story problem using division as a grouping situation. Illustrate this example.

9. How can you help students understand the relationship between basic multiplication and division facts? Describe an activity that you could use for this purpose.

10. Examine three elementary mathematics textbooks. Select two developmental division activities and two practice division activities. Adapt them to meet the needs of a child with a special handicap.

Multiplication and Division Computation

What Do *You* Say?

1. What are the big ideas that guide:
   (a) Multiplication computation?
   (b) Division computation?

2. How would you model the multiplication of two 2-digit numbers?

3. What types of questions and answers might be used in your classroom when introducing the division algorithm?

4. How would realistic problem-solving activities help prepare children for solving division problems?

5. How can computational practice be motivated?

6. How can mastery of multiplication and division computation be assessed?

7. What basic guidelines can be used for remediating computational difficulties with multiplication and division?

Computation: Multiplication

Computation: Multiplication Phase 6

Adequate mastery of the basic multiplication facts, an understanding of grouping by tens, and renaming skills are all indicators that a child is ready to study the multiplication algorithm. Students should have had experience with concrete base-ten models before instruction on the algorithm is begun.

Consider the examples included in Figure 9.36. They would obviously not be taught in the sequence in which they are listed. How could the examples be rearranged to place them in an appropriate learning sequence? To do this, several questions need to be considered. For example, should activities involving physical or pictorial models necessarily precede those involving abstract symbols only? Should exercises without regrouping precede those with regrouping? Should examples involving easy basic facts precede those with harder basic facts?

An instructional sequence for teaching the multiplication algorithm is examined below. As you study that sequence, think about the preceding questions. Notice also that in each step of the development, one of these big ideas is applied:

1. Multiplication by 10 is easy.

   $6 \times 10 = 60$
   $38 \times 10 = 380$
   $14 \times 100 = 1400 \ (14 \times 10 \times 10)$

2. With bigger numbers, use partial products.

   $6 \cdot 8 = 3 \cdot 8 + 3 \cdot 8$
   $12 \cdot 3 = 10 \cdot 3 + 2 \cdot 3$
   $48 \cdot 25 = 40 \cdot 25 + 8 \cdot 25$

Figure 9.36

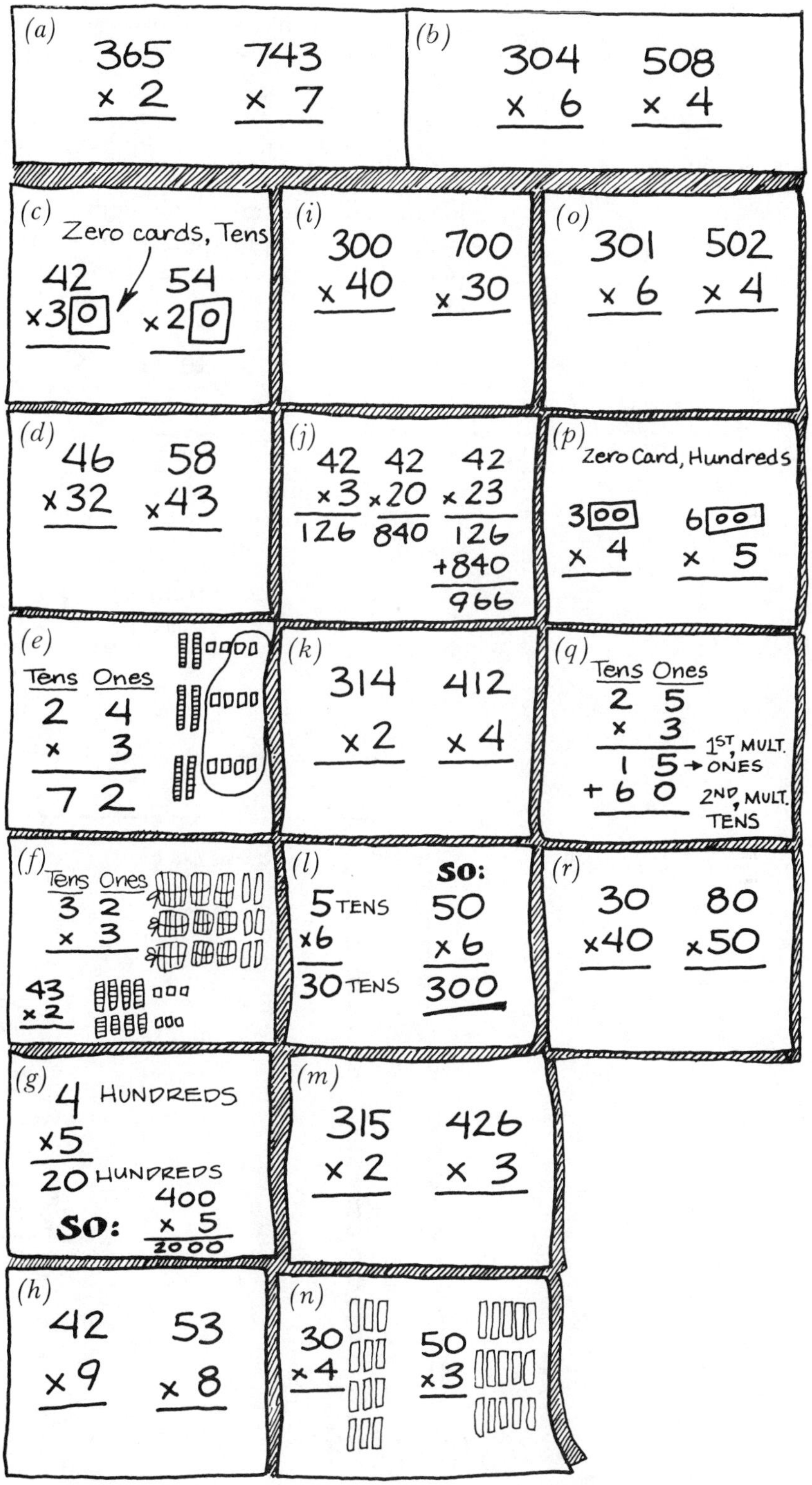

(a)
365   743
× 2   × 7

(b)
304   508
× 6   × 4

(c) Zero cards, Tens
42 ╱ 54
×3⓪   ×2⓪

(i)
300   700
× 40   × 30

(o)
301   502
× 6   × 4

(d)
46   58
×32   ×43

(j)
42   42   42
×3   ×20   ×23
126   840   126
            +840
            966

(p) Zero Card, Hundreds
3⓪⓪   6⓪⓪
× 4   × 5

(e)
Tens Ones
  2   4
  ×   3
  7   2

(k)
314   412
× 2   × 4

(q)
Tens Ones
  2   5
  ×   3
  1   5 → ONES   1ST. MULT.
+ 6   0   2ND. MULT.
                       TENS

(f)
Tens Ones
  3   2
  ×   3

  43
  ×2

(l)
5 TENS   SO: 50
×6   × 6
30 TENS   300

(r)
30   80
×40   ×50

(g) 4 HUNDREDS
×5
20 HUNDREDS
        400
SO: × 5
     2000

(m)
315   426
× 2   × 3

(h)
42   53
×9   × 8

(n)
30   50
×4   ×3

Chapter 9   Whole Number Multiplication and Division   295

Keep in mind that the following is but one sequence. A particular textbook series may follow a different sequence than the one presented here.

Step 1: Basic facts are learned first. These are committed to memory and form a basis for later work.

Step 2: Multiplication by 10 is easy. You just "add a zero." Figure 9.37 illustrates how multiplication by 10 might be taught.

Step 3: Multiplication of one-digit numbers by multiples of 10 (e.g., 3 x 40) uses basic facts and multiplication by 10—which is easy.

$$3 \text{ x } 40 = \underbrace{3 \text{ x } 4}_{\substack{\text{basic} \\ \text{fact}}} \text{ x } \underbrace{10}_{\substack{\text{multiplication} \\ \text{by } 10}}$$

Step 4: Multiplication of one-digit numbers by multiples of 100, 1000, and 10,000 helps to extend and generalize the concept in step 3.

$$8 \text{ x } 4000 = \underbrace{8 \text{ x } 4}_{\substack{\text{basic} \\ \text{fact}}} \text{ x } \underbrace{10 \text{ x } 10 \text{ x } 10}_{\substack{\text{3 multiplications} \\ \text{by } 10 \text{ (add three} \\ \text{zeros)}}}$$

Color cueing sometimes helps children to see this pattern.

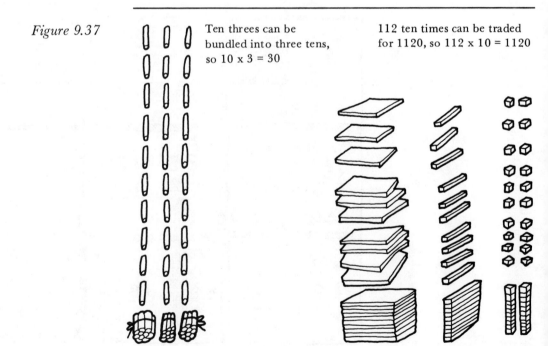

Figure 9.37

Ten threes can be bundled into three tens, so 10 x 3 = 30

112 ten times can be traded for 1120, so 112 x 10 = 1120

$$7 \times 3\underline{00} = 21\underline{00}$$
$$5 \times 7\underline{000} = 35\underline{000}$$
$$4 \times 5\underline{000} = 20\underline{000}$$

Step 5: Multiplication of a one-digit number by a two-digit number involves the previously introduced skills along with the idea of partial products. A physical or pictorial model is useful at this point to help the child understand partial products. Figure 9.38 illustrates the use of such models.

Step 6: The concept of partial products can be extended to multiplication of one-digit numbers by three-digit numbers.

$$246 \times 7 = 200 \times 7 + 40 \times 7 + 6 \times 7$$

Figure 9.38 $24 \times 6 = 20 \times 6 + 4 \times 6$

The 6 must be multiplied times the 20 and the 4.

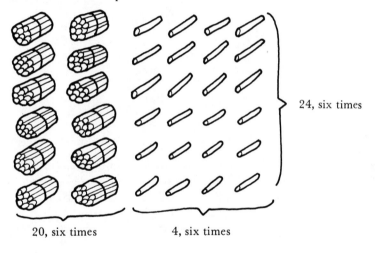

20, six times   4, six times   24, six times

- -

$35 \times 4 = 30 \times 4 + 5 \times 4$

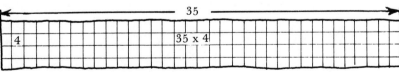

35

4   35 x 4

The rectangle can be broken into two parts giving us two partial products which together equal the total product.

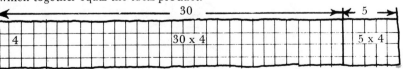

30   5

4   30 x 4   5 x 4

Step 7: At this point in the development it is appropriate to develop the skill of regrouping (or carrying) as it is applied in multiplication (Figure 9.39).

Step 8: Multiplication of a two- or three-digit number by a multiple of 10 comes next. In this step we apply the preceding skills with multiplication by 10.

$$27 \times 60 = \underbrace{27 \times 6} \times \underbrace{10}$$

I already   This is easy.
know how
to do this.

$$572 \times 30 = 572 \times 3 \times 10$$

Step 9: Multiplication of a pair of multi-digit numbers comes next. On completion of this step, the sequence is complete and the child should be able to multiply any pair of whole numbers. We apply, once again, the notion of partial products.

$$42 \times 38 = \underbrace{42 \times 30} + \underbrace{42 \times 8}$$

I already   I can also
can do this. do this.

```
    376
  x  43
   1128  ←— 376 x 3
  15040  ←— 376 x 40
  16168  ←— Add the partial products to get the total product.
```

Since the ideas of multiplication by 10 and partial products reoccurred constantly throughout the development, they should receive special emphasis as the big ideas of multiplication. Such an emphasis will help give meaning to the procedures as well as make the learning of those procedures easier for the child.

Figure 9.39   Instead of   36   the recording might look like this.

```
        36
       x  4
        24
       120
       144
```

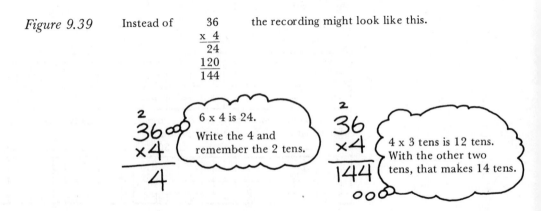

Some elementary mathematics texts present transitional algorithms for computation to help children function at their current level of conceptual understanding (see Figure 9.40). Experience has suggested that many transitional forms are not effective with some children who have learning difficulties in mathematics. The teacher of such children may find it helpful, instead, to:

1. Use physical models such as popsicle sticks or base-10 graph paper pieces to illustrate multiplication with one-digit multiplier.

2. Use column labels (hundreds, tens, ones) to record multiplication work.

3. Use grid paper to record the problem, work, and answer since the grid helps the child to keep columns aligned properly.

4. Sometimes use color cueing to highlight patterns.

5. Mark through each regrouping number after it has been used whenever an example requires more than one regrouping.

$$
\begin{array}{r}
\cancel{3} \\
\cancel{4} \\
58 \\
\times\ 46 \\
\hline
348 \\
2320 \\
\hline
2668
\end{array}
$$

Figure 9.40

$$
\begin{array}{ccccc}
16 & 10 & 6 & 16 \\
\underline{\times\ 4} & \underline{\times\ 4}^{+} & \underline{\times\ 4} & \underline{\times\ 4} \\
 & 40 & 24 & 24 \\
 & & & 40 \\
 & & & \underline{} \\
 & & & 64
\end{array}
$$

**Activity File:
Multiplication
Computation**

Activity 1: Do It With Pieces
   Need: Graph paper tens and ones in envelopes or divided container.
   Procedure:

1. Pose a multiplication problem (one-digit multiplier), e.g., 3 x 24.

2. Ask a child to picture the problem (show 24 three times) with graph paper pieces (Figure 9.41).

3. The child then combines ones (Figure 9.42).

4. Any time there are 10 ones, go to the bank and trade for 1 ten piece. Place it with the other tens.

5. Count the tens. Do it the fast way: three twos is 6. Then add the extra ten. Record 3 x 24 = 72.

6. Children should work similar examples. When they have grasped the process, have them record each step as they go (Figures 9.43 — 9.45). Instruct the children to strike out each "carry" digit as it is used. This will minimize errors when multiple renamings occur with two-digit multipliers. When children are comfortable with the process, they can move to doing problems without the aid of graph paper pieces.

Variation: Write problems on cards. Have children work in pairs, taking turns illustrating and computing the problem cards they draw.

Activity 2: So Many Rows!

Need: Square-centimeter graph paper, crayon, ruler, scissors.

Procedure: Have children mark off (cut out) rows on the graph paper to illustrate given problems. Some models can already be cut out, and the children then can select the appropriate models (Figure 9.46).

Variation: Write problems on cards. Have children work in pairs, taking turns illustrating and computing the problem cards they draw

Note: It is important, when two-digit multipliers are used, that children see the problem as a combination of two easier problems

Figure 9.41   The problem:

$$\begin{array}{r} 24 \\ \times\ 3 \\ \hline \end{array}$$

Figure 9.42   After the trade:

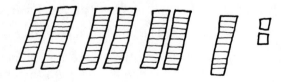

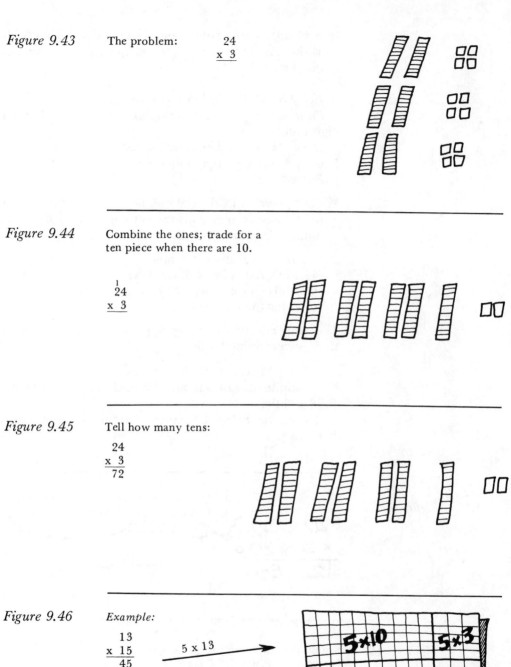

Figure 9.43   The problem: 24
                                    x 3

Figure 9.44   Combine the ones; trade for a
               ten piece when there are 10.

               $\overset{1}{2}4$
               x 3

Figure 9.45   Tell how many tens:

               24
               x 3
               ——
               72

Figure 9.46   *Example:*

               13
               x 15
               ——
               45
               130
               ——
               175

               5 x 13

               10 x 13

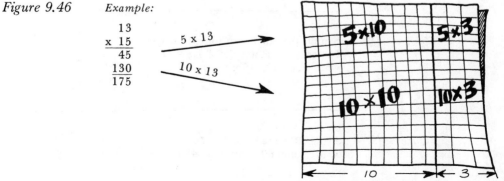

they already know how to do. Placing these two problems on a worksheet aside the two-digit multiplier problem emphasizes this point (Figure 9.47).

Activity 3: Roll, Find Products and Sum
   Purpose: To practice multiplying a one-digit number by a two-digit number.
   Need: Two number cubes—a red number cube labeled 3, 4, 5, 6, 7, 8 and a white number cube labeled 30, 40, 50, 60, 70, 80.
   Procedure:

1. First player rolls the red number cube, multiplies that number by 8, records, then rolls the white number cube, multiplies that number by 8, records, then adds the two products.

2. Second player does the same.
   (a) Rolls red cube, x 8, records;
   (b) Rolls white cube, x 8, records;
   (c) Adds the two products.

The game continues until each player gets four turns. Players keep their own running totals.

   Winner: Player with the highest total wins.
   A sample play of one turn for each player is shown in Figures 9.48 and 9.49.
   Variation: Select a number rather than 8 to multiply each number by.

Figure 9.47

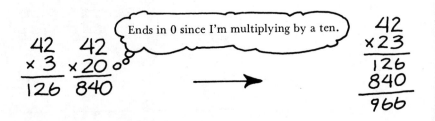

Figure 9.48   Sample play of 1 turn for each player:
Player A: red cube 7,   7 x 8 = 56   56
white cube 40, 40 x 8 = 320   +320
                                          376

Figure 9.49   Player B;   red cube 4,   4 x 8 = 32   32
white cube, 60,   60 x 8 = 480   +480
                                          512

Background Information

Jeff, a learning disabled child, is 11 but is functioning at a fourth-grade level in mathematics. Because of the severity of Jeff's spatial organization, integrative processing and sequencing difficulties, he is in a self-contained LD room. The teacher is about to introduce multiplication problems involving two-digit factors (of the type 47 x 23). Jeff has already worked with and mastered computation for problems like those shown below (multiplication by a one-digit factor; multiplication by a two-digit multiple of 10).

32	37	302	73
x 4	x 6	x 5	x20

Teacher's plan for Jeff. In conjunction with the standard presentation of the computational procedure, the teacher prepared dittos using square centimeter boxes and stoplight green-red color coding. Jeff would be instructed to start (go) on green and end (stop) on red, and to write answers into boxes with the same color as the multiplier (Figure 9.50). Colors would be gradually eliminated, as would the boxes.

Figure 9.50

Shaded = green
Nonshaded = red

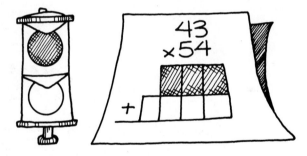

The division algorithm requires that children have certain prerequisite knowledge and skills and understand how they fit together. Knowledge of the following is necessary: numeration ideas; basic facts of subtraction, multiplication, and division; algorithms for subtraction, and multiplication; and estimation skills. Mastery of these prerequisites will minimize difficulties with the complex computation involved in division. These prerequisites are necessary in order to work successfully with the division algorithm and may need to be reviewed as instruction proceeds.

Estimation is an important skill when using the division algorithm. Knowing the thought patterns applicable to finding a range for the quotient help a child solve a division problem more easily.

Estimation skills also help a child check a division answer. "Is my answer reasonable?" "Does my answer make sense?" (See Figure 9.51.)

Developing the Algorithm

Encourage children to use concrete models or materials in working through the steps of the division algorithm. Materials such as bundling sticks, popsicle sticks, tongue depressors, base-ten blocks, graph paper pieces, or copies of dollar bills (ones, tens, and hundreds) can be used. Relate division problems to daily problem-solving situations with which students are familiar.

Two distinct approaches to the development of the division algorithm may be adopted. These approaches arise out of the two different kinds of situations described earlier in this chapter. In each case, we start with a single quantity and separate it into equal parts. For example, we might start with 78 cents and divide it into equal parts as shown in Figure 9.52.

Sometimes, in such a situation, you knew the size of each of the equal parts and need to know the number of parts. Then it would be natural to adopt a successive subtraction approach.

Figure 9.51 For example, 32 ÷ 5 is about 6 **oo o oo** (since 5 x 6 = 30).

32 tens ÷ 5 is about 6 tens.   5)320

Figure 9.52   78¢ can be separated into 3 equal parts.   Each part contains 26¢.

$$78¢$$
$$\underline{-26¢} \leftarrow \text{That's one part.}$$
$$52¢$$
$$\underline{-26¢} \leftarrow \text{That's another part.}$$
$$26¢$$
$$\underline{-26¢} \leftarrow \text{That's three equal parts.}$$
$$0¢$$

If instead, you know the number of parts, a natural way to solve the problem would be to distribute the money in much the same way as you would deal cards to three players. First, distribute the quarters (Figure 9.53).

Then, distribute the pennies (Figure 9.54).

Figure 9.53

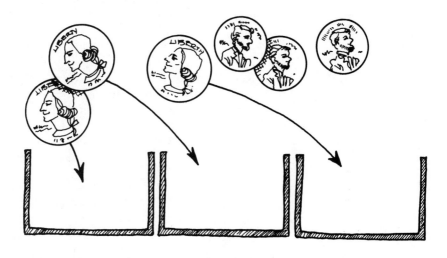

Figure 9.54

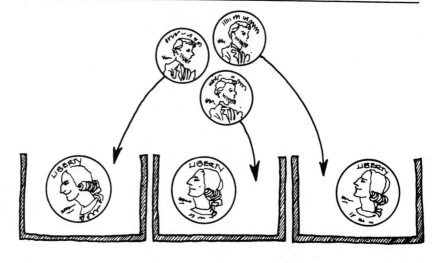

The first of these approaches (successive subtraction) is called measurement division. The second approach (equal distribution) is called partition division. The developmental sequence that follows uses the partition approach. In this development, two important procedural ideas will emerge. The child who learns to do these two things will be able to complete any whole number division problem—provided the prerequisites mentioned earlier have been mastered. Watch for these two big ideas. They should be emphasized.

Step 1: Understanding the symbolism. When a ÷ b = c is written in the form

$$\frac{c}{b)\,a}$$

a is what we start with, b is the number of equal parts, and c (the answer) is the amount that will be placed in each part.

Step 2: The first big idea is: *divide up the largest units first.*

tens	ones
3) 6	9

Represent 69 with a model. We will then separate it into three equal piles (Figure 9.55). The largest unit is tens. We have 6 tens. We can put 2 tens in each part (Figure 9.56).

$(6 \div 3 = 2)$

	tens	ones	
	2		2 tens, 3 times
	3) 6	9	we have taken
9 ones are	−6		3 x 2 = 6 tens
left to be		9	from the original
distributed.			amount.

Now, the largest unit left to be distributed is the ones. We will divide up the ones next. There are 9 ones. We can put 3 ones in each of the three parts (Figure 9.57).

$(9 \div 3 = 3)$

	tens	ones	
	2	3	
	3) 6	9	
	−6		
		9	We took 3 ones, three times.
There is nothing		−9	We have taken 3 x 3 = 9 ones
left to distribute.		0	from the remaining amount.

We have distributed 23 to each of our three parts. 23 is the answer.

Step 3: The second big idea is: *trade remainders for the next-size-smaller unit.* The following example illustrates the use of the big

Figure 9.55

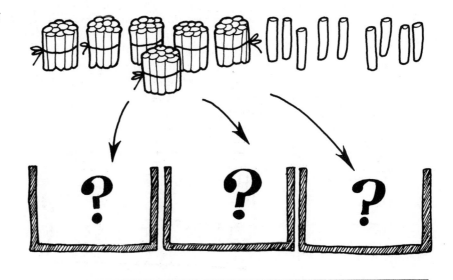

Figure 9.56

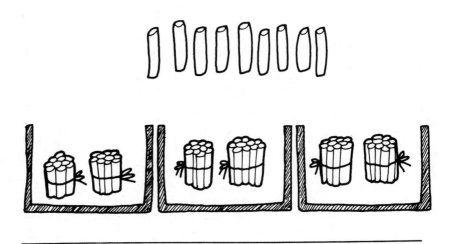

Figure 9.57

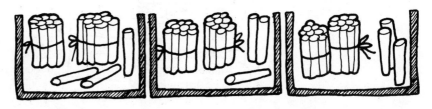

idea (Figure 9.58). The largest unit is hundreds, so divide the hundreds first (Figure 9.59). Now, the largest unit left to be distributed is the tens (Figure 9.60). At this point in the problem, we are left with one ten and two ones. One ten is not enough to distribute four ways, so *we trade for the next-size-smaller unit* which is ones. This gives us 12 ones (Figure 9.61). Now we have 12 ones—enough to distribute four ways. Since 12 ÷ 4 = 3, we will distribute 3 ones to each part (Figure 9.62).

Figure 9.58

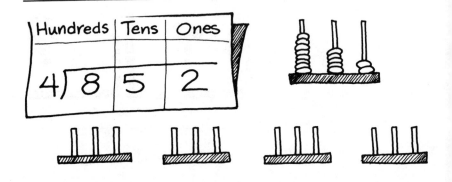

Figure 9.59

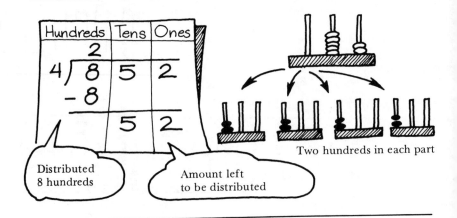

Figure 9.60

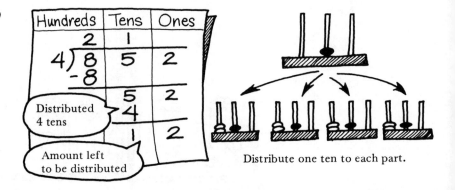

Figure 9.61

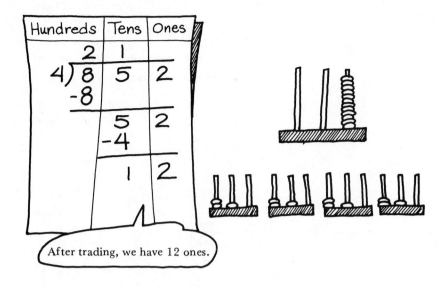

After trading, we have 12 ones.

Figure 9.62

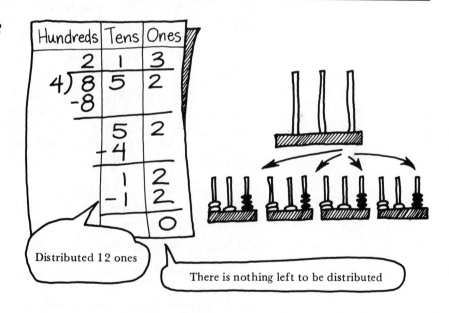

Distributed 12 ones

There is nothing left to be distributed

In the next example, we will not illustrate the use of a model, but think about how you could model the steps of the solution.

$7\,\overline{)\,235}$ We start with 235 and separate it into seven equal parts.

$7\,\overline{)\,\underline{23}5}$ There are only 2 hundreds—not enough for seven equal parts so we trade the two hundreds for tens. We now have 23 tens to distribute.

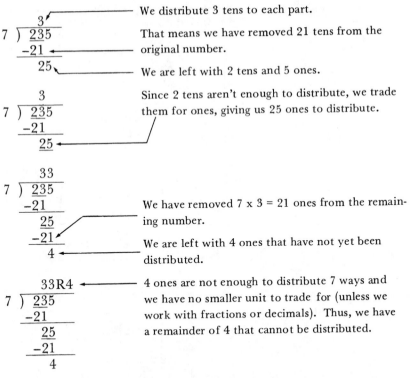

We distribute 3 tens to each part.

That means we have removed 21 tens from the original number.

We are left with 2 tens and 5 ones.

Since 2 tens aren't enough to distribute, we trade them for ones, giving us 25 ones to distribute.

We have removed 7 x 3 = 21 ones from the remaining number.

We are left with 4 ones that have not yet been distributed.

4 ones are not enough to distribute 7 ways and we have no smaller unit to trade for (unless we work with fractions or decimals). Thus, we have a remainder of 4 that cannot be distributed.

Out of work with physical models like that described in the preceding examples, the following pattern should emerge. *Divide, then multiply, then subtract and check; then divide, then multiply, then subtract and check. . . .* Thorough familiarity with this sequence is essential if the algorithm is to be habitualized.

Special Help for Special Problems

There are but two big procedural ideas that must be learned for the division algorithm:

1. Divide largest units first.

2. Trade remainders for smaller units.

Figure 9.63

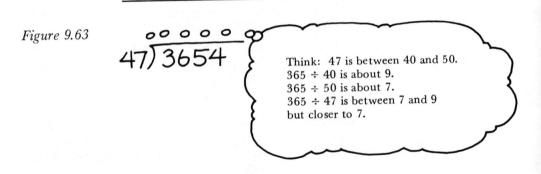

Think: 47 is between 40 and 50.
365 ÷ 40 is about 9.
365 ÷ 50 is about 7.
365 ÷ 47 is between 7 and 9 but closer to 7.

Even though this is the case, there are still two points of difficulty that nearly all children experience, and teachers need to be prepared to deal with them.

The first of these "hard spots" is the occurrence of a zero in the middle of the quotient.

<table>
<tr><td align="center">correctly done</td><td align="center">common error</td></tr>
</table>

```
        207                      2 7
17 ) 3519               17 ) 3519
     -34                      -34
     ───                      ───
     119                      119
    -119                     -119
    ────                     ────
       0                        0
```

The procedure that must be learned is that *whenever you trade for the next smaller unit, you must record something in that position of the answer.*

```
        20                Trade the remaining hundred for 10 tens,
17 ) 3519                 giving 11 tens altogether. This is not enough
     -34                  to distribute 17 ways so there will be 0 tens
     ───                  in each of the 17 equal parts.
     119

        207               Trade the 11 tens for ones. Now there are
17 ) 3519                 119 ones to be distributed 17 ways. Distri-
     -34                  bute 7 ones to each of the 7 equal parts.
     ───
     119
```

The easiest way to convince children that the zero belongs in the answer is to model the problem so they can see where the zero comes from.

The second of the two "hard spots" in long division is the estimation of partial quotients when the divisor is large (two digits or greater). There are various techniques that might be taught. Most of them involve some combination of rounding procedures with division by multiples of 10, 100, 1000, and so forth (Figure 9.63).

A teacher should not expect success with such estimation procedures unless adequate practice with exercises like the following is provided just prior to instruction on estimation of partial quotients.

Mental multiplication practice:

$$40 \times 7 \qquad 300 \times 8$$
$$6 \times 90 \qquad 4 \times 5000$$

Followed by mental division practice:

$$2800 \div 700 \qquad 720 \div 80$$
$$400 \div 50 \qquad 36000 \div 4000$$

And finally followed by practice on division with remainders:

$$272 \div 30 \qquad 60 \overline{)\ 430}$$

$$40 \overline{)\ 295} \qquad 500 \overline{)\ 3972}$$

DIVISION DOLLARS

Background Information. Ed, a new transfer student to a seventh grade classroom, needs special help with long division. He is just basically confused by the algorithm and is unsuccessful with the two-digit division problems being assigned as review work.

Teacher's plan for Ed: Use money to model division problems. In the past this approach has been particularly successful with older students. To begin, the teacher used problems with one-digit divisors and play money to illustrate the equal distribution.

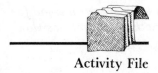

Activity File

Sample Problem:

$456 to share among three boys. They are most interested in getting the HUNDRED dollar bills first (Figure 9.64a).

$$3 \overline{)\ \$456}$$

There are four hundreds. Each gets 1 one hundred dollar bill. This uses up three of the four one hundred dollar bills (Figure 9.65b).

$$3 \overline{)\ \begin{array}{r} \$1 \\ \$456 \\ -3 \\ \hline 156 \end{array}}$$

Only 1 hundred left. Not enough to distribute again. Trade for 10 tens. Now there are 15 tens to divide among the three boys (Figure 9.65c).

$$3 \overline{)\ \begin{array}{r} \$1 \\ \$456 \\ -3 \\ \hline 156 \end{array}}$$

Each boy gets 5 ten dollar bills. No tens left to distribute (Figure 9.65d).

$$3 \overline{)\ \begin{array}{r} \$15 \\ \$456 \\ -3 \\ \hline 156 \\ -15 \\ \hline 6 \end{array}}$$

Only 6 one dollar bills. Each gets two. In all, each boy gets 1 hundred, 5 tens, and 2 ones, or $152. (Figure 9.65e).

$$3 \overline{)\ \begin{array}{r} \$152 \\ \$456 \\ -3 \\ \hline 156 \\ -15 \\ \hline 6 \\ -6 \\ \hline 0 \end{array}}$$

Figure 9.64

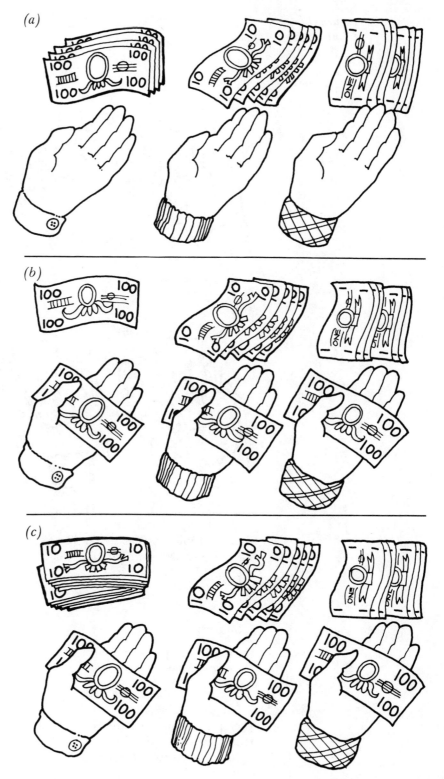

Figure 9.64
(continued)   *(d)*

(e)

Similar problems were worked and the general flow to the computation was noted and written on a file card for ready reference:

1. Divide

2. Multiply

3. Subtract

4. Check: are there enough left over to distribute more bills of this denomination?

The use of play money was dropped during follow-up work with two-digit divisors. The storyline and file card summary continued to help Ed remember the flow of the algorithm. The storyline also helped Ed determine in advance the number and placement of digits in the answer (Figure 9.65).

Figure 9.65   *Sample Problem*   *Ed's Thinking*

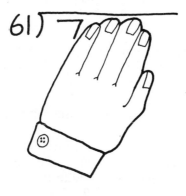

$7265 to share
with 61 people

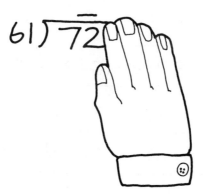

Not enough thousand
dollar bills to share
with 61 people. Trade
down to hundreds.

72 hundred dollar
bills. We'll divide
up the hundreds.

. . . then the tens
and the ones.
My answer will
have three digits
(be in the hundreds).

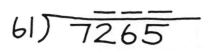

Activity File:
Long Division

Activity 1: Estimate And Calculate
   Need: Gameboard (similar to Figure 9.66), calculators, chips.
   Procedure: In turn, two players select a problem on the gameboard and state the number of tens. The player must tell what digit
will be in the tens place of the quotient. Then, check it on the calculator. If correct, the player may cover that block on the gameboard with a chip.

Winner: The first player to cover three consecutive blocks in the gameboard (Figure 9.66).

Variation: Players could be asked to find the hundreds digit of a quotient. Problems on the blocks may be changed on one gameboard, or make five different gameboards having different division problems on them.

Note: This activity, which involves the child taking one step at a time, gradually helps develop estimation skills.

Activity 2: Layout

Purpose: To have children work with base-ten blocks or strips of paper to develop a concrete understanding for working with division. Using the concrete models will help children visualize the division algorithm.

Need: Set of problem cards with division problems written on them.

Procedure: Teams of children work together, with each child needing to explain each step of the process. A point is given for each correct explanation with the use of the base-ten materials. An example showing $562 \div 3$ appears in Figure 9.67.

Figure 9.66

$142 \div 8$	$174 \div 8$	$231 \div 5$	$253 \div 6$	$382 \div 4$
$346 \div 5$	$358 \div 9$	$250 \div 4$	$342 \div 7$	$420 \div 6$
$128 \div 8$	$137 \div 4$	$524 \div 7$	$458 \div 5$	$631 \div 7$

Figure 9.67

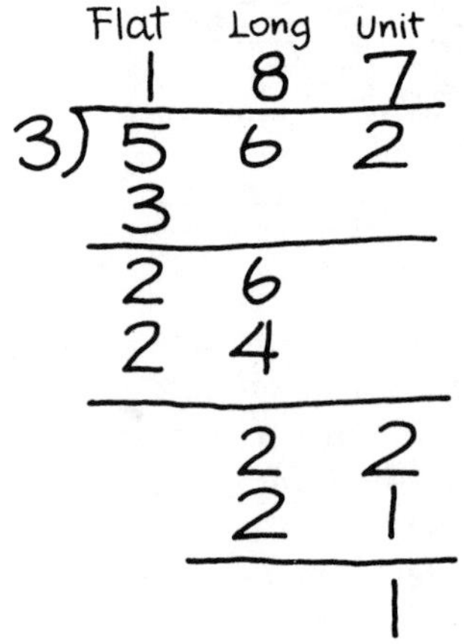

Teaching Mathematics to Children with Special Needs

Action:

1. Take 5 flats, 6 longs, 2 ones.

2. Place 1 flat in each of 3 groups, leaving 2 flats, 6 longs, and 2 ones.

3. Exchange 2 flats for 20 longs.

4. Place 8 longs in each of 3 groups.

5. Exchange 2 longs for 20 units.

Play continues similarly.
   Winner: First player to obtain ten points.

Activity 3: Practice Mental Division With A Friend
   Need: Number cube labeled 1, 2, 3, 4, 5, 6.
   Procedure:
1. The first player names a two-digit number and rolls the cube.

2. The other player divides (mentally) the number mentioned by the number on the cube.

3. The player doing the division scores a number of points equal to the remainder.

Variation: Extend this activity by having the player name three-digit numbers. Prohibit the use of prime numbers as dividends.

Trouble Shooting

Trouble Shooting
   Teachers as trouble shooters need to give attention to potential trouble spots in multiplication and division. Errors that are found early can be corrected and reteaching can take place before a bad habit is set. The following errors are common:

Multiplication

1. Mastery of basic multiplication facts is lacking (Figure 9.68).

2. Digits in the first product are reversed in the regrouping process (Figure 9.69).

Figure 9.68

```
  6 2
x   3
-----
1 9 6
```

Figure 9.69

```
  75
x 23
-----
 261
```

3. The regrouping digit is added before the second indicated multiplication is carried out (Figure 9.70).

4. Zero difficulties.
   (a) The product of a number p x 0 is given as the number p (Figure 9.71).
   (b) The sum of a number p + 0 is given as 0 (Figure 9.72).

5. Multiplying by only one digit when the problem has a two-digit multiplier (Figure 9.73).

Figure 9.70
$$\begin{array}{r} 36 \\ \times\ \ 4 \\ \hline 184 \end{array}$$

(4 + 2) x 3

6 x 3

Figure 9.71
$$\begin{array}{r} 307 \\ \times\ \ 4 \\ \hline 1268 \end{array}$$
$$\begin{array}{r} 803 \\ \times\ \ 6 \\ \hline 4878 \end{array}$$

4 x 0 = 4, so 4 + 2   6 x 0 = 6, so 6 + 1

Figure 9.72
$$\begin{array}{r} 504 \\ \times\ \ 7 \\ \hline 3508 \end{array}$$
$$\begin{array}{r} 409 \\ \times\ \ 6 \\ \hline 2404 \end{array}$$

0 + 2 = 0   5 + 0 = 0

Figure 9.73
$$\begin{array}{r} 63 \\ \times\ 57 \\ \hline 441 \end{array}$$
$$\begin{array}{r} 58 \\ \times\ 45 \\ \hline 290 \end{array}$$

Figure 9.74
$$\begin{array}{r} 32 \\ \times\ 46 \\ \hline 192 \\ 128 \\ \hline 320 \end{array}$$
omits 0 in product

6. Alignment difficulties (Figure 9.74).

7. Confusing addition with multiplication (Figure 9.75).

Division

1. Wrong placement of digits in quotient (Figure 9.76).

2. Zero difficulties.
   (a) Middle zero is omitted (Figure 9.77).
   (b) Final zero is omitted (Figure 9.78).

3. Incomplete division, remainder is greater than divisor (Figure 9.79).

Figure 9.75

$$
\begin{array}{r}
36 \\
\times\ \ 6 \\
\hline
190
\end{array}
\qquad
\begin{array}{r}
63 \\
\times\ \ 5 \\
\hline
308
\end{array}
$$

Figure 9.76

$$
\begin{array}{r}
45 \\
3\,\overline{)\,162} \\
15 \\
\hline
12 \\
12 \\
\hline
\end{array}
\qquad
\begin{array}{r}
56 \\
4\,\overline{)\,260} \\
24 \\
\hline
20 \\
20 \\
\hline
\end{array}
$$

Figure 9.77

$$
\begin{array}{r}
12 \\
6\,\overline{)\,612} \\
6 \\
\hline
12 \\
12 \\
\hline
\end{array}
\qquad
\begin{array}{r}
27 \\
4\,\overline{)\,828} \\
8 \\
\hline
28 \\
28 \\
\hline
\end{array}
$$

Figure 9.78

$$
\begin{array}{r}
58\ \ \text{r } 2 \\
4\,\overline{)\,2322} \\
20 \\
\hline
32 \\
32 \\
\hline
2 \\
\end{array}
$$

Figure 9.79

$$
\begin{array}{r}
95 \\
4\,\overline{)\,387} \\
36 \\
\hline
27 \\
20 \\
\hline
7 \\
\end{array}
\qquad
\begin{array}{r}
14\ \ \text{r } 9 \\
5\,\overline{)\,79} \\
5 \\
\hline
29 \\
20 \\
\hline
9 \\
\end{array}
$$

4. Dividing the first digit of the divisor only, ignoring the second digit of the divisor (Figure 9.80).

5. Failure to recognize when quotient digit is too small (Figure 9.81).

6. Failure to recognize when quotient digit is too large. (Children "reverse" subtraction or otherwise treat the problem their own way.) See Figure 9.82.

Special learning problems with multiplication and division algorithms require special treatment. Sample problems and solutions follow.

Multiplication
   Problem 1. Trading and making exchanges of tens and ones:
(a) Use concrete physical models.
(b) Label column headings with *tens*, *ones*.
(c) Use graph paper pieces.

Figure 9.80

$$
\begin{array}{r}
3282 \\
37\overline{)\,9846} \\
9 \\
\hline
8 \\
6 \\
\hline
24 \\
24 \\
\hline
6 \\
6 \\
\hline
\end{array}
\qquad
\begin{array}{r}
716\ \ r\ 3 \\
43\overline{)\,2867} \\
28 \\
\hline
6 \\
4 \\
\hline
27 \\
24 \\
\hline
3
\end{array}
$$

Figure 9.81

$$
\begin{array}{r}
299 \\
21\overline{)\,6473} \\
42 \\
\hline
227 \\
189 \\
\hline
383
\end{array}
$$

Figure 9.82

$$
\begin{array}{r}
6 \\
63\overline{)\,3455} \\
378 \\
\hline
47
\end{array}
$$

(d) Relate multiplication to addition in terms of needing to regroup or rename with ten or more (Figure 9.83).

Problem 2. Zero difficulties:

(a) Provide a variety of examples of multiplication with zero in different positions within a multiplication problem.

(b) Provide practice in multiplication of zero by a number and then adding a number, such as (4 x 0) + 3 (Figure 9.84).

Problem 3. Difficulties with alignment:

(a) Use graph paper and have children write a digit per box.

(b) Label columns and stress the meaning of the position of the digits, such as ones, tens, etc.

(c) Encourage estimation in checking to see if the answer is reasonable.

(d) In multiplying by a two-digit number, emphasize that the second partial product will end in zero since the multiplication is by ten or a multiple of ten in that product (Figure 9.85).

Figure 9.83

45
x 3

Tens	Ones
4	5
x	3

Figure 9.84

40	206	203	510
x 3	x 8	x 15	x 30

Figure 9.85

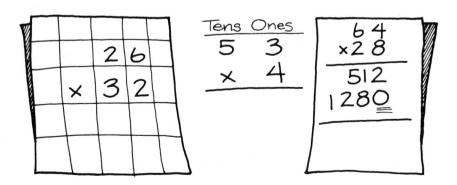

2 6
x 3 2

Tens	Ones
5	3
x	4

6 4
x 2 8
———
5 1 2
1 2 8 0

Problem 4. Confusing addition and multiplication in a multiplication problem.
(a) Encourage the use of concrete physical models with one-digit
multipliers.
(b) Encourage estimation. Use a story problem situation to explain
the meaning of the multiplication problem.
(c) Emphasize the use of the distributive property informally with
two-digit multipliers, e.g., 53 x 24 = 53 x 4 + 52 x 20; two
partial products (Figure 9.86).

Division
Problem 5. Placing digits in the proper position in the quotient.
(a) Stress where the first digit in the quotient will go; provide
worksheets where the child simply indicates the first quotient
digit.
(b) Label columns.
(c) Provide extra practice with estimation skills. Is this a reasonable answer? See Figure 9.87.
(d) Use the division dollar storyline suggested in the case study of
the Division section of this chapter.

Figure 9.86

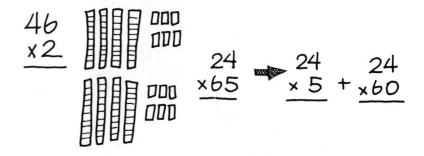

Figure 9.87

How many 5's in 34?
5 x _____ = 30

Quotient Search: Emphasize estimation.
Where is the ten's digit?
Where is the one's digit?

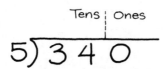

Problem 6. Zero difficulties. Final zero or middle zero is omitted in the quotient.
(a) Encourage children to estimate the quotients and determine how many digits will be needed.
(b) Make boxes in the quotient's place-value positions as reminders that a digit will belong in each position from the leftmost to the ones position (Figure 9.88).
(c) Use square-centimeter paper. Encourage estimation. Will the answer be a one-digit, two-digit, or three-digit quotient? See Figure 9.89.
(d) Refer to the suggestions given in the Division section of this chapter, "Special Help for Special Problems."

Problem 7. Incomplete division, where remainder is greater than divisor.
(a) Encourage children to estimate the quotient.
(b) Use story problem situations.
(c) Review the meaning of the words *remainder* and *division*.
(d) Use the division dollar storyline as you focus on the remainder. Refer to the case study in the Division section of this chapter.

Figure 9.88

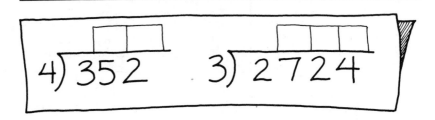

Figure 9.89

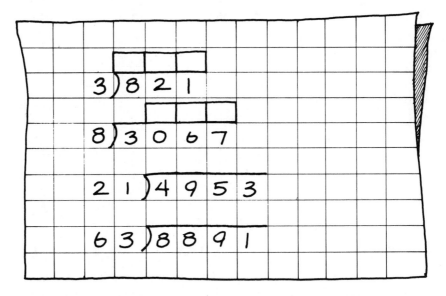

Questions and Activities

1. Examine each set of exercises to see if you can find an error pattern. Check yourself by doing the last two problems using that error pattern as the child would. Then develop a sequence of related developmental and practice activities to help a child overcome this error (Figure 9.90).

2. Do the following exercises, activities and investigations.
   (a) Create developmental activities or worksheets based on some of the hints given in this chapter for multiplication computation.
   (b) Create developmental activities or worksheets based on some of the hints given in this chapter for division computation.
   (c) Use information from Chapter 3 and construct a test to assess strengths and weaknesses in multiplication computation.
   (d) Use information from Chapter 3 and construct a test to assess strengths and weaknesses in division computation.
   (e) Use two elementary mathematics textbooks from different textbook series. Find four learning activities for multiplication computation. State whether each activity is for development, drill and practice, or enrichment.
   (f) Use two elementary mathematics textbooks from different textbook series. Find four learning activities for division computation. State whether each activity is for drill and practice, development, or enrichment.
   (g) Think of a child who has a sequential memory difficulty. Select for this child an appropriate learning activity involving multiplication or division computation.

Figure 9.90

1.
```
  5 3        2 7        3 1        5 4        7 3
x   6      x   8      x   4      x   7      x   5
-----      -----      -----
  3 0 8      1 6 6      1 2 4
```

2.
```
  2 5        2 7        3 1        5 3        8 3
x   7      x   8      x   4      x   7      x   4
-----      -----      -----
  4 2 5      5 6 6      1 2 4
```

3.
```
    3 7        2 9        5 7        3 8        4 2
  x 2 6      x 4 5      x 2 9      x 3 4      x 5 3
  -----      -----      -----
    2 2 2      1 4 5      5 1 3
  1 0 4      1 2 6      1 6 4
  -------    -------    -------
  1 2 6 2    1 4 0 5    2 1 5 3
```

4.
```
  4 1 7        4 2 6        7 2 4        5 1 7        2 8 5
x     5      x   3 8      x 4 5 1      x   2 6      x 4 6 3
-------      -------      -------
2 0 8 5      1 3 0 8      2 9 0 4
```

5.
```
      1 3 3          2 3 1          4 1 2
  3 ) 3 9 1      3 ) 7 1 3      4 ) 5 1 8      4 ) 8 2 4      2 ) 1 7 6
```

6.

```
      23            67            19
   3)96          3)228         8)728        6)516        4)164
```

7.

```
      49          63 r 5        83 r 2
   7)2863        6)3623        7)583        8)3219       9)5426
     28            36            56
       63            23            23
       63            18            21
                      5             2
```

References

Barnard, Janet, elementary and junior high mathematics teacher. Division dollars were used by Janet Barnard in her mathematics classes to help children with the division algorithm.

Bley, Nancy, learning disabilities specialist. Case study based on personal experience with learning disabled children at Park Century School in California. Relayed in unpublished communication with authors.

Tucker, Benny F., "The division algorithm," *The Arithmetic Teacher*, December 1973, pp. 639-646.

10 DECIMALS

Teacher Background

Modern school mathematics curricula are beginning to focus on decimals much earlier than in the past. Previously the development of decimals depended on fraction prerequisites. There is now a gradual shift to numeration emphasis like that used to introduce whole numbers and computation with whole numbers. Indeed, a trend to treat decimal computation before fraction computation is emerging.

The development of basic decimal concepts and skills generally moves through three phases in elementary school instruction. The first phase focuses on helping children to conceptualize decimals and to read and write them correctly. Physical and pictorial models are used to develop decimal numeration meanings. The same models are used in the next two phases to help children compare and order decimals, and to develop computational skills. Big ideas introduced for whole-number operations are now extended and applied to decimals. In decimal addition one adds like units and makes a trade whenever there are ten or more of a kind (Figure 10.1). Big ideas for subtraction are similar. Subtract like units. If there are not enough, make a trade. These basic ideas are extended to decimal multiplication and division.

Figure 10.1

The three-phase development of basic decimal concepts and skills is presented next, along with ideas for anticipating and avoiding many common difficulties. In addition, troubleshooting sections are included to help teachers handle special problems.

What Do *You* Say?

1. How are each of the following teaching aids used to help children develop basic decimal meanings?
   (a) squares and strips
   (b) hundreds square
   (c) money
   (d) meter stick
   What is the "1" in each model?

2. How can teachers help children read and write decimals correctly?

3. What common difficulty do children experience when they compare decimals like .18 and 0.4? How might teachers help children avoid this problem?

4. What big ideas introduced for whole-number addition and subtraction apply to decimal addition and subtraction?

5. How can visual cueing be used to build decimal meanings?

Basic Concepts and Models: Phase I

As in whole number numeration, instructions for decimal numeration should include the following:

1. After hearing the decimal spoken, the student should be able to use a model to show the decimal and write it correctly (Figure 10.2).

2. When shown the model representation of a decimal, the student should be able to both say and write the decimal (Figure 10.3).

3. When seeing a decimal written, the student should be able to say the decimal and show it with a model (Figure 10.4).

Instruction in decimal numeration is complete if, when given a decimal in any of these three forms (the model, spoken, or written), the child can give it back in the other two forms. Since young children's thinking is typically concrete rather than abstract, the real

---

Figure 10.2

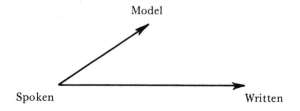

Figure 10.3

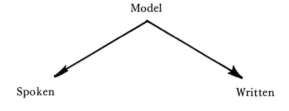

Figure 10.4

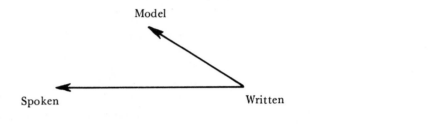

Figure 10.5

Figure 10.6

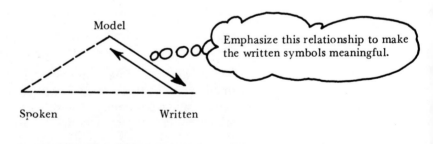

Figure 10.7

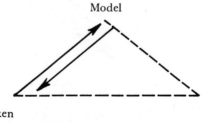

meaning for spoken or written decimals will come from their experiences with the physical or pictorial model.

In general, if you want to give meaning to what's spoken, let the child *see* what you are talking about (Figure 10.5). And, if you want to give meaning to what's written, let the child *see* what is written (Figure 10.6).

Four models for illustrating decimals are presented in this section: squares and strips, the hundreds square, money, and the meter stick. Most school textbooks rely on hundreds-square illustrations in early developmental sessions and use other models to reinforce where appropriate. Activities with squares and strips prepare for meaningful use of the hundreds-square model. Activities with squares and strips emphasize the relationship illustrated in Figure 10.7, and this is where instruction normally begins.

Squares and Strips. Large squares are used to represent the ones (Figure 10.8). A large square can be separated into ten equal strips which are used to represent tenths, and a strip can be cut into ten small squares to represent hundredths (see photo).

Figure 10.8

Ones

Tenths

Hundredths

The Hundreds Square. The hundreds square represents 1. It is subdivided into ten rows of ten equal squares. Any row or column can be shaded to represent one tenth. Any of the small squares can be shaded to represent one hundredth (Figure 10.9).

Money (Dollars, Dimes, Pennies). It is easy to use money to represent decimals since children are exposed to it daily. The dollar is used to represent one, the dime one tenth, and the penny one hundredth (Figure 10.10).

Meter Stick. One meter can be used to represent 1 (Figure 10.11). The decimeter would then be a tenth and the centimeter would be a hundredth. The length measured corresponds to a decimal number.

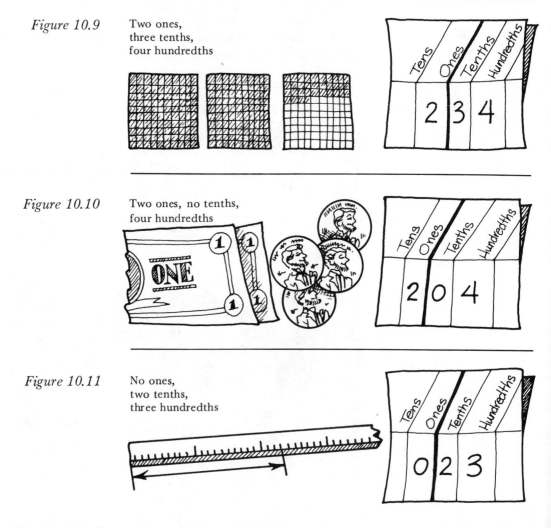

Figure 10.9 Two ones, three tenths, four hundredths

Figure 10.10 Two ones, no tenths, four hundredths

Figure 10.11 No ones, two tenths, three hundredths

The Decimal Point

You will note that in the preceding illustrations no effort was made to introduce the decimal point. Rather, *the focus at this early stage* was on *developing the mental imagery for the decimal units*. When a child hears a decimal spoken or sees one written, the child should have a mental picture of what that number looks like.

Another goal of early instruction is establishing the *relationship between units*. The notion of *what constitutes even trades is important*. Two ones can be traded for 20 tenths. Ten hundredths can be traded for 1 tenth. When even trades such as these are made, the decimal name will change but the decimal amount (the *quantity*) will remain the same (Figure 10.12).

Another important idea that should emerge from early decimal instruction is standard position for the units. *Let students see the symmetry of the system* (Figure 10.13). As they become familiar with the standard positions for the units, students should be able to identify the various place values if they are given any one of them. For example, if the 2 in Figure 10.14 represents the number of tenths, what does the 5 mean? The 8? The 0?

Figure 10.12

 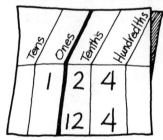

Two names for the same amount

Figure 10.13

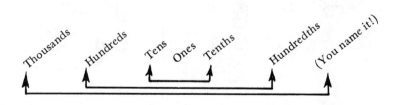

Figure 10.14

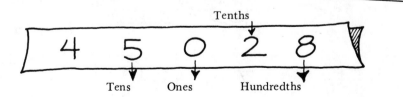

The decimal point is always placed to the right of the ones digit. And, since the decimal point tells us the location of the ones, we can figure out all the rest (Figure 10.15). Of course, we should note that the decimal point also happens to separate the numeral into a whole-number part and a fraction part.

Reading Decimals. Since there are many ways to represent a given decimal amount, it follows that each of those representations might be named and read differently. In the example of Figure 10.16, the decimal 2.13 is represented and named in two ways. Either name is, of course, equally correct. However, when reading decimals, the second name is less cumbersome and thus preferred (Figure 10.17).

Figure 10.15

Figure 10.16   Two ones, one tenth, three hundredths   Two and thirteen hundredths

Figure 10.17

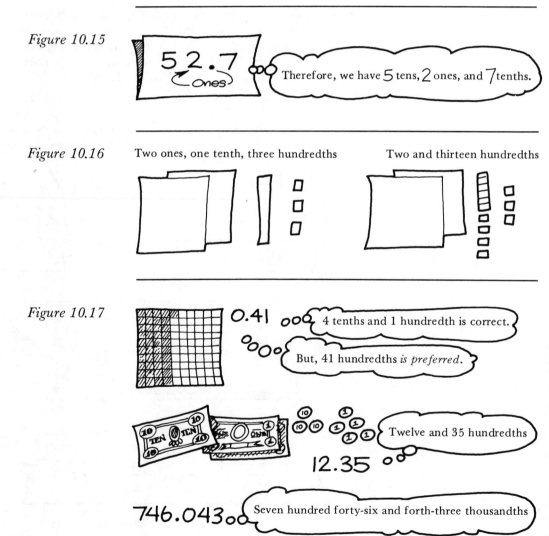

Children need many informal, exploratory experiences so they can read, write, visualize, and illustrate the decimals they use. Suggestions follow for activities along this line. Other ideas for introducing or reinforcing decimal concepts and skills are included in the Math Lab Activities of Appendix C.

Activity File: Basic Decimal Concepts—Reading and Writing

Activity 1: Draw 'N' Shade. (Developmental activity for independent study; for small or large groups)

Need: For each child, a sheet of paper stamped with hundred squares (Figure 10.18); crayons; problem deck for tenths and hundredths (six of each); envelope for cards.

Note: Rubber stamps, available from school-supply companies can be used to prepare the pupil sheets.

Procedure: Cards are mixed and placed in an envelope. One player/teacher serves as caller. A card is drawn, shown to all, and read aloud. Refer to the example of Figure 10.19. Players shade their square to illustrate the decimal named, then turn the card over to check.

Figure 10.18

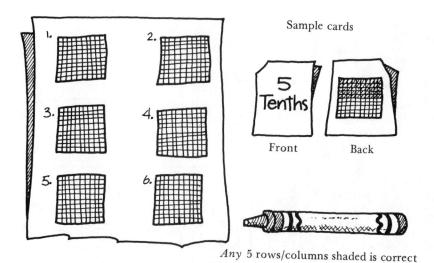

Sample cards

Front   Back

Any 5 rows/columns shaded is correct

Figure 10.19

Example: Caller says "62 hundredths."

Note: For larger groups the number words may be written on the chalkboard.

Variations: (For individuals or small groups)

1. Draw a card and write the decimal (Figure 10.20).

2. Use the hundreds-square side of the cards. Draw the cards and write the decimal (Figure 10.21).

3. Use the hundreds-square side of the cards. Draw card, and name aloud a decimal to describe the shading (Figure 10.22).

Activity 2: Make 1. (Developmental activity for two or three to play)

Need: A "bank" of squares and strips (Figure 10.23). A die marked .1, .2, . . . , .6, and a die marked .01, .02, . . . , .06.

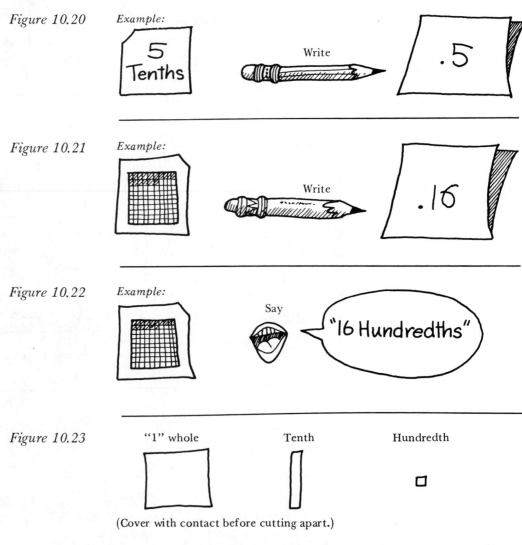

Figure 10.20   Example:

Figure 10.21   Example:

Figure 10.22   Example:

Figure 10.23   "1" whole   Tenth   Hundredth

(Cover with contact before cutting apart.)

Procedure: Players each roll the "one-tenth" die. High point goes first. Then, in turn, players roll both dice and take squares or strips to match (Figure 10.24). Whenever a player has ten of one piece, he or she trades with the bank for the next larger piece.

Winner: First player to get one large square.

Activity 3: Draw It. (Developmental activity for individuals; for small or large groups)

Need: Meter stick, pencil.

Procedure: Draw line segments to represent given decimals. Work in pairs. Check with your neighbor. Do your drawings match? (See Figure 10.25.)

Activity 4: Calculator Punch. (Practice activity for two to play)

Need: Hand calculator for each child; problem cards for tenths and hundredths like those of Figure 10.26.

Procedure: Cards are mixed and placed in an envelope. In turn, players draw a card. Both players punch their calculator to show the decimal on the display (see Figure 10.27). They check with

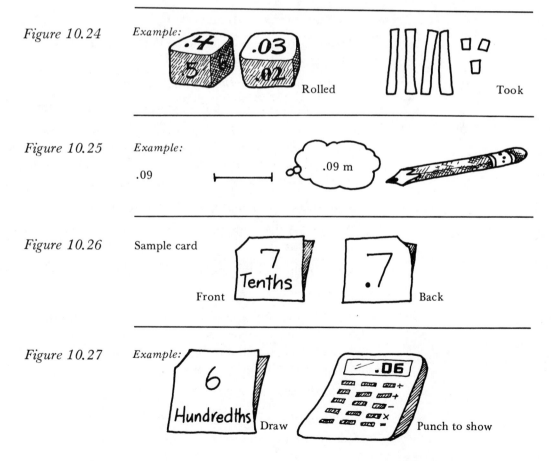

Figure 10.24   *Example:*   Rolled   Took

Figure 10.25   *Example:*   .09   .09 m

Figure 10.26   Sample card   7 Tenths   Front   .7   Back

Figure 10.27   *Example:*   6 Hundredths   Draw   .06   Punch to show

each other, then turn the problem card over. Does it show what the calculators do?

Extension: Add thousandths to the problem deck when children are ready.

Naming and Writing Decimals

Case Study:
Kim

Background information. Kim, a learning disabled seventh grader with difficulties in abstract reading, auditory discrimination, and association, is working two years below level in mathematics. At the beginning of this study she was having difficulty naming and writing decimals. Conventional approaches did not trigger the necessary associations. Kim's teacher (Bley, 1979) developed the following sequence to help her with this problem. The sequence relies on Kim's strong visual-association abilities. Within it, writing decimal and money amounts are interrelated to build understandings and skills in both areas. The five major steps to the sequence are:

1. Providing kinesthetic and visual stimuli to cue correct writing of decimals.

2. Relating decimals to money.

3. Interrelating money and decimal models, emphasizing visual associations.

4. Writing decimals *without* visual cueing.

5. Associating decimal number words with numeric forms.

Prior to the sequence Kim had participated in activities with squares and strips. She could verbally name the "one" square as well as the tenths and hundredths pieces. She *could* show you, using the squares and strips, how:

1. Ten strips (tenths), put together, form a "one" square

2. A strip (tenth) could be cut into ten small squares called hundredths

3. One hundred of these small squares filled the "one" square.

Step 1

Providing kinesthetic and visual stimuli to cue correct writing of decimals. Kim's teacher used squares and strips with worksheets like that of Figure 10.28. Red color cueing was used for all decimal points and underscores on the sheets. For each problem Kim:

1. Used the squares and strips to picture the problem

2. Told and then recorded the number of pieces pictured

3. Wrote the decimal form (with teacher assistance).

Underscoring on the worksheet, along with use of squares and strips, provided the necessary cueing for Kim to write the decimals correctly. In particular, the underscoring prompted the association between:

1. "One zero" (for 1_0_ths) and "one decimal place" (e.g., 4 1_0_ths → 0.4);

2. "Two zeros" (for 1_00_ths) and "two decimal places" (e.g., 26 1_00_ths → 0.2_6_).

Step 2

Relating decimals to money. Once the idea of decimal place value was established to hundredths, decimals were related to money. At first real money was used, with worksheets like that of Figure 10.29. Writing money amounts was reviewed while

Figure 10.28

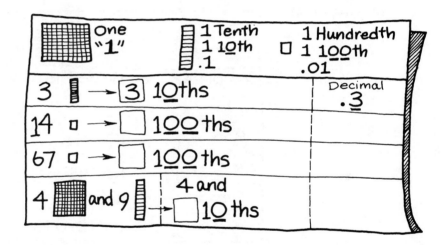

Figure 10.29

focusing especially on the decimal part. Eventually, the dollars were eliminated and only pennies were used. Oral discussion which accompanied the example at the bottom half of the page emphasized that "we have six of the 100 pennies in a dollar." Underscoring was again used on this part of the page.

Step 3

Interrelating money and decimal models, emphasizing visual associations. To make use of Kim's strong visual-association abilities, she was given concrete aids and worksheets which stress the relationship between writing money and decimals (Figure 10.30). Worksheets containing problems like those of Figure 10.31 were also used. Kim could use real money or the squares and strips whenever this seemed helpful.

Figure 10.30

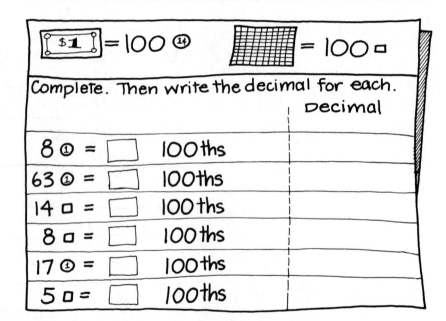

Figure 10.31

Step 4

Writing decimals without visual cueing. For the next set of worksheets, pictures were eliminated except at the top of each page (Figure 10.32). Kim was encouraged now to turn away from models and show she could read and write decimals just by thinking about what they meant.

Step 5

Associating decimal number words with numeric forms. Kim could read but had an auditory learning difficulty. To help her function to her full potential and make it possible for her to use the calculator and other activities of the preceding section, it was necessary to help Kim read correctly the decimal number words and associate them with the numeric forms. Worksheets like that of Figure 10.33 were used to accomplish this goal. All underscored letters on the worksheet of Figure 10.33 are printed RED.

Figure 10.32

```
6  100ths = Six hundredths  or .06
9   10ths = nine tenths    or .9
Complete, then write the decimal for each:
four tenths = _____ or  [  ]
eighteen hundredths= _____ or  [  ]
Seven hundredths= _____ or  [  ]
Write the decimal:
              nine hundredths    =  [  ]
              sixteen hundredths =  [  ]
              Five tenths        =  [  ]
```

Figure 10.33

```
[$1] = 100 ⑩      [▥] = 100 □
Write the decimal for:
    4  10ths :
   23  100ths:
    7  100ths:
Read the decimal. Tell what it means.
   .9  =  [  ]  10ths
  .03 =
  .18 =
  .16 =
```

Note: The most critical aspect of this sequence was the repeated association of the written decimal form with a model for the decimal or with its verbal form. In early stages the underscoring also helped.

Examples: 6 10ths → 0.6   27 100ths → 0.27

Comparing and Ordering Decimals: Phase 2

The overlearning promoted by these procedures helped Kim learn to read and write decimals correctly despite her abstract reasoning and auditory difficulties. Many children relate 0.42 and 0.9 to the whole numbers 42 and 9. Mistakenly, they infer that since 42 is greater than 9, then 0.42 is greater than 0.9. The use of shaded regions to picture each of the decimals can help remedy the problem (Figure 10.35). (This technique is best used during early teaching sessions so the problem will not appear at all.) Since a big idea of comparison is that we compare like units, children can also be cued to add zeros to even out decimal places to the right of the decimal point (Figure 10.36). Activity suggestions such as those below reinforce these ideas.

Figure 10.34

Circle to tell which is greater:

.42   .9   .34   .81

.6   .8   .21   .8

Figure 10.35

Example of the shading technique:

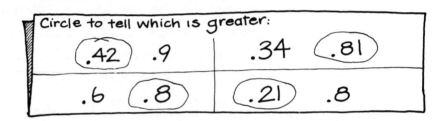

.42   .9   .9 is a lot *more.*

Figure 10.36

Example:

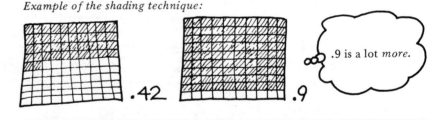

8 tenths is the same as 80 hundredths

.37   .80

So 80 hundredths is greater than 37 hundredths.

**Activity File:
Comparing and
Ordering Decimals**

Activity: Decimal War. (Developmental activity for two to play)
   Need: A "bank" of squares and strips; deck of problem cards
for tenths and hundredths (see Figure 10.37).
   Procedure:

1. Cards are mixed and placed face down between players.

2. Each player turns up one card and pictures the number on that
card with squares and strips.

3. Player with the greater decimal keeps both cards. Any ties
(e.g., 0.5 and 0.50) are played off in the traditional War manner.

   Winner: Player with the most cards when the deck is used up.
   Note: Later, as a practice activity, omit the squares and strips
and use only the card deck.

Figure 10.37 Sample cards

**Decimal Addition
and Subtraction:
Phase 3.1**

   Computational procedures for adding and subtracting decimals
are basically the same as for whole numbers. As in whole-number
addition, there are only two big ideas to master for decimal addition. First, children must understand that you only add like units
(ones to ones, tens to tens, tenths to tenths, and so on, as in Figure
10.38). Second, they must learn that when there are too many of

Figure 10.38

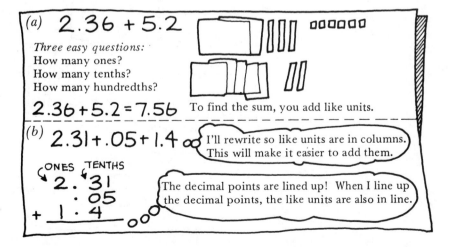

some unit to write down in one position, a trade must be made (Figure 10.39).

These same two ideas were emphasized when whole-number addition was taught, so it should be relatively easy to reteach them as they are used in decimal addition. Similarly, decimal subtraction is almost identical to whole-number subtraction. Always subtract like units. If there are not enough of some unit to do the subtraction, then a trade must be made (Figure 10.40).

Figure 10.39

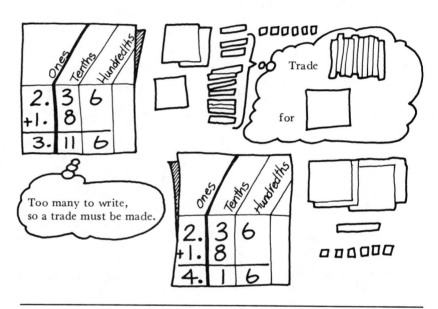

Figure 10.40

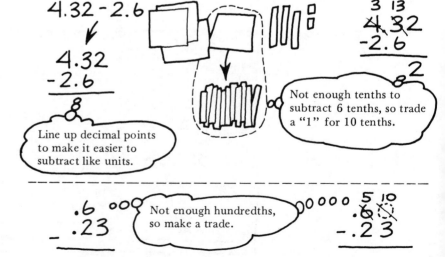

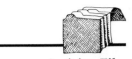

**Activity File:
Phase 3.1
Adding and
Subtracting Decimals**

Activity 1: Do It But Prove It. (Developmental activity for two or three to play)

Need: "Bank" of squares and strips; twenty-four problem cards like those shown in Figure 10.41; envelope for card deck.

Procedure:

1. Mix cards and place them in the envelope.

2. To begin a round, each player draws one card, copies and solves the problem on paper, then uses squares and strips to "prove" to the others that the answer is correct (Figure 10.42).

Figure 10.41

Front   Back

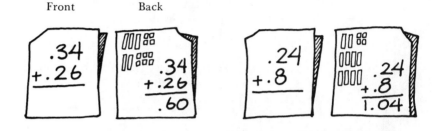

Figure 10.42

Example:

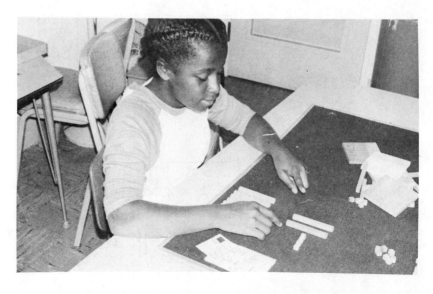

3. Players keep a tally of *rights*.

Variation: Use subtraction problems (Figure 10.43), or mixed addition and subtraction problems on cards.

Activity 2: Three Wins. (Practice activity for two or three to play)
    Need: Problem deck of decimal cards, two each of the following: .1, .2, . . . , .9 and .01, .02, . . . , .09—one each of .001, .002, . . . , .009; paper and pencil. A four-column recording sheet may also be provided for each player.
    Procedure:
1. Cards are mixed and placed between players.

2. Take turns drawing cards. Each player keeps a running sum of decimal numbers drawn.

3. If deck runs out, mix and use the cards again.

Winner: First to get three or more as a sum. (Other players must agree with the computations made.)

Figure 10.43   *Example:*

1. Prepare, as your instructor directs, answers to the major questions on page 327.

2. Consult the sample curriculum guide given in Appendix A.
   (a) Use the list of objectives to identify the grade level at which decimals are commonly introduced.
   (b) Are students asked to compare and order decimals at that same level? Are they introduced to decimal addition and subtraction at that level?

3. Choose a teacher's guide from a current elementary-school mathematics series. Select and share with classmates one developmental, one practice, or one application activity for each of the following:
   (a) Basic decimal concepts
   (b) Comparing or ordering decimals
   (c) Addition or subtraction of decimals

4. Rewrite the directions of the Make 1 activity on page 336. Reverse the procedures. Have children start with one large square and *give away* pieces.

5. A decimal pocket chart like that illustrated in Figure 10.44 is a useful aid for many students.

   The chart can be made from a half-sheet of poster board and 3 x 5 index cards. The numerals 0-9 should each be written on several index cards. The cards can be inserted into the proper columns of the chart to show how a number with decimal places should be written. Since every slot between the decimal point and the last digit must be filled, the need sometimes to insert a zero is emphasized (refer to the preceding example). Another useful function of the chart is helping children correctly *read* decimals.
   Review the Draw 'N' Shade activity and its variations described on page 335. Rewrite the activity (variations as well), using the pocket chart where appropriate. Make the chart, if you can, and try out your activity with children.

6. Ordering decimals is sometimes difficult for students.
   (a) Show how you can use copies of the hundreds square to order the following numbers from least to greatest.
       0.01   0.1   0.18   0.8   1   0.81
   (b) Describe how you could use money to help a student order the same set of numbers.
   (c) How might the number line be used to illustrate the ordering?
   (d) Clearly, you would not use all three models (hundreds square, money, number line) within a lesson on ordering. Choose *one* of the three aids. Then outline a sequence of developmental and practice activities which uses that aid, at least in the early part of the sequence, to help a child learn to order decimals.

7. Tell how you would use money to demonstrate:
   (a) 0.5 − 0.07 = 0.43
   (b) 0.36 = 0.3 + 0.06

8. Adapt the activity Three Wins (p. 344) to provide practice *subtracting* decimals through thousandths. Give your activity an interesting title.

9. Refer to the Trouble Shooting section at the end of Chapter 8. Near the end of this section seven problems and sample solutions are presented. Use ideas from that section to outline *four* problems, with solutions, for decimal addition or subtraction.

10. Examine the computational errors that appear in Figure 10.46. For each error describe:
    (a) What the child has done wrong
    (b) What remedial experiences you would provide

Figure 10.44

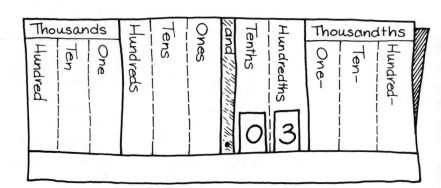

Figure 10.45

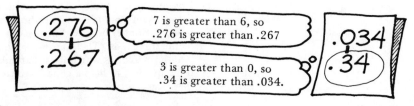

Figure 10.46

(1) .6
    +.7
    ‾‾‾
    .13

(2) .6
    .23
   +.06
   ‾‾‾‾
    .95

(3) .13−.1=.12

(4) 2.6
    −.73
   ‾‾‾‾
    1.93

What Do *You* Say?

1. How can multiplication and division of decimals be taught *without* reference to fractions?

2. How can models be used to develop the "rule" for placing the decimal point in decimal products?

3. What role does estimation play in checking decimal multiplication and division answers?

4. Which problem type for decimal division is easier for students: whole-number divisors or decimal divisors?

5. What types of problems should be included on a diagnostic/evaluative instrument to assess a child's grasp of decimal concepts and skills?

Decimal Multiplication: Phase 3.2

Decimal multiplication is much like whole-number multiplication. When computing, one multiplies the smallest unit first. If there are too many of some unit to write down in one position, a trade must be made. After adding any partial products, however, the problem is not complete until the decimal point is correctly placed. Helping children see where to place it is the major objective of this section.

Like addition and subtraction, multiplication of decimals can be taught without relying on prerequisite fraction concepts and skills. However, since many current textbooks still use fraction ideas to develop decimal multiplication, we have included an outline of that development in the discussion that follows.

Area as a model. Students ready to begin decimal multiplication normally have already studied the topic of area of rectangular regions. This idea, then, can be used to introduce multiplication of decimals.

Area of a rectangular region can be found by multiplying the lengths of its sides. It follows, then, that two names for the shaded region of Figure 10.47 are 4 x 7 and 28. Similarly, two names for the shaded region of Figure 10.48 are 0.1 x 0.1 and 0.01; so 0.1 x 0.1 = 0.01.

Figure 10.47

Figure 10.48

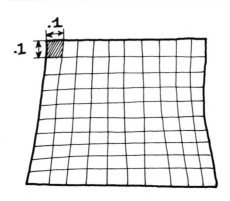

Figure 10.49

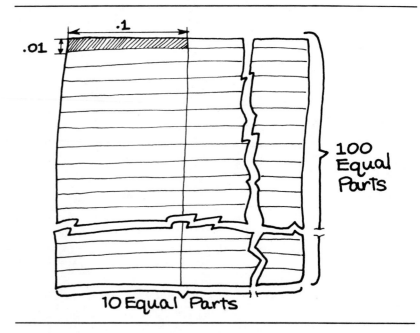

Figure 10.50

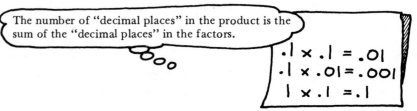

The number of "decimal places" in the product is the sum of the "decimal places" in the factors.

.1 × .1 = .01
.1 × .01 = .001
1 × .1 = .1

The region in Figure 10.49 is separated into 100 rows of ten equal parts, or 1000 equal parts. *One* of these parts is 0.001; so 0.01 x 0.1 = 0.001.

Examples like these can be used as the basis for an important generalization about multiplying decimals (Figure 10.50). Note that underscoring the decimal places often helps the child focus on this pattern.

Mathematically speaking. Children could apply the associative and commutative properties to obtain the same generalization for multiplying any two decimals. Study the analysis given in Figure 10.51.

Results like this can be easily verified by going back to the model (Figure 10.52). This same procedure could be used for any decimal multiplication problem (Figure 10.53).

Figure 10-51   A problem like .7 x 1.6
Involves multiplying 7 "tenths" by 16 "tenths." That is,
.7 x 1.6 = (7 x .1) x (16 x .1).
 (Tenths)   (Tenths)

Or, rearranging and regrouping: .7 x 1.6 = (7 x 16) x (.1 x .1).
 (Multiply whole numbers.)   (Two decimal places.)
                          then
And, since 7 x 16 = 112, ⟶ .7 x 1.6 = 1.12.

Figure 10.52   .7 x 1.6 = .7 + .42
                        = 1.12.

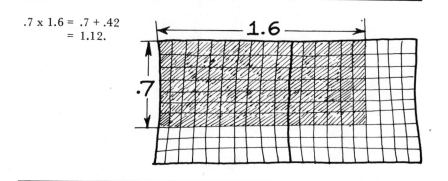

Figure 10.53   *Example:*

.4 x 2 = (4 x .1) x 2   (rewrite)
        = (4 x 2) x .1   (rearrange and regroup) = .8.

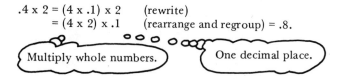
 (Multiply whole numbers.)   (One decimal place.)

The model in Figure 10.54 shows this area, 0.8. Of course, it is easy to see the decimal places without rewriting. Figure 10.55 illustrates how we might think.

Using fraction concepts. The fraction development illustrated in Figure 10.56 appears to be far simpler—and it is, *provided all the fraction concepts and skills have been mastered.* A teacher should expect to have to teach, reteach, or review the necessary fraction ideas before using this approach.

Figure 10.54

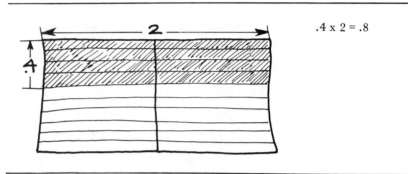

.4 x 2 = .8

Figure 10.55

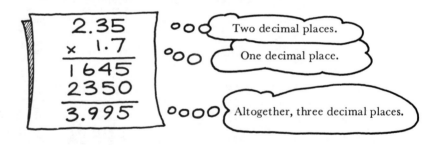

Figure 10.56

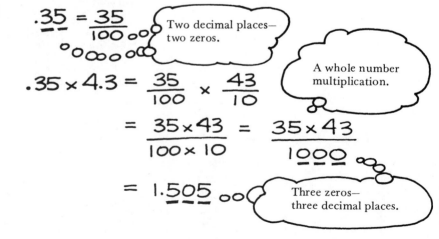

Teaching Mathematics to Children with Special Needs

MULTIPLYING
DECIMALS

Case Study:
Sarah

Background Information
   Sarah, a seventh grader, is at a fifth-grade level in mathematics. She is strong in basic facts and her whole-number computational skills are good. Because of experience with students like Sarah, the teacher decided to develop a unit on decimals, including computation, *before* turning to computation with fractions. Like others in her class, Sarah progressed quickly from early introductory work with decimals through addition and subtraction of decimals. She enjoyed the unit and found it rather easy. The following activities, which involve estimation techniques along with the use of models to check, outline the teacher's plan for developing decimal multiplication with Sarah's class. The basic steps to th teacher's plan were:

1. Estimate to place the decimal point in given problems.

2. Use models to check.

3. Examine completed problems to identify the pattern for placing the decimal point in decimal products.

4. Use the pattern to complete other decimal multiplications.

Before beginning the four-step plan, the teacher reviewed finding areas of rectangles.

Step 1

Estimate to place the decimal point in given problems.
   Sarah's teacher introduced an activity called Place the Point and the chart of Figure 10.57, with product digits given. Sarah and her classmates were asked to estimate each product and place the decimal point to complete each problem correctly. The teacher did the first example, then called on others in the class for the remaining problems. Sample responses are illustrated in Figure 10.58.

Step 2

Use models to check.
   After each estimation the teacher gave students a set of cards like that of Figure 10.59. Could they find the card that pictured their problem? Did their answer check with the area shown by the model? (The teacher could also have allowed students to use calculators to check.)

Figure 10.57

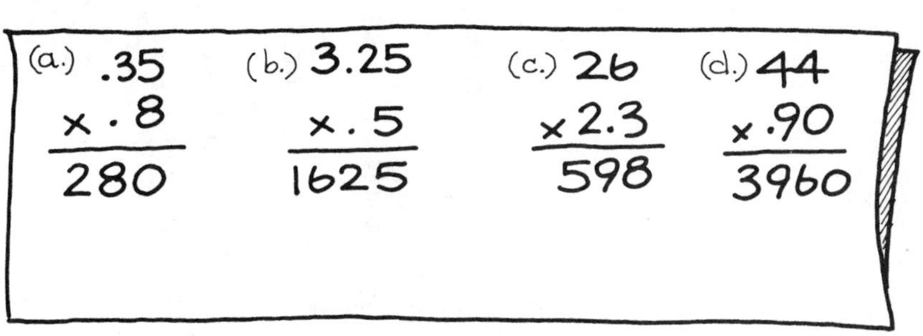

(a.) .35
 × .8
─────
 280

(b.) 3.25
 × .5
─────
 1625

(c.) 26
 ×2.3
─────
 598

(d.) 44
 × .90
─────
 3960

Figure 10.58

(a)
.35
× .8
.280

A little less than 1.35.

Then it must be .28. 2.8 or 288 wouldn't do at all.

(b)
3.25
× .5
1.625

About half of 3.

That's 1.6. .16 is too small. 162 is too big.

(c)
26
× 2.3
59.8

A little more than twice as much.

Not 5, not 598, but 59.

(d)
44
×.90
39.60

Almost 1 × 44.

39.6 is about that. 3.9 or 296 won't do.

Figure 10.59

Sample card:

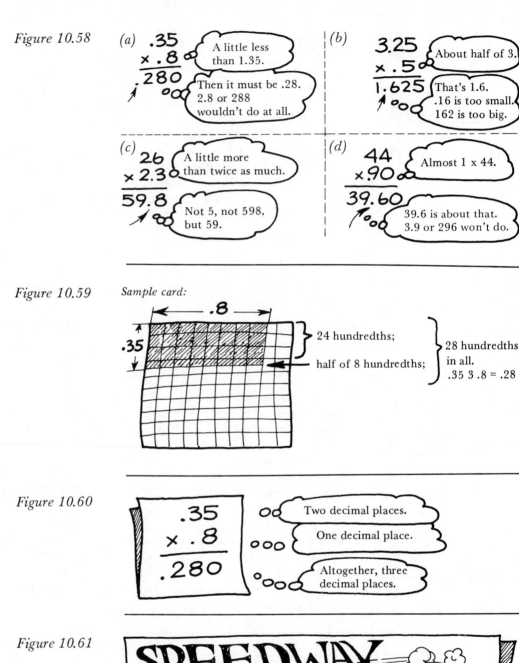

.8

.35

24 hundredths;

half of 8 hundredths;

28 hundredths in all.
.35 3 .8 = .28

Figure 10.60

.35
× .8
.280

Two decimal places.

One decimal place.

Altogether, three decimal places.

Figure 10.61

SPEEDWAY

.52 × .02 104	.003 × .23 69	.25 × .7 175	.103 × .04 412
72	13	47	51

Step 3

Step 4

**Activity File:
Decimal
Multiplication
Phase 3.2**

Examine completed problems to identify a pattern.

When all problems had been completed, the teacher asked Sarah and her classmates to study the chart of completed problems. For each problem, the students underscored the "decimal places" in the product. The teacher led them to compare this number with the sum of the "decimal places" in the factors. The students noted that these numbers are always the same (Figure 10.60). The students then used the estimate-model check procedure with other examples. The pattern of decimal places always held.

Use the pattern to complete other decimal multiplications.

The teacher introduced the following Speedway activity so children could use the pattern for other decimal multiplications.

Activity: Speedway. (Practice activity which each student in the group could play)

Need: Ditto of Speedway mat like that of Figure 10.61 for each child; a decimal multiplication problem (product digits given) in each square.

Procedure: At Go, children use the pattern they discovered to place the decimal point in each product. When time is up, children check their work against a key and count the number of problems they completed correctly.

Variation: Several mats, each containing different problems, can be laminated, placed in a file folder (Figure 10.62), and stored in the classroom learning center. Children who do the activity independently can be encouraged to keep a log of the time taken to complete each card correctly. This log allows the teacher to monitor a child's progress through the card set.

Follow Up: Problems from the students' mathematics books can be assigned. For each, children should perform the multiplication and then place the decimal point in each product.

Figure 10.62

Sides taped shut

**Decimal Division:
Phase 3.3**

Computational procedures for division with decimals closely parallels that for whole numbers. Once the decimal point is adjusted as needed, there are only two big ideas to remember. Divide biggest units first. Second, if ever there are leftovers, trade for the next smaller unit.

The new aspect of the computation is handling the decimal point, and that is the major objective of this section. The following discussion focuses on introducing decimal division without relying on fraction concepts or skills. However, since many current mathematic series still refer to fractions in explaining decimal division, an outline of that development is also included.

Easy problems first.

Division with decimals usually examines easy problems first— *those with counting numbers as divisors.* Two examples illustrating this division are presented in Figures 10.63a and 10.63b.

Note: Dividing money amounts is a common occurrence in daily life. It is relatively easy to do. Decimal division frequently starts with examples like that just given.

Children can use money or squares and strips to illustrate decimal division having counting numbers as divisors. In each case children can be cued to note how the decimal point in the quotient is *right above* that of the dividend (Figure 10.64).

Figure 10.63

(a) *Example 1:* $.64. Share with 4 people.

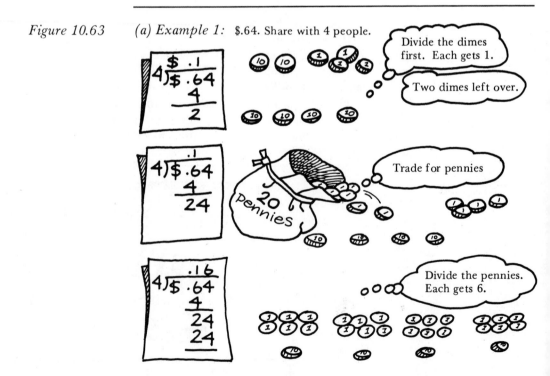

Figure 10-63

(b) Example 2: Use squares and trips. Divide 5.35 into 5 piles.

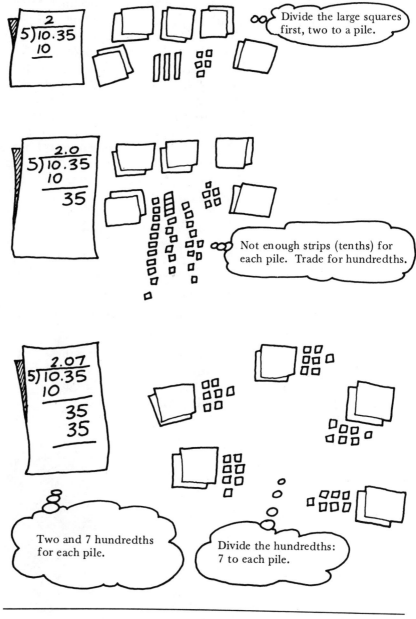

Figure 10.64

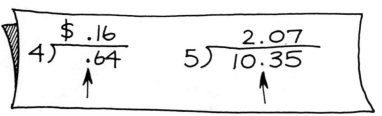

Decimal divisors.

When problems have decimal rather than counting number divisors children can use easier, equivalent problems to compute (Figure 10.65). These "easy" problems have counting numbers as divisors and the children know how to handle them.

The procedure for forming easy problems can be developed in several ways. We shall describe two methods. The first is a more informal, intuitive approach. It asks children to use a hand calculator to verify that given problems have the same answer. It then challenges them to look for a pattern linking the problems.

The second development uses fractions. Since it assumes that certain fraction concepts and skills have been mastered, teachers may have to review, teach, or reteach before using this approach.

Method 1: Using a calculator and finding a pattern. To begin, ask children to use a hand calculator to find answers to problem sets like those listed in Figure 10.66. Children will find, of course, that the two problems in each set have the same answer. The next step is to discover the pattern linking each pair of problems (Figure 10.67). In general, children can discover that:

Multiplying both dividend and divisor by ten or a power of ten gives a counting number divisor.

They know how to handle these divisions. Children should be encouraged to test this rule with problems they themselves create. A calculator check will show that, each time, the two problems do have the same answer.

Figure 10.65   *Given problem*   *Equivalent "easy problem"*

.5) 21.5   .5) 21.5

.02) 26.843   02) 26.84 3

Figure 10.66   *Problem Sets*

(a)   .5) 21.5   5) 215

(b)   2.6) 24.   26) 240

(c)   .02) 26.846   2) 2684.6

(d)   .82) 434.6   82) 43460.

Figure 10.67   (a)

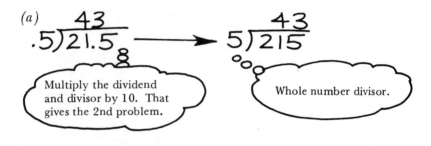

(b)

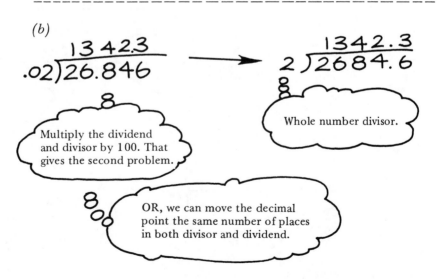

Method 2: Using fractions. Three basic steps are involved. They are illustrated in Figures 10.68a and 10.68b. Is it always necessary to rewrite into fractional form? Children will discover, with teacher guidance as needed, that:

Multiplying both dividend and divisor by ten or a power of ten gives a counting number divisor.

It is not really necessary to rewrite into fractional form each time. The multiplication will immediately give an equivalent, "easy" problem they know how to do.

Division with decimals.
    The hand calculator has replaced the need for many paper-and-pencil calculations. In day-to-day living, one may only occasionally hand compute a decimal division. Many students, however, need to pass minimal competency exams in mathematics in order to receive grade- or high-school diplomas. In their case computational skills, including those for decimals, must be mastered. Certainly any student preparing for high-school algebra must capably handle decimal computations.

Trouble Shooting

Figure 10.68   *(a) Example 1:*

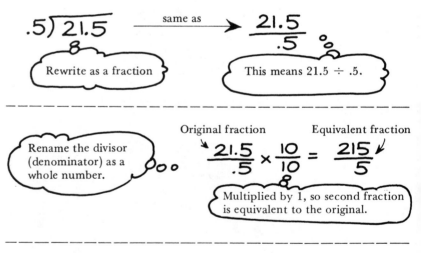

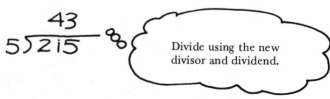

Figure 10.68   (b) Example 2:

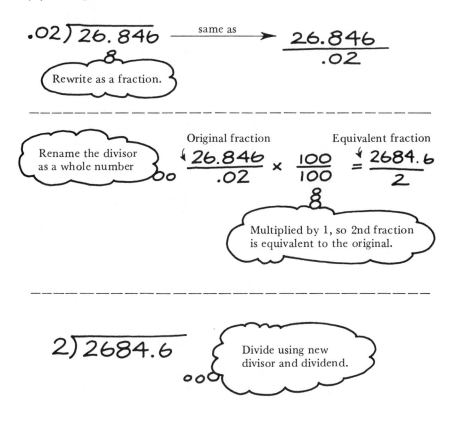

For children with learning difficulties in mathematics, some techniques promote skill mastery better than others. One such technique is taking things *one step at a time*. Focusing only on *one* difficult step of a computation often provides practice where it is needed most. The examples of Figure 10.69 suggest ideas for applying this one-step technique to decimal divisions.

The last example presents a topic that requires special attention. Many school mathematics texts require that decimal answers be carried to the nearest tenth or hundredth. When rounding techniques for whole numbers are mastered, it is easier to apply those skills to decimals. The same basic rule applies.

When rounding, look at the digit <u>after</u>. If less than 5, round <u>down</u>. If 5 or greater, round <u>up</u>.

Example: Round to the nearest tenth. Examine the digit *after* that in the tenth's place (Figure 10.70).
(a) The "2" is less than 5. Round *down* to 43.3.
(b) The "8" is greater than 5. Round *up* to 43.4 (Figure 10.71).

Figure 10.69   (a)   *Example 1:*

Given problem   Child just places decimal point.

$$.03\overline{)6.963}^{\,2\,3\,2\,1}$$ $$.03\overline{)6.963}^{\,2\,3\,2.1}$$

- -

(b)   *Example 2:*

Given problem   Child quickly estimates (or actually
                                 multiplies) to check answer.

$$.5\overline{)35.25}^{\,7\,0.5}$$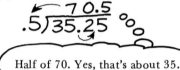

                                 Half of 70. Yes, that's about 35.

- -

(c)   *Example 3:*

Given problem   Child just makes the divisor a whole
                                 number.

$$.03\overline{)24}$$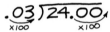

Note: Children often hesitate when the dividend is a whole number.
They forget the decimal point belongs after the one's place and frequently make errors because of this.

- -

(d)   *Example 4:*
(Round to nearest tenth)

Given problem   Child just adds zero in hundredths place.

$$.03\overline{).626}$$ $$.03\overline{).6260}$$

                                 I need hundredths in the
                                 quotient so I can *then*
                                 round to the nearest tenth.

- -

Figure 10.70

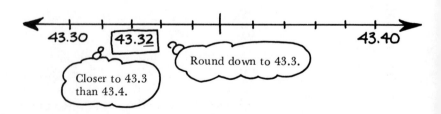

Figure 10.71

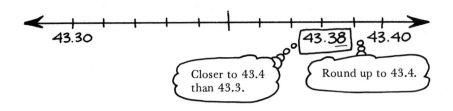

The reasons behind children's difficulties with decimal computation for multiplication and division are many. Three prerequisites,
often lacking, are the following:

1. Ability to visualize decimals

2. Mastery of basic facts

3. Skills for multiplication or division of whole numbers

Unless these prerequisites are in hand, it is premature to even begin
decimal computation.

When prerequisites are mastered, children still have difficulty
occasionally. Reasons for problems include:

1. Children miss out on important developmental work due to illness, emotional upsets in the home that lead to inattention during class, or other similar reasons.

2. Children are careless.

3. Children "forget" and need both reteaching and additional
   practice.

4. Children fall into error patterns that mushroom in intensity
   when they are not noticed.

5. Children have special handicaps which make it difficult for them
   to keep up with the pace of instruction or learn by the techniques used.

A teacher's knowledge of and sensitivity to each child will help
sort out underlying causes and plan appropriate remediation.

Problems like those of Figure 10.72 are characteristic of those
children experience. When children are cued to check by estimating,
they frequently realize when they have made an error (Figure 10.73).
Sometimes the estimation process helps children independently to
correct their errors. More often, teacher guidance and reteaching
are necessary. The suggestions of the last two sections should prove
helpful in this regard.

When specific learning difficulties are present, as in the following

Figure 10.72

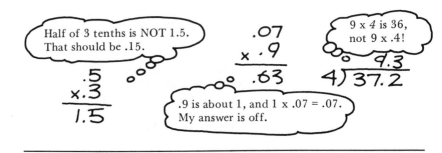

Figure 10.73

Half of 3 tenths is NOT 1.5.
That should be .15.

$$\begin{array}{r} .5 \\ \times .3 \\ \hline 1.5 \end{array}$$

$$\begin{array}{r} .07 \\ \times .9 \\ \hline .63 \end{array}$$

.9 is about 1, and 1 x .07 = .07.
My answer is off.

9 x 4 is 36,
not 9 x .4!

$$4\overline{)37.2}^{\;9.3}$$

---

case, then more specialized techniques must be employed. The case study illustrates one important technique: color cueing.

Case Study: Wendy

Special help for special needs.

Wendy, a seventh grader, is learning disabled. Her disabilities are: auditory sequencing, auditory discrimination, and abstract reasoning. A major strength, helpful in mathematics, is her visual association abilities. Color cueing techniques were used by the teacher (Bley, 1979) in conjunction with the estimation approach of the multiplication section to help Wendy with decimal multiplication (Figure 10.74). Wendy had already mastered the basic facts and computational skills for whole numbers. The high visual cueing of written exercises such as that illustrated, coupled with discussion of completed work, led Wendy to discover and use successfully the rule for placing the decimal point in decimal products.

Figure 10.74

Sample worksheet problems given Wendy during early work with decimal multiplication:

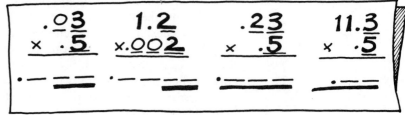

The bold letters and heavy underscores are green. Other underscores are red.

A minimal evaluation of a child's grasp of decimal concepts and skills would include the following tasks.

1. Decimal concepts
   (a) Give the child a decimal in written form (e.g., 0.63). Have the child read the decimal and picture it with squares and strips.
   (b) Orally read a decimal to the child. Have the child write it and picture it with squares and strips.
   (c) Show the child squares and strips that represent a decimal you have in mind. Have the child orally name and then write the decimal.

2. Comparison
   Give the child three decimals (written form) to order. Include both tenths and hundredths. Include a comparison such as the following in case children *mistakenly* think "36 > 6, so 0.36 > 0.6."

   0.36   0.6   0.3

3. Computation
   Give the child written computational problems to complete. Include:
   (a) Addition or subtraction problems in horizontal form. See if the child aligns decimal points so as to *add like units* when recopying to compute.
   (b) Ragged columns for addition and subtraction (Figure 10.75). Does the child "bring down" the 4 in the first example? Does the child correctly add digits in each column of the second example?
   (c) Multiplication problems for which a zero must be inserted into the product (Figure 10.76).

Figure 10.75   *Example for b:*

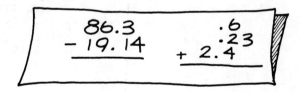

Figure 10.76   *Example for c:*

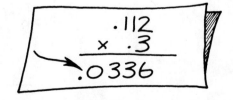

(d) Division problems with whole *and* decimal divisors.
(e) Division problems with whole number dividends (Figure 10.77).
(f) Division problems that require rounding to the nearest tenth or hundredth.

Figure 10.77   Example for e:

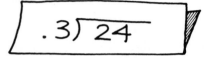

Figure 10.78

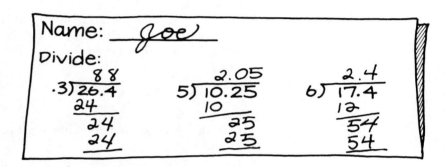

Figure 10.79

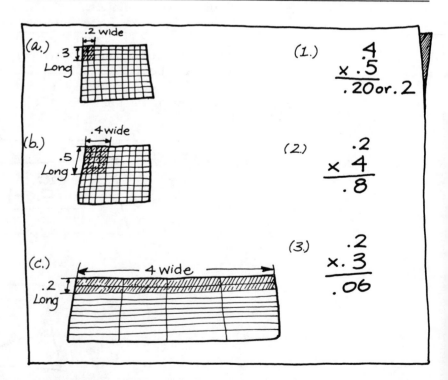

Your Time: Activities, Exercises, and Investigations

1. Prepare, as your instructor directs, answers to the major questions on p. 347.

2. What skills should be mastered *before* a child begins multiplication and division of decimals?

3. Refer to the curriculum guide given in Appendix A. At what grade level is a child normally expected to master multiplication and division of decimals?

4. Refer to Sarah's case study (p. 351). Suppose an additional practice activity were needed to help Sarah and her friends master multiplication of decimals. What would you suggest?

5. Refer to the suggestions for dividing with decimals on pp. 354-358 Use these ideas and those you may find in other sources (e.g., teacher's editions of school mathematics texts). Create a sequence of developmental and practice activities to help Joe, whose work is shown in Figure 10.78.

6. Refer to example 4 of Figure 10.69. An idea for helping children round decimal quotients is given: just focus on one step of the problem.
   (a) What instructional activities should *precede* an assignment like that suggested in the example?
   (b) What *next step* should be focused on during instruction?
   (c) Can you suggest a practice activity to reinforce rounding skills for decimal quotients?

7. Review Wendy's case study (p. 362). How might color cueing be used to help Wendy with decimal *division*?

8a. Match each of the diagrams of Figure 10.79 with the correct multiplication problem.

8b. Children may have difficulty in placing the decimal point when estimating. How might a visual-manipulative activity based on the idea illustrated in (a) help children in early developmental work for decimal multiplication?

8c. What division problems are also illustrated by each of the diagrams in (a)?

9. Refer to the assessment section on p. 363. Use the suggestions. given to construct a diagnostic/evaluative test for decimals. If possible, try out your test on an elementary-school child who has studied the topics you are assessing.

References

Bley, Nancy, learning disabilities specialist. Case study based on personal experience with learning disabled children in Park Century School in California. Relayed in unpublished communication with authors, 1979.

11 FRACTIONS

Teacher Background

With the increased availability and use of hand-held calculators and the emergence of the metric system, questions are being raised concerning students' need to compute with fractions. Children and adults still do use fractions and speak in terms of fractional parts of items. Fractions and the application of fractions are needed in many different situations. For example, fractions are needed in home economics and industrial arts classes, auto repair shops, carpentry work, cooking, sewing, stock markets, algebra classes, and vocational area classes. The need for fractions is evident. Students need both fraction concepts and fraction computation skills. Even when calculators are used in computation, there is still a need to understand how to change from common fractions to decimal fractions.

Fractions are used in many everyday situations to express a part of something, yet children often develop misconceptions about them. "I want the biggest half." or "How much of the candy bar do you have?" "Half" (regardless of the actual portion). Preschool and primary children often use the language of fractions without understanding the words.

This chapter focuses on concrete models and techniques for developing basic fraction concepts and skills. It presents ideas for helping children represent, name and write fractions; find equivalent fractions; compare and order fractions; and compute with fractions. The chapter contains sections with ideas for teaching conversion of fractions to decimals; ratio, proportion, and percent; and concludes with a trouble-shooting section.

The elementary school curriculum corrects and extends children's concepts of "fraction." Early fraction activities involve only the more common fractions (e.g., halves, thirds, fourths, and tenths). At this level children's experiences are founded in the opportunity to use real-life objects and concrete aids to represent fractions. Three basic questions permeate activities of these early sessions:

1. What equals the unit *one*?

2. How many parts are in the unit *one*?

3. Are the parts the same size?

For example, a whole unit with two equal-sized parts is divided into halves; a whole unit with three equal-sized parts is divided into thirds; a whole unit with four equal-sized parts is divided into fourths, and so on (Figure 11.1). The ordinal number names help children learn the fraction names.

Activities which encourage children to use correct *oral* names to describe "parts of one" are necessary in early developmental work with fractions. Fraction concept development experiences should be exclusively concrete, using manipulative aids and visual materials. Having children *talk* to describe the manipulation comes second. Only later do we ask children to read and write fraction symbols. Gradually, children approach the more difficult tasks of fraction comparison and computation.

The teaching-learning triangle is a helpful guide for teaching and for assessing a student's grasp of fraction concepts and skills. After hearing a verbal name for a fraction, a student should be able to use a model to show the fraction and correctly write it (Figure 11.2).

Figure 11.1

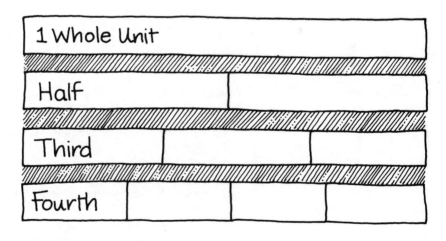

Figure 11.2

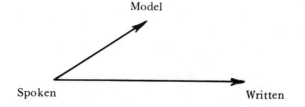

When shown the model representation of a fraction, a student should be able to both say and write the fraction (Figure 11.3). Seeing a fraction written, a student should be able to say the fraction, name, and show it with a model (Figure 11.4). Young children usually think in terms of physical objects or concrete models, and the meaning for spoken or written fractions is learned most easily from their experiences with physical or pictorial models. Given a fraction in any one of the three forms (model, spoken, or written), a child should be able to give the other two forms following fraction instruction.

Figure 11.3

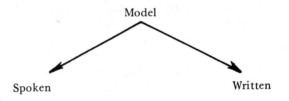

Figure 11.4

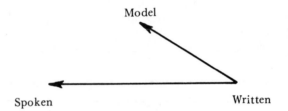

What Do *You* Say?

1. How might manipulative aids like the following be used to help children develop a basic understanding of fractions?
   (a) fraction strips
   (b) fraction chart

2. How can a teacher introduce the numerator and denominator to children and help them understand and write what each represents in an example?

3. How might examples from real life or other illustrations be used to help children understand that the following are true:
   (a) $\dfrac{1}{2} = \dfrac{3}{6}$
   (b) $\dfrac{3}{4} = \dfrac{6}{8}$

4. How might teachers help children determine the greater of two given fractions (e.g., $\dfrac{1}{2}$, $\dfrac{2}{4}$)?

**Basic Concepts
—Fractions:
Phase 1**

A variety of concrete manipulative aids and activities is necessary in the early stages of the development of fraction concepts. Using concrete materials has a great impact on children's conceptual understanding of fractions.

The following diagram highlights the sequence underlying early fraction activities with young children (Figure 11.5).

Region models including rectangular bars, strips, squares, and circles are commonly used to illustrate fractions in very early work with fractions. Research shows that it is easier for children to grasp fraction concepts illustrated by geometric regions. More sophisticated fraction models such as set models or the number line can be introduced later.

Geometric regions are used to represent fractions in the activities that follow. The use of rectangular regions is emphasized throughout the activities in most developmental work with fractions. The choice of these regions over circular regions is a pragmatic one. Dividing a circle into equal-sized pieces is just more difficult for most children. Rectangular region units can be constructed by children or the teacher from cardboard, tagboard, construction paper, graph paper, or other sheets of paper.

In early work, colored rods can be used as region models instead of paper or cardboard fraction strips. The colored rods have lengths of 1 cm, 2 cm, 3 cm, ..., 10 cm (Figure 11.6).

Figure 11.5

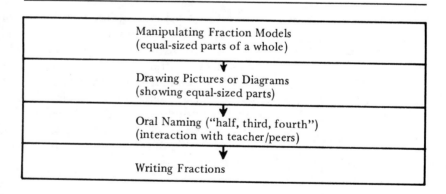

Figure 11.6

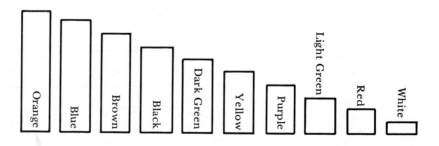

The fraction $\frac{1}{2}$ can be represented by one of the two colored rods if one rod is twice the length of the other rod. Since the purple rod is twice the length of the red rod, it can be shown that the two red rods each represent the fraction $\frac{1}{2}$ if the purple rod has the value 1. See Figure 11.7 for other examples that show $\frac{1}{2}$.

Discussion: Teachers can use activities such as the one just given to develop meanings for fractions. Here are basic questions that can be repeated to guide children's thinking:

1. What represents the number *one*?

2. How many parts is *one* divided into?

3. Are the parts all the same size and shape?

4. Into how many equal-sized pieces is the region separated?

5. How many parts are shaded?

6. What fraction names the shaded part of the region?

7. What fraction names the part unshaded?

Manipulation and discussion of fraction models gives an excellent way to provide answers to these questions. As the child relates models to written fractions, and in the discussion of that relationship, the major concepts can be emphasized and the vocabulary of fractions eventually developed.

In this development one typically emphasizes that the bottom number of the fraction indicates the number of equal-sized parts into which the unit has been divided. The top number indicates how many of those parts are being considered. At a later time the words *denominator* and *numerator* would be introduced. The word *denominator* (namer) can be related to the word *denote*. It should be understood that the denominator tells the size of the fractional unit (fourths, halves, thirds). The word *numerator* (numberer) can be related to numerating or counting the number of pieces or fractional units.

Figure 11.7

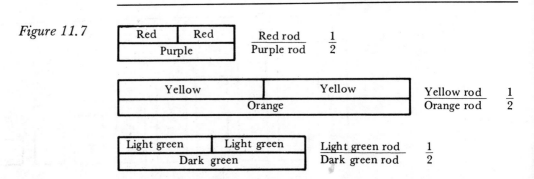

**Activity File:
Basic Fraction
Concepts**

Activity 1: Cut And Match
  Purpose: Emphasize the concept and naming of *equal-sized parts* of a unit.
  Need: Paper and scissors.
  Procedure: Children can cut (or fold) rectangular regions into equal-sized parts and learn the names of the fractional parts. Children can check to see that each part of the region will fit exactly on top of other parts.

Activity 2: Roll A Strip (three or four players)
  Purpose: Identify fractional parts.
  Need: Fraction chart (halves to tenths) with fractional pieces (velcro on back) and number cubes with fractions marked on them:

$$\frac{1}{2}, \frac{1}{3}, \frac{1}{4}, \frac{1}{5}, \ldots, \frac{1}{10}.$$

  Procedure:
1. The fraction chart is placed on the table or floor.

2. Players take turns rolling a cube at a time. The player sticks on the fractional piece that the fraction on the cube indicates.

3. Play continues with the next player rolling the cube.

4. The player who turns over the last remaining fractional piece of a rectangle strip takes the entire rectangular strip.

  Winner: Player with the most rectangular strips.

Activity 3: Give And Take
  Purpose: To help the child conceptualize fractions and develop mental imagery for fractions.
  Need: Two fraction pie sets with halves, thirds, fourths, and eighths. Spinner with the fractions

$$\frac{1}{2}, \frac{1}{3}, \frac{2}{3}, \frac{1}{4}, \frac{3}{4}, \frac{1}{8}, \frac{3}{8}, \frac{5}{8}, \frac{7}{8}.$$

  Procedure: Give each of two children a set of fraction pies. The children take turns spinning the spinner then taking the indicated amount from the other player. After ten turns, the one with the greater amount is the winner.

Activity 4: Fractional Numbers And An Egg Carton
  Purpose: To present the use of fractions to represent part of a set.
  Need: Egg cartons, various common small objects.
  Procedure: An egg carton could be used to present a set. Children could place various objects in the different sections to represent various fractions and state orally the name of the fraction. Egg cartons could also be cut into smaller sections (e.g., 2 x 4,

2 x 2) for working with sets other than twelve. Children can be helped to realize that the size of the denominator is dependent on the size of the set. Some examples of egg carton fractions are shown in Figure 11.8.

The application of fractions to sets can be shown with practical problems after children have had many experiences with regional models. For example, fractions may be used to represent the number of students out of a larger group who eat in the school cafeteria, or the number among those voting who cast their ballot for a particular candidate. Conceptually, the idea of "equal-sized parts of a region is no longer applicable. Now the focus is on the number of persons or objects which are part of a larger group. When dealing with groups or sets in this way, it is important to guide children's thinking with the following questions.

1. What is equal to one?

2. How many pieces (or subsets) are there in the set?

3. What is a fair share (equal size subsets)?

Figure 11.8

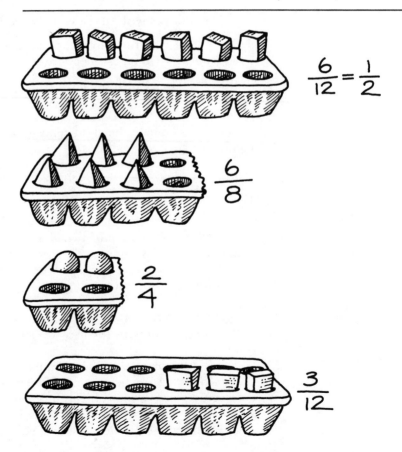

Background Information

Terry is learning disabled. He was sent to a resource teacher for mathematics during his first three years at school. This year he is being mainstreamed into fourth-grade mathematics. Terry's learning disabilities involve difficulty with spatial organization, auditory association, and auditory memory. Visual memory is intact, and is, for Terry, a definite learning strength. The teacher has been sensitive to Terry's special learning needs, and has used simple adaptive techniques when necessary. Because of this, Terry has been able to maintain a C grade or better in mathematics.

At the present time the class is starting a unit on fractions. The teacher is certain Terry has an adequate grasp of the basic *concept* of fraction, but notices that Terry makes many mistakes *writing* fractions to describe situations modeled or pictured. A major problem is Terry's difficulty with spatial organization. For the first time now, Terry needs to write numerals "on top of each other." Besides organizing his writing to control the size of the numerals, Terry must think carefully and position the correct numeral on top. Terry sometimes writes the reverse of what should appear. Just *telling* him what digit should be written in the numerator (and *why*) has not been enough. Apparently Terry's difficulty with auditory association and memory interfere and call for another approach.

The Teacher's Approach to Terry's Problem of Writing Fractions

Terry's teacher prepared colored cards (for whole class review of fractions) and followup worksheets based on examples like that shown in Figure 11.9. On the cards, the teacher outlined the geometric shapes and the partitioning lines with Elmer's glue. The shaded part(s) of each region were also covered with glue. When dry, the glue gave a textured surface for finger tracing. The teacher's goal was to help Terry during a whole-class review of fractions, thereby minimizing the time necessary for one-on-one instruction. Terry's teacher had learned that, with a little thought, one can often deal with handicaps within whole class instruction, to the benefit of all. In this case, since Terry was a poor auditory learner, the teacher

Figure 11.9

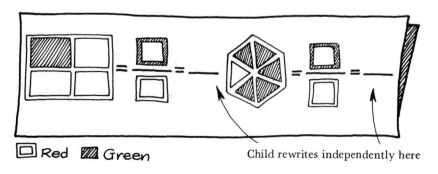

☐ Red ▨ Green   Child rewrites independently here

sought a visual-kinesthetic approach to the problem. Familiar stop-light colors were used. Green meant *Go* or *start here*.

"How many *equal-sized parts* are colored green? . . . Write this number in the green box." "How many *equal-sized parts* in all are outlined in red? . . . Write this number in the red box."

During class several students, including Terry, would be invited to finger-trace the cards before responding. Experience has shown this approach often triggers the necessary associations, giving children a *feel* for the fraction they are expressing. The color-coding scheme cues both the correct writing and naming of the fraction, numerator first. Besides using the cards, the teacher also planned to use green and red chalk on board drawings and follow up with the worksheets. The color cue would be used as long as necessary for Terry, then systematically be faded.

Equivalent Fractions: Phase 2

In order to compare fractions with unlike fractional units we must be able to rename them so that they are named with the same fractional unit. Let's look now at the process of renaming fractions with different fractional units. Using the region model, as in Figure 11.10, we see that the denominator (or unit size) corresponds to the size of the pieces. When the pieces are cut into smaller pieces, the same fractional amount can be named using a different denominator.

Examples like those pictured in Figure 11.10 can be used to help children to see the basic principle of equivalent fractions. Notice that in the first case $\left(\frac{2}{3} = \frac{4}{6}\right)$, we cut all the pieces in half. There were then twice as many pieces and twice as many shaded pieces.

2 x 2 ⟵ twice as many shaded pieces
3 x 2 ⟵ twice as many pieces

Figure 11.10

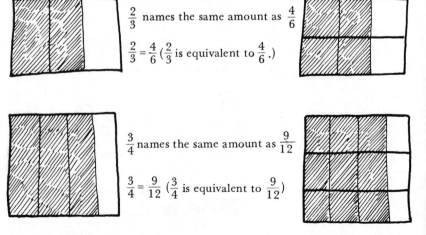

$\frac{2}{3}$ names the same amount as $\frac{4}{6}$

$\frac{2}{3} = \frac{4}{6}$ ($\frac{2}{3}$ is equivalent to $\frac{4}{6}$.)

$\frac{3}{4}$ names the same amount as $\frac{9}{12}$

$\frac{3}{4} = \frac{9}{12}$ ($\frac{3}{4}$ is equivalent to $\frac{9}{12}$)

This produced a new name for the same fractional amount.

$$\frac{2 \times 2}{3 \times 2} = \frac{4}{6}$$

In the second example $\frac{3}{4} = \frac{9}{12}$, we have three times as many shaded pieces and three times as many pieces altogether.

$\frac{3 \times 3}{4 \times 3}$ ← 3 times as many shaded pieces

  ← 3 times as many pieces

We then have a new fraction name equivalent to $\frac{3}{4}$.

$$\frac{3}{4} = \frac{3 \times 3}{4 \times 3} = \frac{9}{12}$$

These examples, and others like them, can be used to help children to generalize the basic principle of equivalent fractions, namely, that *if you multiply both numerator and denominator by the same number, the resulting fraction will be equivalent to the original.*

$$\frac{a}{b} = \frac{a \times n}{b \times n}$$

The only time this will not work is if you multiply by zero. This would produce a denominator of zero which is meaningless.

Reducing Fractions

Sometimes it is helpful to rename a fraction using a smaller numerator and denominator. This renaming process, called reducing fractions, relies heavily on the basic principle of equivalent fractions mentioned earlier. If we can multiply the numerator and the denominator of $\frac{2}{3}$ by 5 to get the equivalent fraction $\frac{10}{15}$, we can reverse the process. Dividing numerator and denominator of $\frac{10}{15}$ by 5 will produce the *reduced* fraction, $\frac{2}{3}$.

The problem for the child then becomes deciding what number to divide numerator and denominator by. Textbook programs include various techniques for determining what number to divide by. In general, we want children to understand that to reduce a fraction "completely" it is necessary to divide by the greatest number possible (the greatest common divisor of the numerator and denominator).

Activity 1: Fold And Fold Again
   Purpose: To practice finding equivalent fractions.
   Need: Sheets of paper or construction paper (all uniform in size).
   Procedure: Give each child a sheet of paper.

Ask children to: Fold the sheet in half. Open the sheet. Shade one of the halves. Discuss this fraction. $\left(\frac{1}{2}\right)$

Ask children to: Fold the sheet of paper again. Make sure that all parts are the same size and shape.

1. How many parts are now on the sheet of paper?   (4)

2. How many of them are shaded?   (2)

Discuss this new fraction $\left(\frac{2}{4}\right)$ in relation to $\left(\frac{1}{2}\right)$. Emphasize in a discussion that the number of pieces in all are doubled and also the number considered is doubled.

Ask children to: Fold the sheet of paper again.

1. How many parts are now on the sheet of paper?   (8)

2. How many of them are shaded?   (4)

Discuss this new fraction $\left(\frac{4}{8}\right)$ in relation to $\left(\frac{1}{2} \text{ and } \frac{2}{4}\right)$. Try this activity again. Repeat these same questions or ask a leader in the room to ask these questions.

Activity 2: Using A Fraction Strip Chart

A fraction aid helpful in developing understanding of many concepts is a fraction strip chart. To make the chart, children can start with ten pre-cut strips of paper about 3 cm x 20 cm, a piece of construction paper, glue and a marker.

The first strip represents 1 and is labeled as shown in Figure 11.11. The second strip is folded in half and labeled using Figure 11.12. Folding (with teacher assistance) and labeling continues for thirds, fourths, fifths, sixths, eighths, ninths, tenths, and twelfths.

Figure 11.11

Figure 11.12

The strips are then glued in order of denominator size onto a sheet of construction paper to make a chart as shown in Figure 11.13. Emphasize that accuracy in folding and gluing is important.

Children can use the chart to identify equivalent fractions, nonequivalent fractions, sets of equivalent fractions, fractions in lowest terms and common denominators. Children can also compare fractions and add or subtract those with like denominators.

Figure 11.13

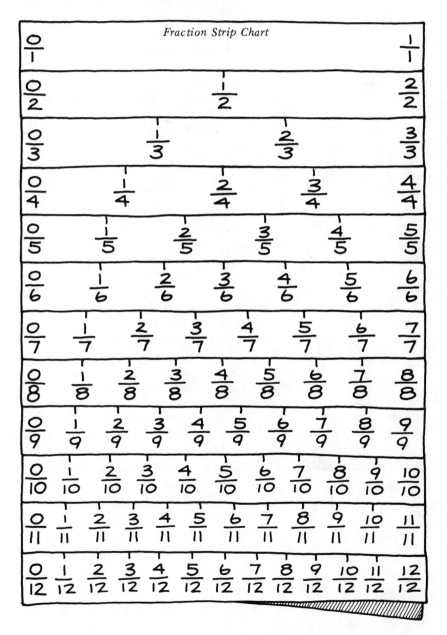

Have children use a ruler and markers to identify sets of equivalent fractions. For example, $\frac{1}{2}$, $\frac{2}{4}$, and all other fractions equivalent to these fractions might be color-highlighted in red. The fractions $\frac{1}{3}$, $\frac{2}{6}$, and all other fractions equivalent to these fractions might be color-highlighted in blue. Having children handle and fold the fraction strips is extremely helpful. Of course, the fraction strip chart can be simply a dittoed sheet of paper from which children read. However, there are greater positive results with children who can make their own personal fraction strip charts. A classroom fraction chart can be made from larger strips. Extend the chart to study mixed numerals and improper fractions, too.

Activity 3: Cover-Up!
 Purpose: To practice recognizing equivalent fractions.
 Need: Deck of 48 cards: four sets of 12 cards numbered 2, 3, 4, 5, 6, 7, 8, 9, 10, 12, 15, 16.

Gameboard: 5 x 5 grid with the following fractions—

$$\frac{1}{2}, 1, \frac{9}{12}, \frac{1}{3}, \frac{10}{10}, \frac{6}{8}, \frac{9}{10}, \frac{1}{4}, \frac{1}{5}, \frac{5}{15}, \frac{4}{8}, \frac{8}{10}, \frac{5}{6}, \frac{4}{5}, \text{ etc.}$$

Forty chips: twenty each of two colors
 Procedure: Each player is given twenty chips of one color. To start, cards are shuffled and placed face down and five cards are dealt to each player. In turn players try to create a fraction equal or equivalent to a fraction on the board by using two cards from those in hand—one card for numerator and one for the denominator. When a fraction is formed, the player places a chip on that fraction on the board. Used cards are placed on a discard pile and the player draws two new cards from the deck. Each fraction on the board can only be used once.
 Winner: Player who has the most chips on the gameboard after all cards have been drawn or no moves can be made.

Comparing Fractions: Phase 3

The comparison and ordering of fractions is easy when they have like denominators. If we first represent the fractions with physical or pictorial models, the comparison is much like the comparison of two whole numbers (Figure 11.14). Indeed, children very quickly see that since the denominators just tell us the size of the fractional unit, all we need to compare are the numerators (which tell us the number of units).

$\frac{7}{13}$ is greater than $\frac{4}{13}$ because 7 is greater than 4.

It is slightly more difficult to compare fractions which have different denominators but the same numerators. It is important that children develop mental imagery for fractions. If they can visualize fractions, it is easy to see that $\frac{3}{8}$ is less than $\frac{3}{5}$ because the pieces (the fractional units) are smaller (Figure 11.15).

Figure 11.14

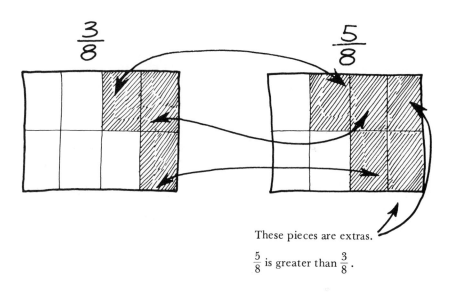

$\frac{3}{8}$ $\frac{5}{8}$

These pieces are extras.

$\frac{5}{8}$ is greater than $\frac{3}{8}$.

Figure 11.15

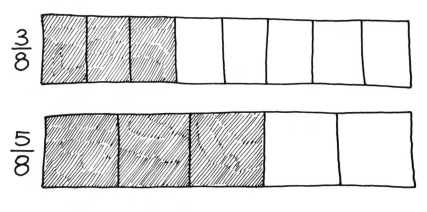

$\frac{3}{8}$

$\frac{5}{8}$

$\frac{3}{5}$ is obviously greater than $\frac{3}{8}$, because "fifths are greater than eighths."

Comparing Unlike Fractions

The comparison of fractions which have both numerators and denominators different is more complex than when either the numerators are the same or the denominators are the same. It is hard to tell if 75 km is more or less than 50 miles because they are named with different units. Consider the comparison of 7 feet and 98 inches. If we rename 7 feet using inches as the new unit, we can then tell, quite easily, that 98 inches is greater than 7 feet. 7 feet = 84 inches. 98 inches is more.

In much the same manner, we can compare $\frac{3}{4}$ and $\frac{7}{8}$. If we multiply the numerator and denominator of $\frac{3}{4}$ by 2, we get an equivalent fraction that can be compared easily to $\frac{7}{8}$ $\frac{3}{4} = \frac{3 \times 2}{4 \times 2}$ $= \frac{6}{8}$ $\times \frac{7}{8}$ is greater.

Unfortunately, we cannot always use one of the two denominators as a common unit for comparison. For example, if we wish to compare $\frac{5}{7}$ and $\frac{3}{4}$, we cannot conveniently rename $\frac{5}{7}$ as fourths. Nor can we rename $\frac{3}{4}$, as sevenths. However, we can name both fractions as twenty-eighths.

$$\frac{5}{7} = \frac{5 \times 4}{7 \times 4} = \frac{20}{28}$$

$$\frac{3}{4} = \frac{3 \times 7}{4 \times 7} = \frac{21}{28} \longleftarrow \frac{3}{4} \text{ is greater than } \frac{5}{7}.$$

Notice that if we multiply numerator and denominator by 2, 3, 4, 5, 6, 7 and so forth, we get many new names for $\frac{5}{7}$.

$$\frac{5}{7} = \frac{10}{14} = \frac{15}{21} = \frac{20}{28} = \frac{25}{35} = \frac{30}{42} = \frac{35}{49} = \boxed{\frac{40}{56}} = \frac{45}{63} \dots$$

Similarly, we can find many fraction names for $\frac{3}{4}$.

$$\frac{3}{4} = \frac{6}{8} = \frac{9}{12} = \frac{12}{16} = \frac{15}{20} = \frac{18}{24} \dots \frac{36}{48} = \frac{39}{52} = \boxed{\frac{42}{56}} = \frac{45}{60} \dots$$

The denominator 56 could also be used as the common denominator for comparison, as could 84, 112, 140, and an infinite number of other numbers.

Which of these denominators should be used as the common denominator for our comparison? Of course, any of them would be satisfactory.

An efficient technique for comparison of fractions that is usually included for older children is the "cross product" technique.

$$5 \times 3 = 15 \quad \frac{5}{7} \times \frac{2}{3} \quad 7 \times 2 = 14$$

15 is greater than 14, so $\frac{5}{7}$ is greater than $\frac{2}{3}$.

The cross product technique is really a disguised version of the common denominator technique discussed earlier.

$$\frac{5 \times 3}{7 \times 3} = \frac{5}{7} \times \frac{2}{3} \qquad \frac{2 \times 7}{3 \times 7}$$

We know that the denominators will be the same, allowing us to first compare numerators. Hence, we don't bother to compute the denominators.

$$\frac{a \times d}{b \times d} = \frac{a}{b} \times \frac{c}{d} = \frac{c \times b}{d \times b}$$

The denominators are the same.

We just compute numerators and compare them.

$$a \times d \quad \frac{a}{b} \qquad \frac{c}{d} \quad b \times c$$

$$\frac{a}{b} > \frac{c}{d} \quad \text{if } a \times d > b \times c$$

$$\frac{a}{b} = \frac{c}{d} \quad \text{if } a \times d = b \times c$$

$$\frac{a}{b} < \frac{c}{d} \quad \text{if } a \times d < b \times c$$

**Activity File:
Comparing and
Ordering Fractions**

Activity 1: Call It

Purpose: To provide practice in identifying which fraction is greater.

Need: Card deck of thirty-eight cards with two sets of each of the following fractions:

$$\frac{1}{2}, \frac{1}{3}, \frac{2}{3}, \frac{1}{4}, \frac{3}{4}, \frac{1}{5}, \frac{2}{5}, \frac{3}{5}, \frac{4}{5}, \frac{1}{6}, \frac{5}{6}, \frac{1}{10}, \frac{3}{10}, \frac{7}{10},$$

$$\frac{9}{10}, \frac{1}{12}, \frac{7}{12}, \frac{11}{12}.$$

Fraction strips for each of these fractions.

Procedure: The cards are shuffled, and all dealt out face down among the players. Simultaneously, each player flips over a card. The children verify with the fraction strips which fraction is the greater one. For example, a fraction showing $\frac{1}{3}$ can be placed directly beside a fraction strip showing $\frac{1}{4}$ (Figure 11.16).

Figure 11.16

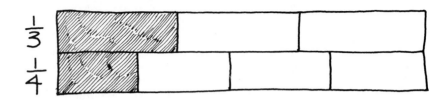

It is obvious at a glance which fraction is greater. The player who has the fraction card with the greater value says out loud, "$\frac{1}{3}$ is greater than $\frac{1}{4}$," and takes both cards.

Winner: Player with the most fraction cards after all have been played.

Activity 2: Shade It All! (For two to play).

Purpose: To help children visualize the comparison of fractions with the same numerator.

Need: Number cube labeled $\frac{1}{2}$, $\frac{1}{3}$, $\frac{1}{4}$, $\frac{1}{6}$, $\frac{1}{12}$, and $\frac{3}{12}$. Two gameboards (sheets of paper divided into twelfths and laminated and glued to posterboard), and two markers.

Procedure: Player rolls the number cube; if $\frac{1}{12}$ is rolled, player shades $\frac{1}{12}$ of gameboard. If $\frac{1}{4}$ is rolled, player finds that $\frac{1}{4}$ is equivalent to $\frac{3}{12}$ and shades $\frac{3}{12}$ of gameboard. Players take turns. Children can use the fraction strip chart to help them find the equivalent fractions if necessary.

Winner: First player to completely shade the entire gameboard.

Variation: Use a different set of fractions.

Addition and Subtraction of Like Fractions: Phase 4

Children should have developed the meaning of the concept of fractions through physical manipulation and the use of oral names and written symbols before addition or subtraction of fractions is introduced. When two or more fractions have the same denominator, they are called *like fractions* (e.g., $\frac{5}{8}$ and $\frac{1}{8}$). The same physical materials that were used to develop the concept of a fraction can be used to illustrate addition or subtraction of like fractions.

To introduce the operations at this level, the teacher might ask children use strips and orally name the new fraction formed by the combining or comparing the shaded sections of two strips the same color.

A next step, before moving on to writing symbols for fractions, would be to use word-form examples like the following:

        3 fifths   or   4 sixths
       +1 fifth   −1 sixth
        ? fifths   ? sixths
       (4 fifths)   (3 sixths)

Writing the denominator as a word rather than as a numeral will help to emphasize how two like fractions are added. The word form also emphasizes the importance of the size of the fractional pieces (the fractional units). It helps to name the denominators in this way, as there is otherwise a tendency to naturally want to add the denominators, too.

A teacher can suggest these steps for adding like fractions:

1. Are the fractional units the same?

2. If not, then we cannot solve it yet.

3. If yes, add (subtract) the numerators and write the fraction that tells the number of fractional units you have altogether (left after the subtraction).

At first, do not insist that results be given in "simplest form." Gradually, standard formats are adopted and used for carrying out the computation. For example,

$$2\frac{3}{8}$$
$$+1\frac{1}{8}$$
$$\overline{3\frac{4}{8}}$$
$$\qquad \frac{7}{8} - \frac{4}{8} = \frac{7-4}{8} = \frac{3}{8}$$

The two big ideas which guided work with addition and subtraction of whole numbers again emerge here.

Figure 11.17

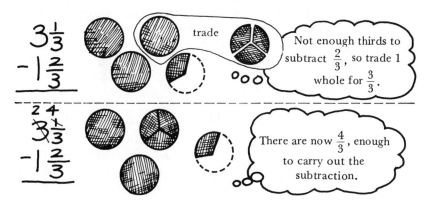

1. Add (subtract) like units—in this case fractions to fractions, whole numbers to whole numbers.

2. If there's too many (not enough) of a unit, make a trade.

As with whole numbers, modeling each of these ideas is important in early work with fraction computation. Presenting visuals to illustrate renaming when it occurs in addition or subtraction is particularly critical. Figure 11.17 suggests how this might be done for subtraction.

**Activity File:
Adding and
Subtracting Like
Fractions**

Activity 1: Fraction Strips—Hold 'Em And Add 'Em
  Purpose: To understand addition of like fractions.
  Need: Fraction strips from fraction strip chart; chips.
  Procedure: Have children work in pairs. One player will hold up two fraction strips of the same denominator. The other player will add the strips and state the sum. If correct, the player will receive a chip. Each player gets three consecutive turns. Then the roles change.
  Winner: Player with the most chips when time is called.

Activity 2: Subtracting Likes
  Purpose: To provide practice in subtracting like fractions.
  Need: Three dice, chips, paper and pencil.
  Procedure: First player rolls one die and that number is the denominator. Second player rolls other two dice and those numbers are the numerators. On paper, each player subtracts the smaller fraction from the larger. The first player to finish the problem correctly gets a chip.
  Winner: The player with the most chips when time is called.
  Variation: Add instead of subtracting the two fractions.

**Fractions Greater Than
or Equal to One:
Phase 5**

Improper Fractions
  There are some situations where fractions are equal to or greater than 1. Such fractions where the numerator is greater than the denominator are called improper fractions (Figure 11.18). Improper fractions may be written as mixed numbers. The definition of a mixed number is that it is the sum of a whole number and a proper fraction. In Figure 11.18, we can see that $\frac{8}{6} = 1 + \frac{2}{6}$, which would be written without the plus sign as $1\frac{2}{6}$.

  Figure 11.19 illustrates the procedure for renaming an improper fraction as a mixed number. If the procedure is reversed, a mixed number can be renamed as an improper fraction (Figure 11.20). In fraction computation, it frequently is necessary to convert improper fractions in answers to their mixed number equivalents. For multiplication and division the reverse procedure, changing a mixed number to an improper fraction, is often carried out to obtain a solution.

Figure 11.18

Figure 11.19

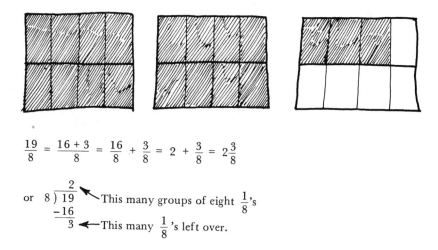

$$\frac{19}{8} = \frac{16 + 3}{8} = \frac{16}{8} + \frac{3}{8} = 2 + \frac{3}{8} = 2\frac{3}{8}$$

or $8\overline{\smash{)}19}$ $\leftarrow$ This many groups of eight $\frac{1}{8}$'s
  $\underline{-16}$
  3 $\leftarrow$ This many $\frac{1}{8}$'s left over.

Figure 11.20

$$4\frac{2}{3} = 4 + \frac{2}{3} = \frac{12}{3} + \frac{2}{3} = \frac{14}{3}$$

(This many $\frac{1}{3}$'s in 4) (This many more $\frac{1}{3}$'s)

or $4\frac{2}{3} = \frac{(4 \times 3) + (2)}{3} = \frac{12 + 2}{3} = \frac{14}{3}$

Unlike fractions are those with different denominator numbers. Two fractions such as $\frac{2}{3}$ and $\frac{1}{2}$ cannot be added until a *common denominator* is found. As is, the fractional units are not the same, so there is no sensible way to express a sum or difference. The fractions will have to be renamed to have like denominators. Then they may be added or subtracted. In this case,

$$\frac{1}{2} = \frac{3}{6} \quad \text{and} \quad \frac{2}{3} = \frac{4}{6}.$$

Now that the two fractions are expressed having like units, they may be added:

$$\frac{3}{6} + \frac{4}{6} = \frac{7}{6} \quad \text{or} \quad 1\frac{1}{6}.$$

Analyzing the preceding problem, we see that skill in two areas is prerequisite to adding or subtracting unlike fractions: (1) the ability to find equivalent fractions by renaming fractions and (2) the ability to add fractions with like denominators. Then skill finding the appropriate common denominator must be developed or reviewed.

Although *any* common denominator may be used, it generally is to the student's advantage to use the *least* common denominator. This allows them to work with smaller numbers.

Formal instruction for finding the least common denominator is included in most elementary school textbook series. The most common of these approaches is that described earlier, in the Comparing Fractions section of this chapter. The approach is based on the idea that the least common multiple (LCM) of the denominators will be the least common denominator (LCD) of the fractions. The following summarizes the process that can be used to find the LCD:

1. Check to find whether one denominator is a factor of the other. If it is, then the larger denominator is the LCD. For example,

3 is a factor of

6 so 6 is the LCD

$\Rightarrow$

$$\frac{1}{3} \times \frac{2}{2} = \frac{2}{6}$$

$$\frac{1}{6} \qquad = \frac{1}{6}$$

$$\frac{3}{6} \quad \text{or} \quad \frac{1}{2}$$

2. If neither denominator is a factor of the other, try successive multiples of the larger denominator. Use the first number that is also a multiple of the other factor. For children who cannot mentally skip count to identify the needed multiple, the multiplication basic fact chart provides a visual reference (Figure 11.21). Rows of the chart can be cut apart and glued to tagboard strips or tongue depressors. During early work, some children profit from

having their own personal set of multiple sticks to use for solving problems. For example,

5   10   (15)   20 . . . multiples of 5

$$\begin{aligned}
\left.\begin{array}{l}
\dfrac{1}{3} \\[4pt]
\text{5 is the larger} \\
\text{denominator} \\[4pt]
+\dfrac{2}{5}
\end{array}\right\} \rightarrow
\end{aligned}
\quad
\begin{aligned}
\dfrac{1}{3} \times \dfrac{5}{5} &= \dfrac{5}{15} \\[6pt]
+\dfrac{2}{5} \times \dfrac{3}{3} &= \dfrac{6}{15}
\end{aligned}$$

15 is the first number of the multiples for 5 that is also a multiple of 3. This can be verified by comparing the multiple sticks for 3 and 5, the two denominator numbers (see Figure 11.22).

Figure 11.21

X	1	2	3	4	5	6	7	8	9	10
1	1	2	3	4	5	6	7	8	9	10
2	2	4	6	8	10	12	14	16	18	20
3	3	6	9	12	15	18	21	24	27	30
4	4	8	12	16	20	24	28	32	36	40
5	5	10	15	20	25	30	35	40	45	50
6	6	12	18	24	30	36	42	48	54	60
7	7	14	21	28	35	42	49	56	63	70
8	8	16	24	32	40	48	56	64	72	80
9	9	18	27	36	45	54	63	72	81	90
10	10	20	30	40	50	60	70	80	90	100

Figure 11.22

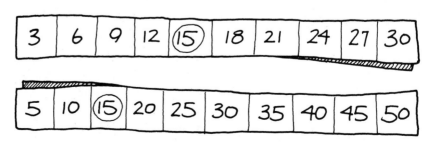

Children can use three steps to identify the least common multiple of the denominator numbers:

1. Think of the multiples of each denominator.

2. Find the common multiples.

3. Then find the least common multiple.

Note: For children with learning difficulties, *writing* the multiples for each denominator may be necessary, with possible reference to multiple sticks, as an intermediate step to memorizing them.

This three-step approach helps children better understand why 15 was selected in the example of Figure 11.22. This approach also helps children grasp the meaning of the LCD or LCM terminology.

After trying other similar examples, some children may be able to formulate a generalization for adding fractions. When the denominators have no factors in common, they are relatively prime. In each of the instances, *the least common denominator is the product of the two denominators.* Problems of this type need to be examined and discussed. The above process results in the mathematical definition for adding fractions:

$$\frac{a}{b} + \frac{c}{d} = \frac{a}{b} \times \frac{d}{d} + \frac{b}{b} \times \frac{c}{d}$$

thus

$$\frac{a}{b} + \frac{c}{d} = \frac{ad + bc}{bd}$$

The definition for subtraction of fractions is similar.

Figure 11.23   (a)

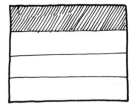

$\frac{1}{4}$ (horizontal bars)
The units are fourths.

(b)

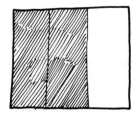

$\frac{2}{3}$ (vertical bars)
The units are thirds.

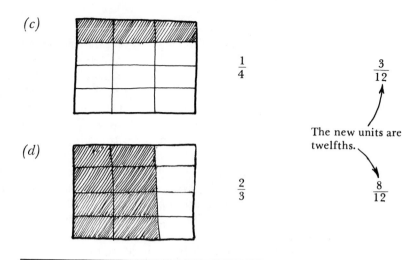

(c)

$\dfrac{1}{4}$

$\dfrac{3}{12}$

The new units are twelfths.

(d)

$\dfrac{2}{3}$

$\dfrac{8}{12}$

Pictures may be used to help children visualize addition or subtraction of fractions with unlike denominators. For example: For example:

1. Represent each fraction in a diagram (Figures 11.23 a and b).

2. Make the pictures look "alike."

Subdivide the squares so that both fractions are shown using the same size units (twelfths) (Figures 11.23 c and d).

The above pictures show that:

$$\dfrac{1}{4} = \dfrac{3}{12}$$

$$+\dfrac{2}{3} = \dfrac{8}{12}$$

In all → $\dfrac{11}{12}$

Analyzing this and similar problems should lead children to note how, in each instance, when the denominator numbers are relatively prime, *the least common denominator is the product of the two denominators.* A similar development is possible for subtraction. subtraction.

**Activity File:
Adding and
Subtracting
Unlike Fractions**

Activity 1: Roll And Subtract

Purpose: To find the difference when subtracting two fractions.

Need: Fraction strips from fraction strip chart. Two number cubes, each labeled $\dfrac{1}{2}, \dfrac{1}{3}, \dfrac{1}{4}, \dfrac{2}{3}, \dfrac{3}{4}, \dfrac{5}{8}$.

Procedure: Player rolls the two cubes; identifies the larger fractional number; subtracts the smaller by placing the larger fraction strip underneath the smaller one. The difference can then be stated and "proven" with the fraction strips. Players take turns and check each other. One point is given for each correct response.

Winner: Player with the most points when time is called.

Activity 2: Estimate And Order

Purpose: To estimate answers rapidly for computation examples that involve addition and subtraction of fractions.

Need: Forty cards: each card has a different fraction addition or subtraction problem.

Procedure: Shuffle the cards and deal out five cards to each player, face down. When the dealer says go, the players arrange their cards by estimating the answers so that the answers are in order, from smallest to largest.

Winner: The first player to order the cards correctly wins the hand. That player is the new dealer.

Activity 3: Book One (For three or four players)

Purpose: To provide opportunity for practice with addition and subtraction of fractions whose sum is one.

Need: Fifty-two cards with common fractions on them. A suggested list follows: three each of $\frac{1}{2}, \frac{1}{3}, \frac{1}{4}, \frac{1}{6}, \frac{5}{6}, \frac{1}{8}, \frac{3}{8}, \frac{7}{8}, \frac{1}{10}, \frac{9}{10}$; two each of $\frac{2}{3}, \frac{3}{4}, \frac{1}{5}, \frac{2}{5}, \frac{3}{5}, \frac{4}{5}, \frac{5}{5}, \frac{3}{10}, \frac{7}{10}$; one each of $\frac{1}{12}, \frac{5}{12}, \frac{7}{12}, \frac{11}{12}$.

Procedure: Each player is dealt seven cards. The rules for regular rummy are used and the object is to make "books" of cards that total one. As these are formed, they are placed face up on the table in front of the player.

Winner: The first player out of cards.

Activity File: Multiplying Fractions

What Do *You* Say?

1. What kinds of intuitive experiences could a teacher provide for the students for multiplying fractions before introducing the rule?

2. What types of concrete models might be used in teaching multiplication and division of fractions?

3. How might investigating patterns help children understand why division of fractions means to multiply by the reciprocal?

Region models can be used to represent multiplication of fractions. This approach uses the concept of area. Background review on finding the area of rectangles may be needed before using this development with children. For example, children should recognize that the area of the rectangle in Figure 11.24 is a x b.

The steps in the procedure for finding the product of two fractions using the region fraction model are illustrated in Figure 11.25.

There are two names for the area of the shaded rectangle. First, the shaded rectangle has sides $\frac{2}{3}$ and $\frac{1}{4}$, so its area must be $\frac{2}{3} \times \frac{1}{4}$. Second, the unit square has area 1 which has been divided into 12 equal parts. Two of those 12 equal parts are in the shaded rectangle, so another name for the area of the rectangle is $\frac{2}{12}$. We have two names for the area.

$$\frac{2}{3} \times \frac{1}{4} = \frac{2}{12}$$

Figure 11.24   Side a x Side b = Area of rectangle

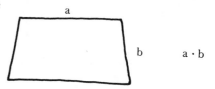

a · b

Figure 11.25

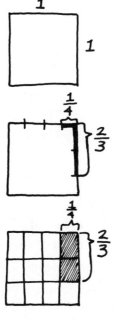

Start with a square that is 1 unit on a side. Its area is also 1.

Subdivide the sides of the square to represent $\frac{2}{3}$ and $\frac{1}{4}$.

Subdivide the unit square and shade the rectangle that is $\frac{2}{3}$ on one side and $\frac{1}{4}$ on the other.

We can show the product of other pairs of fractions in the same way. In Figure 11.26 we use the same process to find $\frac{2}{5}$ x $\frac{3}{4}$.

If children see enough examples like those shown above, they can be led to see that the product can be found by multiplying numerators and denominators.

$$\frac{a}{b} \times \frac{c}{d} = \frac{a \times c}{b \times d} .$$

If the correctness of this "easy way to get answers" is then verified by checking more examples with the model, most children are convinced that the procedure is correct as well as easy.

The development just used is an example of the laboratory development described in Chapter 3. We followed the following series of steps.

1. Experience, observe.

2. Gather data.

3. Organize data / identify patterns / see relationships.

4. Hypothesize.

5. Test hypothesis.

6. Communicate findings.

Experience has shown this laboratory approach is an effective way to teach multiplication of fractions.

Figure 11.26

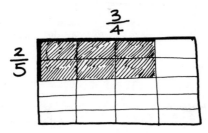

$$\frac{2}{5} \times \frac{3}{4} = \frac{6}{20}$$

Multiplication and Division of Fractions

Phase 7

Activity 1: Fold And Shade

Discuss examples using the word *of* and its meaning in multiplication of fractions. Beth ate $\frac{1}{3}$ of the twelve pieces of candy in the sack. How many pieces did she eat? Tom asked for $\frac{1}{6}$ of the whole pie. The pie is cut into six pieces. He will get one piece of pie. After introductory discussion has taken place, this activity may begin.

Teaching Mathematics to Children with Special Needs

Purpose: To use the word *of* in terms of multiplication of a fraction by a fraction.

Need: Sheets of paper (8½ x 11).

Procedure: Ask the children to fold a sheet of paper in half. Shade one-half of the sheet of paper. Unfold the sheet of paper. Now ask the children to fold the sheet of paper into thirds. Shade two-thirds of the sheet of paper with a different shading than used before. This example illustrates the problem $\frac{1}{2}$ x $\frac{2}{3}$. How many sections do you notice on the sheet of paper? (6) How many of those sections are shaded with cross-hatching or double shading? (2 out of 6 sections). What fraction does that represent? $\frac{2}{6}$

Activity 2: SEO—Shuffle, Estimate And Order

Purpose: To practice multiplication of fractions.

Need: Thirty cards with the following kinds of multiplication of fraction problems: fraction x whole number, fraction x fraction, mixed number x fraction, mixed number x mixed number.

Procedure: Dealer shuffles the cards and deals out five cards to each player, face down. When the dealer says to begin, players turn over their cards and arrange them so that the answers are in order, from smallest to largest.

Winner: The first player to order the cards correctly wins the hand. Two points can be given to the winner of each hand. Continue play until time is called. The winner of a hand is the next dealer.

DIVISION OF FRACTIONS

Early work with division of fractions often begins by dividing a whole number by a unit fraction. A physical or pictorial model is useful in this development (Figure 11.27).

Figure 11.27

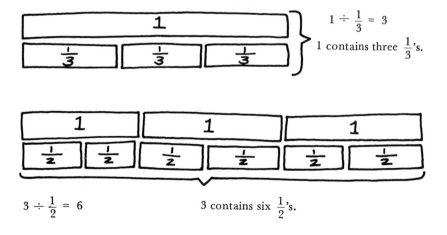

$1 \div \frac{1}{3} = 3$

1 contains three $\frac{1}{3}$'s.

$3 \div \frac{1}{2} = 6$ 3 contains six $\frac{1}{2}$'s.

From examples like those in Figure 11.27 students can begin to see that dividing by a fraction produces the same result as multiplying by its reciprocal. This generalization can be reinforced by carefully chosen problems like those illustrated in Figure 11.28.

The generalization that is most often developed as the procedure for dividing one fraction by another is *invert the divisor and multiply*. We know from previous work with whole numbers that division results can be checked by multiplication.

$$24 \div 3 = 8$$

We know that 8 is the correct answer because 8 x 3 = 24. We can extend this idea to include division with fractions.

$$\frac{3}{7} \div \frac{2}{5} = \frac{15}{14} \; .$$

We know that $\frac{15}{14}$ is the correct answer because $\frac{15}{14}$ x $\frac{2}{5}$ = $\frac{3}{7}$.

Figure 11.28

$$6 \div \frac{2}{3} = 9$$

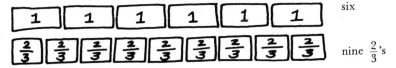

six

nine $\frac{2}{3}$'s

6 contains nine $\frac{2}{3}$'s.

$$6 \div \frac{2}{3} = 9$$

Also: $\frac{6}{1} \cdot \frac{3}{2} = 9.$

Multiply by the reciprocal of $\frac{1}{6}$ to get the same answer.

$$\frac{2}{3} \div \frac{1}{6} = 4$$

1

$$\frac{2}{3}$$

$\frac{1}{6}$ $\frac{1}{6}$ $\frac{1}{6}$ $\frac{1}{6}$ $\frac{1}{6}$ $\frac{1}{6}$

$\frac{2}{3}$ contains four $\frac{1}{6}$'s.

$$\frac{2}{3} \div \frac{1}{6} = 4$$

Also: $\frac{2}{3} \cdot \frac{6}{1} = 4.$

Multiply by the reciprocal of $\frac{2}{3}$ to get the same answer.

In general, we have the following.

$$\frac{a}{b} \div \frac{c}{d} = \frac{a}{b} \times \frac{d}{c} = \frac{ad}{bc}$$ (using the procedure developed earlier).

We know that $\frac{ad}{bc}$ is the correct answer because

$$\frac{ad}{bc} \times \frac{c}{d} = \frac{adc}{bdc} = \frac{a}{b}.$$

Division As Comparison

Another approach to the development of fraction division is based on an understanding of division as comparison. If we want to compare two quantities, for example 12 eggs and 4 eggs, we can divide the first quantity by the second to get the comparison.

$$12 \div 4 = \frac{12}{4} = 3 \quad \longleftarrow \quad \left\{ \begin{array}{l} \text{Twelve eggs is 3 times as} \\ \text{much as 4 eggs.} \end{array} \right.$$

Similarly, the comparison of 8 apples to 12 apples is as follows:

$$8 \div 12 = \frac{8}{12} = \frac{2}{3} \quad \longleftarrow \quad \left\{ 8 \text{ apples is } \frac{2}{3} \text{ as much as 12 apples.} \right.$$

We can apply the same technique to divide $\frac{5}{8}$ by $\frac{7}{8}$. We will compare 5 eighths to 7 eighths.

$$5 \div 7 = \frac{5}{7} \quad \longleftarrow \quad \left\{ 5 \text{ eighths is } \frac{5}{7} \text{ as much as 7 eighths.} \right.$$

So, $\frac{5}{8} \div \frac{7}{8} = \frac{5}{7}$.

Such comparisons can only be made if both quantities are named using the same units. For example, 3 feet is not $\frac{1}{2}$ as much as 6 inches, even though $3 \div 6$ is $\frac{1}{2}$. We only compare quantities named in the same units.

When we wish to compare quantities that are not named with the same units, we must first rename one or both of them so that the units are the same.

Compare 3 feet and 6 inches.

3 feet = 36 inches, so the comparison is:

$$36 \div 6 = \frac{36}{6} = 6.$$

3 feet is 6 times as much as 6 inches.

This is one of the basic ideas on which comparison division is based. *We only compare quantities, when they are named in the same units.*

Consider now, the example, $\frac{3}{5} \div \frac{2}{3}$. If we draw a pictorial model for each fraction, we can see how to compare them to get an answer (Figure 11.29).

The laboratory developmental approach, described in Chapter 3, was mentioned earlier in this chapter as a technique for teaching multiplication of fractions. This approach has proved to be equally effective for developing the generalizations and procedures for fraction division. The visuals and the discussion which are part of the laboratory development convinces children that the procedural "rules" which emerge to guide computation are consistent with what they "see" is true.

Figure 11.29 $\frac{3}{5} \div \frac{2}{3}$ $\frac{3}{5}$ compared to $\frac{2}{3}$

Since the units are different, we cannot compare them. We will cut the pieces to get the same size units for both fractions.

     $\frac{3}{5}$ compared to $\frac{2}{3}$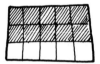

Now the units are the same for both fractions. The comparison is 9 units to 10 units.

So, $\frac{3}{5} \div \frac{2}{3} = \frac{9}{10}$

**Activity File:
Fraction Computation**

Activity 1: Tell A Fraction Story
   Have students tell a story or describe a problem situation illustrating such problems:

$4 \div \frac{1}{2} = 8$ $2 \div \frac{1}{3} = 6$ $6 \div \frac{2}{3} = 9$

Activity 2: Aim: Greatest Quotient
   Purpose: To practice dividing fractions.
   Need: Fifty cards with the common fractions of halves through twelfths. A suggested list follows:

$$\frac{1}{2}, \frac{1}{3}, \frac{2}{3}, \frac{1}{4}, \frac{2}{4}, \frac{3}{4}, \frac{1}{5}, \frac{2}{5}, \frac{3}{5}, \frac{4}{5}, \frac{1}{6}, \frac{2}{6}, \frac{3}{6}, \frac{4}{6}, \frac{5}{6}, \frac{1}{8}, \frac{2}{8}, \frac{3}{8},$$

$$\frac{4}{8}, \frac{5}{8}, \frac{6}{8}, \frac{7}{8}, \frac{1}{9}, \frac{2}{9}, \frac{3}{9}, \frac{4}{9}, \frac{5}{9}, \frac{6}{9}, \frac{7}{9}, \frac{8}{9}, \frac{1}{10}, \frac{2}{10}, \frac{3}{10}, \frac{4}{10}, \frac{5}{10},$$

$$\frac{6}{10}, \frac{7}{10}, \frac{8}{10}, \frac{9}{10}, \frac{1}{12}, \frac{2}{12}, \frac{3}{12}, \frac{4}{12}, \frac{5}{12}, \frac{6}{12}, \frac{7}{12}, \frac{8}{12}, \frac{9}{12}, \frac{10}{12}, \frac{11}{12}.$$

Procedure: Cards are spread out on table face down. Each player takes two cards. The object of the activity is to get the greatest quotient by dividing one of the fractions by the other. Each player has the option of taking another card to improve the result or pass. If a new card is taken, one card must be discarded. So that at any one time a player has only two cards in hand. Two points are awarded to the player who has the greatest quotient with each draw of cards.

Winner: The player who first gets twenty points.

Activity 3: Fraction Cube Toss

Purpose: To review the four basic operations with fractions.

Need: Two cubes labeled:

cube 1: $\frac{1}{2}, \frac{3}{4}, \frac{2}{3}, \frac{1}{6}, 3, 0$

cube 2: $\frac{1}{3}, \frac{5}{5}, 4, \frac{1}{4}, \frac{5}{6}, \frac{1}{2}$

Game board for Fraction Cube Toss: (The board can be laminated. This will permit students to record their results and then erase and reuse the board.)

Toss	Outcomes	Operation	Score
1	_____	_____	_____
2	_____	_____	_____
3	_____	_____	_____
4	_____	_____	_____
Total			_____

Procedure: Each player tosses the cubes and records the outcomes (numbers shown). The player may then choose to add, subtract, multiply, or divide the numbers in any order to get the highest possible result. Each player gets four tosses and the greatest result wins. One point given for each round.

Winner: The first player to reach five points.

FRACTION-DECIMAL EQUIVALENTS

Some fractions are written as decimals very easily because they have denominators which are factors of 10 or some power of 10. For example,

$$\frac{1}{2} = \frac{5}{10} \text{ or } .5 ; \qquad \frac{1}{4} = \frac{25}{100} \text{ or } 0.25.$$

These fractions may be expressed in an equivalent form with 10 or the power of 10 as denominators, and then rewritten as decimals.

For the general case, every fraction can be regarded as a division problem? $\frac{a}{b}$ can be thought of as $a \div b$. When the indicated division is performed, the equivalent decimal is obtained.

$$\frac{3}{8} \Rightarrow 8 \overline{)\begin{array}{r} .375 \\ 3.000 \end{array}}$$

Figure 11.30

Fraction	Decimal	Prime Factorization of denominator
	These terminate.	
$\frac{1}{2}$	.5	2
$\frac{3}{4}$	.75	2 x 2
$\frac{3}{8}$	.375	2 x 2 x 2
$\frac{7}{10}$	.7	2 x 5
$\frac{3}{25}$	.12	2 x 5
$\frac{7}{50}$	.14	2 x 5 x 5
Fraction	Decimal	Prime factorization of denominator
	These repeat.	
$\frac{1}{3}$	$.\overline{3}$	3
$\frac{1}{6}$	$.1\overline{6}$	2 x 3
$\frac{2}{7}$	$.\overline{285714}$	7
$\frac{4}{9}$	$.\overline{4}$	3 x 3
$\frac{2}{3}$	$.\overline{6}$	3
$\frac{2}{15}$	$.1\overline{3}$	3 x 5

Some fractions convert to repeating decimals. Others, in decimal form, terminate. It is interesting to find the decimal equivalent of several fractions. Chart the two types separately, and try to identify the pattern which emerges. Teachers might ask students how it is possible to identify terminating decimals before any conversion to decimals is made. Finding the prime factorization of each denominator may help answer this question. The chart of Figure 11.30, which groups fractions that terminate and that repeat in decimal form, incorporates this suggestion. Note that a horizontal bar is placed above those digits that repeat in a decimal. For example, $\frac{1}{6} = .1\overline{6} = 0.161616\ldots$.

One can see that in the fractions equivalent to terminating decimals, each prime factorization of the denominator contains either all two's, all five's, or a combination of two's and five's. Having such a prime factorization of two's or five's in the denominator of a fraction will guarantee that such a fraction can be renamed as a terminating decimal. Having any other factors such as 3, 7, 11, etc., in the prime factorization of the denominator will indicate that the fraction can be renamed as a repeating decimal.

Any terminating or repeating decimal can be written as a fraction. A terminating decimal can be written with a denominator of some power of ten. For instance, 0.3 is three-tenths or $\frac{3}{10}$, 0.27 is twenty-seven hundredths or $\frac{27}{100}$. A repeating decimal can be renamed to a fraction by (1) identifying the number of digits that repeat; (2) multiplying by that power of ten; (3) subtracting the original decimal; and (4) solving the equation. Examples follow:

$$.\overline{4} = \underline{\hspace{2cm}} \qquad \begin{array}{r} 10n = 4.\overline{4} \\ -\ \ n = \ \ .\overline{4} \\ \hline 9n = 4 \\ \\ n = \dfrac{4}{9} \end{array} \qquad .\overline{45} = \underline{\hspace{2cm}} \qquad \begin{array}{r} 100n = 45.\overline{45} \\ -\ \ \ n = \ \ \ .\overline{45} \\ \hline 99n = 45 \end{array}$$

Activity File: Fraction Decimal Equivalents

Activity 1: Match Fractions With Decimals

Purpose: To practice finding equivalent values between fractions and decimals.

Need: Fifty-two cards with matching sets of cards; some examples follow:

.3 $\qquad \frac{3}{10} \qquad$.$\overline{6}$ $\qquad \frac{2}{3} \qquad \frac{7}{8} \qquad$ .875

Procedure: Players are dealt six cards each. They try to make a match. They can discard an unwanted card and draw from the top of the deck. As a match is made, the cards should be put out in front on the table with the cards being then replaced by another draw. Play continues by repeating this procedure.

Activity 2: Do It Fast (For two to play)

  Purpose: To practice changing terminating decimals to fractions.

  Need: Fifty-two cards containing terminating decimals like those below, answer key giving the fraction equivalent for each decimal listed, paper and pencil, six game chips.

.3   .62   .4   .17

  Procedure: After cards are shuffled, each player takes three, keeping them face down until the signal for turn-over is given. Players take turns saying "Go," upon which students look at their cards and write the fraction equivalent of each decimal. First to correctly complete the set of three wins a game chip.

  Winner: First to win three game chips.

  Note: It should be agreed in advance whether fractions should be reduced to simplest form.

Trouble Shooting

We now describe some common trouble spots for children working with fractions. Possible remedial procedures are also described.

1. Refer to the child's description of the shaded region in Figure 11.31. The student lacks adequate understanding of the concept of fractions, or of the meaning given to fraction symbols. Concrete materials and region models need to be used. The concept of fractions can be developed through the use of region models and sets of objects.

2. Refer to Figure 11.32 which illustrates an example in which " $\frac{1}{4}$ " does not equal " $\frac{1}{4}$." Because the size of the unit you varies so does the size of each fractional part. Conceptually, children *can* understand this apparent discrepancy when physical models or other visuals are used to illustrate.

3. Suppose a student were asked to shade the bar of Figure 11.33 to show $\frac{1}{4}$. If the student shades the incorrect number of

Figure 11.31   The shaded area shows __3/5__ .

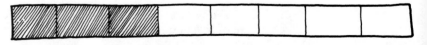

squares, he may not understand the concept of equivalent
fractions; or he may be guessing and shade only one square
since the numerator indicates 1. Fraction strips could be used
to re-develop the idea of equivalent fractions.

4. $\dfrac{1}{3} = \dfrac{2}{6} = \dfrac{?}{12} = \dfrac{?}{15}$.

If the student does not get this problem correct, then the concept of fractions and equivalent fractions needs to be reviewed.
Fraction strips or the fraction family chart could be used. Also,
show the student a sequence of

$$\dfrac{1}{3} = \dfrac{1}{3} \times 1 = \dfrac{1}{3} \times \dfrac{2}{2} = \dfrac{1}{3} = \dfrac{1}{3} \times \dfrac{4}{4} .$$

Figure 11.32

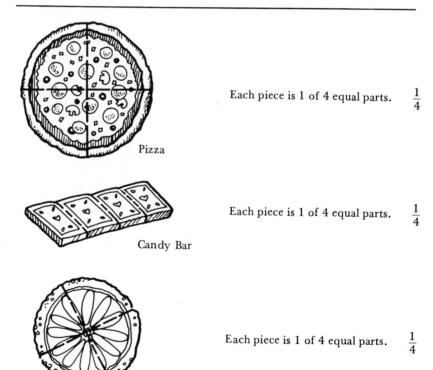

Each piece is 1 of 4 equal parts. $\dfrac{1}{4}$

Pizza

Each piece is 1 of 4 equal parts. $\dfrac{1}{4}$

Candy Bar

Each piece is 1 of 4 equal parts. $\dfrac{1}{4}$

Slice of Orange

Figure 11.33

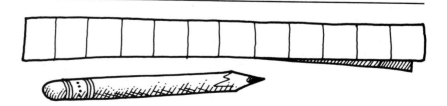

5. $\dfrac{1}{8} + \dfrac{3}{8} =$ _____

If a student misses the problem by adding both of the numerators and then adding both of the denominators, one can use concrete materials or visuals to illustrate the problem. One could also use the word form for the denominator as suggested in the section on Addition and Subtraction of Like Fractions. The child needs to "see" that eighths plus eighths remains eighths.

6.
$$
\begin{array}{r}
\dfrac{3}{4} \\[2mm]
+ \dfrac{1}{2} \\
\hline
\end{array}
\qquad\qquad
\begin{array}{r}
\dfrac{7}{8} \\[2mm]
- \dfrac{2}{3} \\
\hline
\end{array}
$$

An incorrect response to problems like these may mean that the students need help in finding common denominators. First let the child "see" that the incorrect answer is not right, then remediate. A pictorial representation of the problem should be helpful. When the error is recognized, get the child to list fractions in sequence until those with common denominators are found.

$$\dfrac{3}{4} = \dfrac{6}{8} = \dfrac{9}{12} = \dfrac{12}{16}$$

$$\dfrac{2}{3} = \dfrac{4}{6} = \dfrac{6}{9} = \dfrac{8}{12}$$

For the pair $\dfrac{3}{4}$ and $\dfrac{2}{3}$, the common denominator is 12. Multiple sticks, discussed in the section on Addition and Subtraction of Unlike Fractions, can be useful in doing such problems. For each denominator, one can examine the multiples and use the lowest of those that are common as the new denominator for the fraction pair.

7.
$$
\begin{array}{r}
\dfrac{1}{3} \\[2mm]
+ 2\dfrac{1}{6} \\
\hline
\end{array}
$$

Adding a fraction and a mixed numeral may be confusing to the student. Concrete representations or pictorial models may help the student to "see" the sum.

8.
$$
\begin{array}{r}
\dfrac{1}{3} \\[2mm]
+ 2\dfrac{1}{6} \\
\hline
\end{array}
\quad
\begin{array}{l}
= \dfrac{2}{6} \\[2mm]
= 2\dfrac{1}{6} \\[2mm]
\hline
\ \ 2\dfrac{3}{6}
\end{array}
$$

Sometimes the difficulty arises when answers must be reduced to simple form for ease in grading. Posting a chart of numbers 2 through 9 is a simple way to help children who can't readily identify common factors of numerators and denominators. Simply suggest they look at the chart when stuck and ask themselves, "Is 2 the largest number that will divide both numerator and denominator? . . . Is 3? . . ." The chart serves merely to prompt students. It gives them a place to start.

9. $\dfrac{3}{4}$

   $-\dfrac{1}{4}$

   $\overline{}$

Subtraction of fractions can be shown using concrete models or pictorial models by setting up the $\dfrac{3}{4}$ portion of a model and then by taking away the $\dfrac{1}{4}$.

10. $\dfrac{3}{4}$ x $\dfrac{1}{2}$ = _____ .

If students get this problem incorrect, they may have performed a different operation such as addition or subtraction. The basic concept of multiplication of fractions can be visually represented through the use of shading and a sheet of paper, as was illustrated earlier in this chapter. Reviewing this visual interpretation of multiplication should be useful.

11. $3\dfrac{2}{5}$ x 6 = _____ .

A student may have made the common error of multiplying the two whole numbers and then writing the fraction next to the product of the whole numbers. Concrete or pictorial rectangular models would be helpful to the student in trying to visualize the situation.

12. $\dfrac{2}{3} \div \dfrac{5}{6}$ = _____ .

If a student does not have a good understanding of division of fractions, this problem may be done incorrectly. Sometimes the student remembers to invert and multiply, but inverts the wrong fraction. To help this student, demonstrate the relationship between multiplication and division of whole numbers. Then notice the relationship between multiplication and division of fractions.

13. $\dfrac{3}{5} \div 4$ = _____ .

Students may have difficulty with division problems of this type because they are uncertain what the inverse or reciprocal of 4 should be. Finding the reciprocal of a whole number needs to be reviewed.

14. $5\frac{2}{3} \div 3\frac{1}{4} =$ _____ .

A student may make a common error in this type of problem by inverting the fraction before changing the mixed number to an improper fraction. The definition of a reciprocal or inverse must be noted. The reciprocal is that number which when multiplied by its original number yields one. For instance, $3\frac{1}{4}$ does not have as its reciprocal $3\frac{4}{1}$, since $3\frac{1}{4} \times 3\frac{4}{1}$ does not equal 1.

Emphasize that the mixed number first needs to be changed to an improper fraction, and then one can proceed to find the reciprocal.

What Do *You* Say?

1. What everyday experiences can you name that use the idea of a ratio?

2. What types of situations involve the use of percent?

3. What concrete materials might be used in teaching the concept of percent to students?

4. What prerequisite skills do you think are necessary in order for a student to solve a proportion?

MAKING COMPARISONS

Ratio and Proportion

A ratio is a comparison of two numbers which can be written in the form of a fraction indicating division, such as $\frac{a}{b}$. Ratios are used in everyday situations without giving them much thought. For example, 53 hits in 87 times at bat (ratio 53 to 87 or 53:87 or $\frac{53}{87}$), 94 cents for 2 cans of mushrooms (ratio 94 to 2 or 94:2 or $\frac{94}{2}$); 9 out of 14 football passes (ratio 9 to 14 or 9:14 or $\frac{9}{14}$).

Let's examine one example in detail. If there are nine girls in the class to every twelve boys, a ratio of girls to boys could be expressed as $\frac{9}{12}$. Ratios can be written in one of three ways: 9 to 12, 9:12, $\frac{9}{12}$. Ratios can be though of as a type of matching (*a to b*). For example, here, every nine girls are matched to twelve boys. Other matchings are also possible with this example (Figure 11.34). This example indicates also that ratios can be thought of as having more than one name. In this case $\frac{3}{4}$ and $\frac{9}{12}$ are equivalent ratios. Other ratios that can be discussed by the students include: the ratio of students absent to students present, ratio of windows to doors in the room, ratio of students to desks, ratio of girls to all students in room, ratio of school days to days in a month. Encourage students to suggest other common ratios.

A proportion is an equation stating that two ratios are equal. For example, if one roll of film yields twenty pictures, then two rolls will yield forty pictures. So the proportion $\frac{1}{20} = \frac{2}{40}$ is obtained by setting the two ratios equal. One ratio ($\frac{1}{20}$) is equal to another ratio ($\frac{2}{40}$).

Comparisons cannot be made where the quantities' names are in different units (e.g., which is more—18 oranges or 2 sacks of oranges?). To make comparisons, one should first change the ratios into the same units. For example, $\frac{1}{5}$ and $\frac{2}{9}$ have different units, so change to 45ths. The numerators can then be compared to assess the relative value of each ratio.

The cross products method discussed in the section on Comparing Fractions may also be used with ratios. It is another technique for comparing ratios and can also be used to determine whether a

Figure 11.34

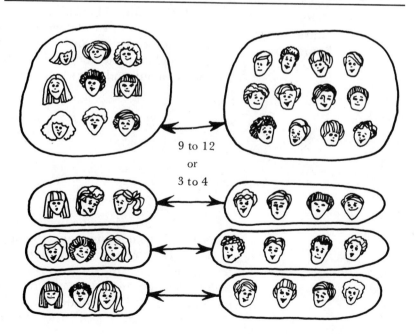

9 to 12

or

3 to 4

proportion is true or false. If the cross products are equal, the proportion is true. For example,

Does $\frac{3}{4} = \frac{12}{16}$?

$3 \times 16 = 4 \times 12$

$48 = 48$

So, this proportion is *true*.

Does $\frac{5}{7} = \frac{30}{35}$?

$5 \times 35 = 7 \times 30$

$175 \neq 210$

So, this proportion is *false*.

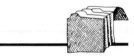

**Activity File:
Ratio and
Proportion**

Activity 1: Understanding Ratio
  Purpose: To use manipulatives as an informal basis for recognizing ratios.
  Need: Red chips, blue chips.
  Procedure: Form two teams, same number on each team. "Extras" can serve as recorders. Give each member of Team A two red chips and each member of Team B one blue chip. Chips can be dropped in a box and recorded. A running total is recorded.

Sample Play

After	*Team A*	*Team B*	*Ratio*
1st drop	1	2	$\frac{1}{2}$
2nd drop	2	4	$\frac{2}{4}$

Activity 2: Sort (For one or two to play)
  Purpose: To determine whether a proportion is true or not true.
  Need: Deck of twenty laminated cards with a proportion written on each one. Students sort the deck into two piles: (1) *Is a true proportion* or (2) *Is not a true proportion* by finding the cross product of each. Students can show their work directly on the laminated cards. (An answer key can be provided.)

Activity 3: Build A Proportion (For two teams to play)
   Purpose: To provide practice with ratios and proportions.
   Need: Set of fifteen cards with the numbers 1-15, one number per card.
   Procedure: Divide the class into two teams. Shuffle the cards. Place six cards on the chalkboard tray. Have two members from one team build as many true proportions as they can using the six numbers. For example, if the cards of Figure 11.35 were displayed, students could build these proportions: $\frac{2}{3} = \frac{6}{9}$ or $\frac{1}{3} = \frac{2}{6}$. Give two points for each true proportion that is named. Students can use the cross product method to check. Reshuffle the cards and select six cards for the opposing team.
   Winner: The team getting the greatest number of points.

Figure 11.35

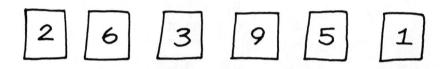

Comparisons Based on One Hundred

   Fractional parts can be represented by fractions or decimals, but they can also be represented by percent. When we want to compare a number to one hundred, we often use percent. The word *percent* means "hundredths" or "per hundred" and is denoted by the symbol %. Percents are fractions with denominators of 100. Some examples include: 25%—can be renamed as $\frac{25}{100}$ or $\frac{1}{4}$; 100%—can be renamed as $\frac{100}{100}$ or 1; 3%—can be renamed as $\frac{3}{100}$. Children should have a good concept of fractions and decimals before working with percents. Since *percent* means hundredths, physical materials that can be used in teaching percent are those that have one hundred units. The following aids can be used in teaching percent: hundred peg board (board with 10 x 10 array of pegs); 10 x 10 square centimeter grid of paper; a meter stick (100 cm); a dollar bill (100 pennies); a hundred board (grid of 100 squares in a 10 x 10 array); and a flat from base-ten blocks (100 units).
   Since a one-dollar bill is equivalent to 100 cents, money can be used to illustrate the relationship between fractions, decimals, and percents. For example, one-half of one dollar is 50 cents, or $.50, or 50% of a dollar. Students can use a 10 x 10 array of pennies to visualize this situation.
   A meter stick can be used to develop the concept of percent. If

A meter stick can be used to develop the concept of percent. If the meter represents one, then 30 centimeters would be $\frac{30}{100}$ or $\frac{3}{10}$ of the meter stick. This measurement could be renamed as 0.30 meter or 30% of a meter. Percents less than 1 percent can be marked on a meter stick by examining the millimeters. For example, 2 milli meters = $\frac{7}{1000}$ since there are 1000 millimeters in a meter; $\frac{7}{1000}$ = 0.007 = 0.7%, or $\frac{7}{10}$ of a percent.

Percents larger than 100 can be expressed in a manner similar to the way in which fractions greater than 1 are written.

Example:

If 1 = 100%, then 2 = 200%, and 1.26 = 126%.

Each of these percents can be modeled using a 10 x 10 grid of square-centimeter graph paper.

It is important that children be aware of the many common uses of percent in day-to-day living. Children can be asked to bring exam ples of percent applications they find in newspapers or magazines. Any percents that are mentioned on the radio or television could also be discussed. The meaning of the percent situation could be reported and displays can be made showing percent applications in sports, consumer business, and other subject areas.

Flowcharts are often helpful for writing decimals as percents or percents as decimals. An example follows.

Write a decimal as a percent:

Decimal	→	Move decimal point two places to the right	→	Write % after numeral
0.50		50		50%

Write a percent as a decimal:

Percent	→	Write numeral without % sign	→	Move decimal point two places to left
25%		25		0.25

Percent problems can be expressed as an equation, which can then be solved. *Of* and *is* are usually translated as "times" and "equals" when writing the equation. Following are some examples:

1. What is 25% of 40? (Let n be the number.)
   Find the number.

   $n = 0.25 \times 40$

   $n = 10$ Find the product.

2. 25% of some number is 10?
   Find the number.

$$0.25 \times n = 10$$
$$n = \frac{10}{0.25}$$ Find the missing factor.
$$n = 40$$

3. What percent of 40 is 10? (Let p be the percent.)

$$p \times 40 = 10$$
$$p = \frac{10}{40} = \frac{1}{4} = \frac{25}{100}$$ Find the missing factor.

$$p = 25\%$$

When the product is missing, the student needs to multiply the two factors. When one of the two factors is missing, the student needs to divide the product by the known factor to solve the equation.

Percent problem-solving often involves buying items or noticing items that are on sale in stores. Students can often be motivated to solve percent problems when they look for sales on items such as minibikes, records, sports equipment, or clothes. Some examples are shown in Figure 11.36.

Figure 11.36

Activity 1: Percent Concentration (For two to six players)
   Purpose: To relate percents to fractions to decimals.
   Need: Deck of 48 cards with twelve cards having *common* percents and other cards with the fractions or decimals equivalent to each percent in the card deck. Sample cards:

25% $\frac{1}{4}$ .25

   Procedure: Have students work in pairs and play by matching equivalent fractions and percents or decimals and percents. Students can fold a hundreds square and count the squares to see the relationships between fractions, decimals, and percents. Some examples follow:

$$\frac{1}{4} = \frac{25}{100} = 0.25 \text{ or } 25\%$$

$$\frac{1}{2} = \frac{50}{100} = 0.50 \text{ or } 50\%$$

$$\frac{1}{3} = \frac{33\frac{1}{3}}{100} = 0.\bar{3} \text{ or } 33\frac{1}{3}\%$$

   Winner: Player with the most cards.

Activity 2: Color-A-Percent
   Purpose: To visualize a percent on graph paper by counting squares out of 100.
   Need: Crayons, graph paper, percent cards.
   Procedure: Have a child mark off a 10 x 10 grid on the graph paper. Count the number of squares to make sure that there are 100. Have children shade squares for five different percents which are listed on the percent card. Color according to the listed color. A sample percent card follows:

(1) 8% (red)

(2) 25% (green)

(3) 3% (yellow)

(4) 50% (blue)

(5) 14% (brown)

Various percent cards can be made in advance for the children to use. Each completed 10 x 10 grid will be completely colored at the end of doing each percent card if it is done accurately.
   Note: Percents greater than 100 can also be considered by using more than one 10 x 10 grid.

Activity 3: Make A Percent Chart

Purpose: A percent chart can be made and used to check the reasonableness of answers for all three types of percent problems. Actual answers to percent problems can be verified by computation.

Need: Piece of string, sheet of centimeter-square graph paper, piece of cardboard.

Procedure:

1. Cut a grid of centimeter-square graph paper as shown in Figure 11.37.

2. Paste it on a piece of cardboard leaving a 5 cm margin.

3. Attach a string through a hole at the top of the percent scale. Make sure that the string is long enough so that it can be stretched tightly to any point on the percent and base scale.

4. Approximate answers to all forms of percent problems can be read directly from the chart.

Example 1: What is 50% of 20?

Stretch the string through a base of 20. Read across the percent 50. The percent is 10.

Example 2: What percent of 75 is 60?

Stretch the string through a base of 75. Read up to a percent of 60. Reading across, the percent is 80.

Example 3: 40% of what number is 20?

Stretch the string so that it passes through the intersection of the horizontal percent line for 40 and the vertical percent line for 20. The base is 50.

Figure 11.37

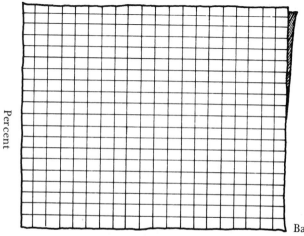

Percent

Base

1. Fraction strips like those illustrated in this chapter have many uses for teaching fractions to children.

   (a) Make a set of fraction strips from construction paper or cardboard, shade in the appropriate fractional parts, and then laminate each for durability. Include halves, thirds, fourths, fifths, sixths, eighths, tenths and twelfths in your set.

   (b) Write two developmental activities based on the fraction strips and try them out with children. Write a brief report on the session(s) during which you used the strips with students. Mention strong points of the activities you created as well as any problem you encountered.

2. Use fraction strips to find two fractions that are equivalent to $\frac{2}{3}$. Illustrate with diagrams how a child might do this problem.

3. Examine the following problems.
   - In what sequence should these problems be taught?
   - Describe or sketch the picture of a concrete aid or model that can be used to help children solve each of these problems.

   (a) $\frac{1}{2} \times \frac{1}{3}$ (g) $\frac{1}{3} = \frac{}{6}$

   (b) $4 \times \frac{1}{2}$ (h) $\frac{1}{3} + \frac{1}{3}$

   (c) $\frac{1}{2} \div \frac{1}{3}$ (i) $\frac{2}{3} - \frac{1}{5}$

   (d) $1 = \frac{}{7}$ (j) $\frac{1}{2} \times 4$

   (e) $7 = \frac{}{7}$ (k) $\frac{1}{2} \div \frac{1}{3}$

   (f) $\frac{2}{3} \times 1$ (l) $1\frac{1}{2} \div \frac{2}{3}$

4. Use a geoboard to illustrate with diagrams why $\frac{1}{2} = \frac{2}{4} = \frac{3}{6}$.

5. Describe two different methods for children to compare the fractions: $\frac{1}{2}, \frac{3}{4}, \frac{5}{6},$ and $\frac{1}{3}$.

6. Show a way to help children rename the following fractions in simplest form:

   (a) $\frac{12}{30}$,   (b) $\frac{49}{56}$,   (c) $\frac{32}{48}$,   (d) $\frac{35}{80}$.

7. Show how to help children order the following fractions from greatest to least. Use a different approach with each set.

   (a) $\frac{3}{8}, \frac{2}{3}, \frac{7}{12}, \frac{5}{6}, \frac{3}{4}$.   (b) $\frac{2}{3}, \frac{4}{5}, \frac{1}{6}, \frac{5}{8}, \frac{1}{2}$.

8. To find the least common denominator of $\frac{3}{10}$ and $\frac{7}{12}$, Mark reasoned that two is the largest number that divides both ten and twelve. Ten divided by two is five. Five times twelve is sixty, so sixty is the least common denominator. Is this method of reasoning correct? Explain your answer.

9. Describe an activity to help a child visualize that writing a fraction in lowest terms does not make the fraction become smaller.

10. Write a practical story problem for each of the following problems so that the fractions expressed will be more meaningful to children:

(a) $2\frac{1}{2} + 3\frac{1}{4}$

(c) $\frac{7}{10} - \frac{3}{10}$

(b) $\frac{1}{2} + \frac{3}{4}$

(d) $4 \times \frac{2}{3}$

11. Describe a sequence of three developmental and three practice activities for teaching *one* of the following operations with fractions:
(a) addition
(b) subtraction
(c) multiplication
(d) division

12. What is the ratio of:
(a) Days in a week to days in a year?
(b) Time in a minute to time in an hour?
(c) Four dimes to three quarters (in value)?
(d) Centimeters in a meter to millimeters in a meter?

13. What types of experiences or situations would you use to teach the concept of ratio to children?

14. Explain how you can meaningfully express the mixed numeral $4\frac{5}{8}$ in the form of the fraction $\frac{a}{b}$. Use a model or diagram to show how you changed the mixed numeral to a fraction.

15. For each of the following sets of problems:
(a) Find the computational errors.
(b) Determine which concept(s) might not be understood by the child.
(c) Briefly describe how you would help the child overcome this error.

(i) 45% of 36 = 162   60% of 32 = 192

(ii) $\frac{4}{5} + \frac{5}{9} = \frac{9}{14}$ $\frac{1}{6}$ of $\frac{2}{3} = \frac{3}{9}$

(iii) $\frac{5}{8} \div \frac{2}{7} = \frac{16}{35}$ $\frac{7}{8} \div \frac{2}{7} = \frac{16}{49}$

(iv) $5\frac{3}{5} - 1\frac{1}{2} = 4\frac{1}{10}$ $6\frac{5}{6} - 2\frac{1}{3} = 4\frac{3}{6}$

(v) $\frac{2}{3} \div \frac{4}{5} = \frac{12}{10}$ $4\frac{1}{8} - 1\frac{1}{2} = 3\frac{3}{8}$

16. Illustrate with a diagram to show what $3/4 \times 5/6$ equals.

17. Solve each of the following percent problems by setting up an equation and then showing all of the steps in solving the equation.
(a) 60% of what is 24?
(b) What percent of 56 is 42?
(c) What is 10.25% of $90,000?

Reference

Bley, Nancy, learning disabilities specialist. Case study based on personal experience with learning disabled children at Park Century School in California. Relayed in unpublished communication with authors, 1979.

12 MEASUREMENT

In order to help mathematics become real to children, it is important that they see mathematics as something that has use outside the classroom.

How much carpet is needed for a room?
How much material is needed for a suit?
Will that new refrigerator fit in the kitchen?
Can I afford that basketball?
How far apart should be beans be planted?
Do I have enough money to write this check?
How much meat is needed for a meal?
How long will it take me to finish this?

Virtually all nonclassroom applications of mathematics involve measurement, except for those who are in technical vocations. It follows, then, that the topic of measurement can provide excellent opportunities for showing the practicality of mathematics. Children should see that the study of measurement is important. It is useful. It is time well spent. In addition, because of the hands-on nature of so many measuring experiences, it is fun!

TEACHER BACKGROUND

The Piagetian concept of conservation is as important in the study of measurement as in the study of early number concepts (see Chapter 6). In layman's language, if a quantity is separated into parts, rearranged, spread out—whatever is done to the quantity, so long as none is removed or added—the total quantity remains the same.

This makes it possible, for example, to recognize that there are many ways to name a given quantity. In Figure 12.1, several names are given for each of two lengths. The "many names" are accepted here because in each case the total length is separated into different-size pieces, or units, to produce each new name. The total length remains the same whatever size of unit is chosen.

The notion of conservation is basic to other considerations as well. In examining properties of length, for example, it is possible to "bend" a ruler around corners to find the perimeter of a rectangle. Or one can measure the sides separately, add the parts, and know that the sum of those parts will be the same as the whole perimeter.

Conservation of area makes it possible to separate a shape into parts, rearrange the parts to get a simpler figure, and be confident that the area of the resulting figure is the same as that of the original. In fact, this is the most common technique used to develop computation formulas for area.

Another consideration in the teaching of measurement is that of precision. How precise should measurements be? How can more precision (in measurements) be gained? Precision depends on the

Figure 12.1

1 meter	= 100 centimeters	3 feet	= 1 yard
	= 10 decimeters		= 36 inches
	= 1,000 millimeters		= 2 half-yards

Figure 12.2

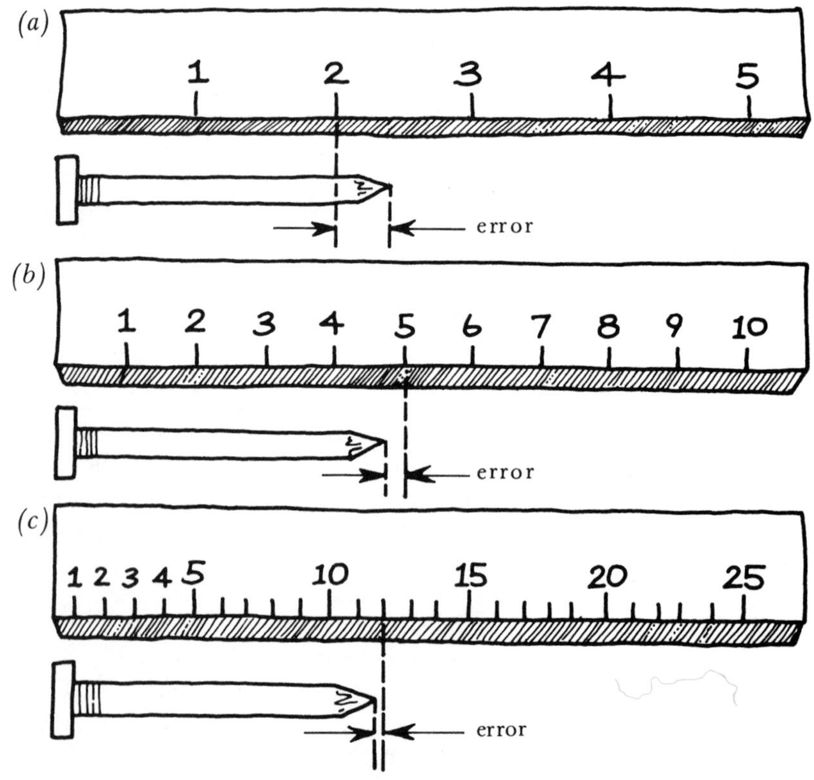

Teaching Mathematics to Children with Special Needs

size of the unit being used. Consider the example given in Figure 12.2a. Using the first ruler, the nail is found to be about two units long (closer to two than to three). Note that there is considerable error.

Measuring again, using a ruler with units that are only half as big, the nail is about five units long. The error is considerably smaller when we use this unit (Figure 12.2b).

Measuring once more, using an even smaller unit, the nail is about twelve units long. The error is even less (Figure 12.2c).

Precision can be improved by making measurements with smaller units.

By examining the ruler it is possible to determine the largest allowable error for any given unit. Observe the three measurements shown in Figure 12.3

Measurements should never have an error of more than half the unit being used. If an object is being measured to the nearest inch, then the maximum allowable error should be one-half inch. If the measurement is being made to the nearest meter, then the answer should be within one-half meter.

Figure 12.3

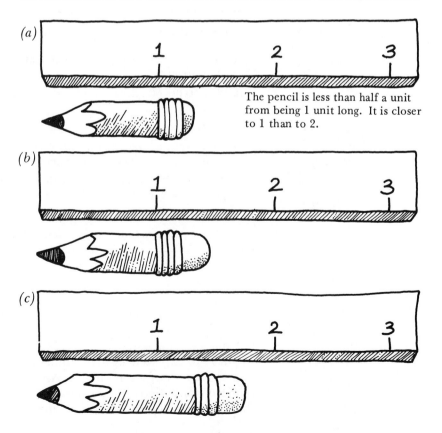

(a)

The pencil is less than half a unit from being 1 unit long. It is closer to 1 than to 2.

(b)

(c)

The maximum allowable measurement error is one-half the unit being used.

The smaller the unit, the smaller the error, and the more precise the measurement will be .

Sometimes very precise measurements are needed. At other times reasonable estimations will do. This chapter suggests ways of preparing children for both types of situations. It focuses on ways of presenting basic measurement ideas related to length, capacity, weight, time, area, volume, and angles. A major goal is to help children become familiar with basic measurement units and to master important measurement concepts and skills through hands-on experiences. In many of the suggested activities children are encouraged to estimate before measuring. This approach not only helps them acquire basic concepts more quickly, but also enhances their ability to use measurement sensibly in daily situations. Other activities require children to make more precise measurements right away. The many facets of measurement are exciting and fun for children. The study usually begins with length.

LENGTH MEASUREMENT

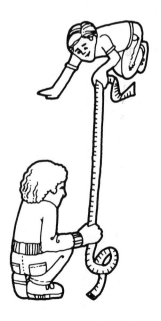

What Do *You* Say?

1. How can teachers help children develop the concept of length measurement?

2. What types of units are used in early measurement activities for length?

3. What system of measurement should be taught?

4. Besides the notion of unit, what other concepts and skills are basic to the study of length measurement?

5. What metric equivalents for length measurement might be presented to intermediate-level school children?

6. What type of activities can help children develop estimation skills?

7. How might area formulas be developed with children.

Instruction in length measurement generally moves through five phases in elementary school instruction. We will now consider these five phases, from early work in the development of appropriate language to experiences that require children to estimate then measure the length of given objects. Although instructional needs may vary from situation to situation, from teacher to teacher, and from child to child, the five phases normally evolve in the order presented.

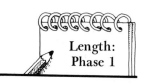

**Length:
Phase 1**

Helping Children Conceptualize Length

Developing the concept of length is largely a matter of language development. The role of the teacher is to help children develop the language of comparison in its proper context—through use. There are many words that mean length (width, height, breadth, depth, and so on). The appropriate word depends on the context in which it is used. Much of this language development grows out of the classification and ordering experiences discussed in Chapter 6.

Although it is usually necessary to use contrived situations in order to provide children adequate experiences related to the concept of length, wise teachers also make constant use of natural situations and ask children to compare objects encountered on the playground, at recess, at lunch, on field trips, or whenever the opportunity presents itself.

**Activity File:
Developing the
Concept of Length**

Activity 1: Sizing

Have three children stand. Have a fourth child arrange them in order from shortest to tallest. Talk about it. Use the words *height, taller, shorter, tallest, shortest.*

Activity 2: Look And See

Choose two objects with obviously different lengths. Have a child pick the longer or the shorter. Use the words *length, long, short, longer, shorter, longest, shortest.*

Activity 3: Strips

Place paper strips of varying lengths about the room (on the wall, tackboard, and so on). Be sure that some are hung vertically, some horizontally, some diagonally. Identify two strips. Have a child choose the longer or shorter of those two strips. Emphasize that length does not depend on direction. Use comparison vocabulary.

Activity 4: Rods

Use graduated centimeter rods (or paper centimeter strips—colored, of different lengths). Identify two rods by color. Have a child choose the longer or shorter of the two. Use appropriate comparison vocabulary.

Activity 5: Circles

Place a number of different-sized round (circular) objects about the room. Identify two of those objects. Have a child choose the wider or narrower of those two objects. Use the words *wide, narrow, wider, widest, narrower, narrowest.*

Most school mathematics programs include workbook exercises which focus on basic vocabulary for length. Examples are given in

Figure 12.4

Figure 12.4. Such exercises can be used to reinforce the experiences of activity sessions.

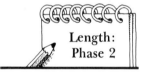

**Length:
Phase 2**

Helping Children Develop the Concept of a Unit of Length

Sometimes it is not possible to compare two objects directly and choose the longer or shorter. There are times when the objects cannot be moved and their lengths are not so obviously different that you can easily see which is longer. In situations like this, both objects can be compared against a third object. This third object, then, serves the function of an informal unit of measurement. The unit is something that is independent of the objects being compared. The unit is something that can be moved about.

At times it is useful to be able to compare objects in a way that gives more information than *longer* or *shorter*. Such questions as "How long?" or "How much shorter?" suggest a need for more than simple comparisons. This need is satisfied as we develop the notion that length can be expressed using numbers. The length of an object can be described as a number of units long.

**Activity File:
Developing the
Concept of a Unit of
Length**

Activity 1: Tall And Small

Have two children who are nearly the same height stand on opposite sides of the room. If the children cannot "look and see" who is taller, ask a third child to stand by the first child, then by the second. After comparing both children against the third child, decide who is taller.

Activity 2: Big And Little

Place about the room a number of objects that are about the same length. Choose two objects to be compared. Give a child a piece of string that is longer than any of the objects. Have the child measure one object with the string, then compare that length of string with the second object.

Activity 3: Stacks

Place two objects about the same height at opposite sides of the room. Have two groups of children stack their math books next to the two objects. How many books high are the objects? Which is taller?

Activity 4: Trains

Using rods of one size from a set of graduated centimeter rods, have children make trains of rods as long as various objects. How many rods long are the objects? (Figure 12.5)

Activity 5: Arbitrary Units

Have children measure an assortment of objects using arbitrary units such as paper clips, pencils, sticks, or lengths of string.

Activity 6: Neat Feet

Trace around a child's foot. Make several copies. Cut them out. Measure things. Measure the teacher's desk. Measure the length of the room. Have Johnny lie down; measure Johnny.

Figure 12.5

(The pencil is about 6 rods long.)

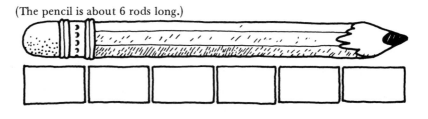

Length: Phase 3

Introducing Systems of Standard Units

While measuring with rods or other arbitrary units, children often become aware of a need for standard units. "Four feet long," for example, is meaningless unless you know whose foot serves as the unit: Four of Julie's feet is not the same as four of Juan's feet.

If people are to communicate meaningfully about measurement, it is necessary for each to know what the other person is talking about. Both must have common (standard) meanings for the terms used. How long is the stick? What size paper clip do you mean? Whose foot?

Once children become aware that there is a need for *standard units*, they can be introduced to those they are likely to encounter outside the classroom.

What System of Measurement Should Be Taught?

Because the metric system is used worldwide in almost all countries outside the United States and because it is superior in many ways to the English system, its official adoption in the U.S. is imminent. For this reason it is necessary that teachers teach the metric system so children will be prepared to use metric units as that system comes into common use.

On the other hand, not even an official Act of Congress will change habits formed over a lifetime. Older generations will continue to use many of the familiar measurement units. Furthermore, our libraries are full of books that use feet, miles, inches, quarts, pints, yards, gallons, acres, and other customary units. And, though business and industry are even now making changes to the metric system on the production line, cost and other practical considerations dictate that such changes must be gradual over a period of years. It seems apparent, then, that teachers will need to continue for a time to teach English units to children. At least one generation will need to be reasonably proficient with both systems of measurement—metric and English.

The teaching emphasis should be on *measurement* concepts and skills, not on metrics. With respect to length measurement, children need to learn *how* to use a ruler, whether it be English or metric; whether the "0" point comes at the end or not (Figure 12.6).

Figure 12.6

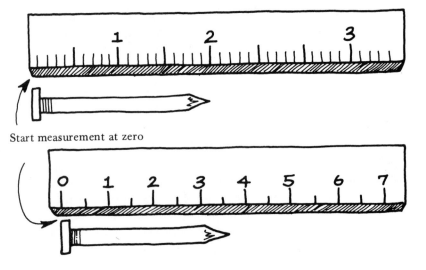

Start measurement at zero

These abilities should be stressed. The child should be able to:

1. Choose an appropriate unit (centimeter—not meter, or inch—not yard) to measure a given object

2. Make quick estimates of lengths and distances in daily situations

3. Round a measurement to the nearest unit or half-unit

Activity File: Introducing Standard Units

The following activities suggest a sequence that might be used to help children:

1. Formalize the need for a standard unit

2. Guide first experiences using common metric and English units.

Activity 1: Stretch
Form two groups of children with the taller children in one group and the shorter children in the second group. There should be one or two more children in the shorter group. Have ach group hold hands and stretch out to form the longest possible line without letting go. How long are the lines? How many people long? Does the largest measurement (number) go with the longest line? (Figure 12.7)

Activity 2: Weird Sticks
Use two sticks, one about a foot long, the other about an inch longer. Have a child use one of the sticks to measure the length of the room. Have the second child use the other stick to check the measurement. Why are the answers different?

Figure 12.7

Activity 3: Feet Feat

Choose two children, one with a large foot, one with a small foot. Have the child with the bigger foot use his feet to measure the length of the teacher's desk. Have the other child use his feet to measure out a piece of string the same number of feet long. Why is the string not as long as the desk?

Activity 4: Big Foot

Write a length that is less than one foot (7 inches, 2 inches, etc.). Have a child tell how much more must be added to be as long as a foot ruler.

Activity 5: Meter Made

Give a length that is less than one meter (30 cm, 8 dm, 600 mm, etc.). Have a child tell how much more must be added for it to be as long as a meter stick.

Activity 6: Concentration

Prepare pairs of cards showing equal lengths (Figure 12.8). Spread the cards, face down. Children take turns turning up pairs of cards. If the cards match, the child takes the cards. If not, the cards are turned back over.

Activity 7: Crazy Race

Prepare a set of cards showing length (Figure 12.9). Mark off a starting line and a finish line about three meters away. Shuffle the cards. Children take turns drawing a card and moving their markers ahead of the given distance. All measurements should be made using centimeter cubes and decimeter rods (Figure 12.10).

Figure 12.8

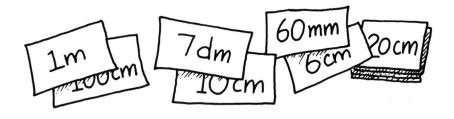

Figure 12.9

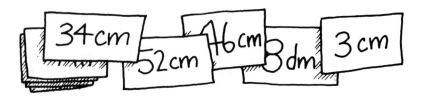

Activity 8: Textbook Exercises

School mathematics textbooks include written exercises designed to help children become familiar with systems of basic units. The examples listed below are fairly typical of what the teacher might find.

1. 1 meter, 52 centimeters = ☐ centimeters

2. 17 centimeters = ☐ millimeters

3. 2.3 meters = ☐ centimeters

4. 321 centimeters = ☐ meters, ☐ centimeters.

Figure 12.10

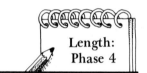

**Length:
Phase 4**

Helping Children Develop Estimation Skills

Activities like those in the previous section that introduce children to basic units also lay the groundwork for estimation activities. Unless children are familiar with a unit, it is impossible to estimate a length using that unit. The activity suggestions that follow may be used to build on previous activities and help children develop estimation skills. Each activity deals with only one unit of measure, but can be easily adapted to other units. Further, since the ability to choose the *most appropriate unit* for making a measurement is important, activities can be modified to require children to select a unit, *then* estimate the length or distance. The child estimates lengths for short things in inches or centimeters and long things in feet, yards, or meters.

**Activity File:
Estimating Lengths
and Distances**

Activity 1: Meet A Meter

Have children find an object that they think is about one meter long. Have them measure the object to see how close to a meter it is.

Note: The following two activities are designed to help children "see" lengths in tens and ones. For example, a length of 36 cm should be "seen" as 3 tens and 6.

Activity 2: Guessing

Have children select an object shorter than a child's arm. Have them guess its length in centimeters. Have them measure to check its length, making the measurement with centimeter cubes and decimeter rods.

Activity 3: How Far?

Have two children sit facing each other across a table. Have them close their eyes, then each place a paper clip somewhere on the table. Have the children look at the paper clips and guess the distance between them. Using centimeter cubes and decimeter rods, measure to check the guesses.

Activity 4: How Big?

Guess distances between objects and lengths of objects. Guess both long and short lengths. Measure to check accuracy.

Developing Skill Using Measuring Tools

Estimating and measuring activities go hand in hand. This was apparent in the "estimate-measure, estimate-measure" activities of the preceding section. The development of measurement skills is an ongoing task. Skills learned now by the child will be extended and refined in later years: how to obtain greater precision when this is needed; how to select appropriate measuring tools; how to make needed measurements as efficiently and accurately as possible; how to use new tools. The climate that nurtures this growth during school years is one that provides children opportunities for measuring *many* things:

- In the room
- In the hall
- In the gym
- On the playground
- At home
- Wherever measurement experiences present themselves

Let children use rulers, tape measures, meter sticks, yard sticks, and other assorted tools to measure things. (The American Printing House for the Blind carries braille rulers if these are needed.) Gradually require greater precision in the measurements made.

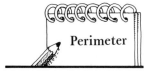

Perimeter: The Distance Around

It was mentioned earlier that there are many names for length. It is appropriate to include at this point two names for length that are frequently used in mathematics. The first of these names is *perimeter*.

The perimeter of a simple, closed plane figure is the total length of the curves and/or segments that make up the figure. In the language of children, the perimeter of a figure is "the distance around" the figure.

The perimeter of a rectangle is the sum of the lengths of the four sides (Figure 12.11).

The perimeter of this figure is the sum of all its sides—both straight and curves sides (Figure 12.12).

If the figure being considered is a circle, then its perimeter is called the *circumference*. Circumference is a special name for the perimeter of a circle.

If one were to experiment, one would discover that the distance around a circle (its circumference) is a little more than three times the distance across the circle (its diameter). By using more and more precise measurement instruments, measuring the distance around (Figure 12.13a), then measuring the distance across (Figure 12.13b), then dividing, perhaps using a hand calculator,

$$\frac{3.14+}{12.8 \,\,\,) \,\, 40.3}$$

a reasonably accurate approximation of π (pi) can be obtained. This approximation was found by dividing the circumference c by its diameter d.

Figure 12.11

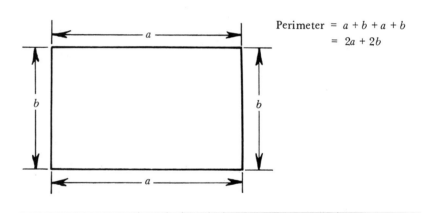

Perimeter $= a + b + a + b$
$= 2a + 2b$

Figure 12.12

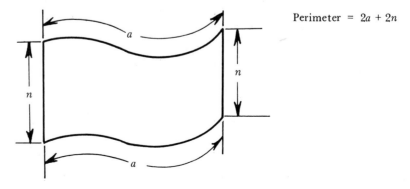

Perimeter $= 2a + 2n$

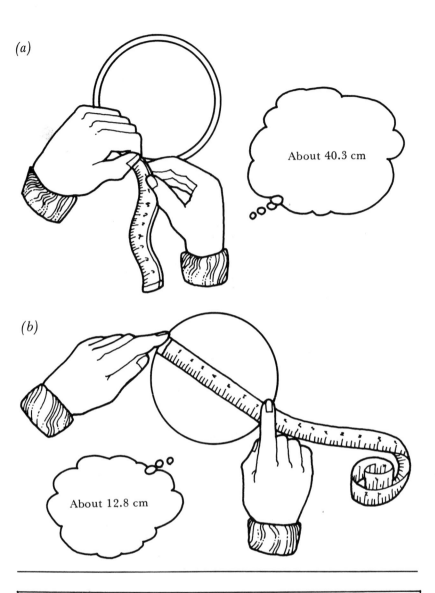

Figure 12.13　(a)

About 40.3 cm

(b)

About 12.8 cm

Since $c/d = \pi$

then $c = \pi d$ Computational formulas for circumference
                         in school math texts
or $c = 2\pi r$

The number π in the above formulas cannot be expressed exactly as a decimal or as a fraction. It is an irrational number. However, for elementary applications of π, either of the approximations, 3.14 or $3\frac{1}{7}$, are close enough. More accurate approximations of π can be used; they are normally developed and discussed in text-books as they are needed.

1. Prepare, as your instructor directs, answers to the major questions on p. 418 of this chapter.

2. Describe a phase-one length activity which is appropriate for a child with a vision impairment. If possible, try your activity with a child.

3. Describe a length activity for phase three which is appropriate for a child who is a weak *auditory* but strong *visual* learner. If possible, try your activity with a child.

4. Name three things you might have a student measure using
   (a) centimeter as a unit and
   (b) meter as a unit.

5. Find the perimeters of Figure 12.14.

6. Find the circumference of each circle in Figure 12.15.

7. Answer the following questions and, for each, describe the procedure you used.
   (a) How much money would you have if you had a stack of quarters one meter high?
   (b) How thick is one sheet of typing paper?

8. Find objects with measurements that fit in the following intervals. Write the name of the object. Measure it and record your answer in three equivalent measures, as in the example following.
   Example: The child's pencil, pictured in Figure 12.16, is 14 centimeters or 1.4 decimeters of 0.14 meters long.

Measurement intervals	Objects	Actual measurements
(a) 100 to 200 cm	_____	___ cm = ___ dm = ___ m
(b) 50 to 100 cm	_____	___ cm = ___ dm = ___ m
(c) 25 to 50 cm	_____	___ cm = ___ dm = ___ m
(d) less than 10 cm	_____	___ mm = ___ cm = ___ dm

9. An old adage is that a person's armspan is equal in length to his or her height. Have a few of your classmates spread their arms for you and estimate whether this appears to be true. Then check your estimates by measuring the armspans and heights.

Figure 12.14

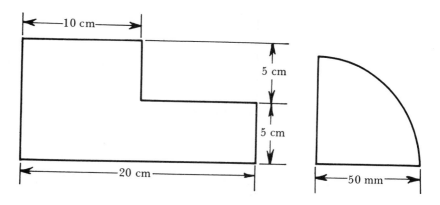

Figure 12.15

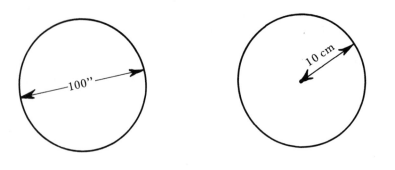

Figure 12.16

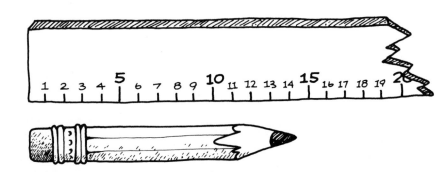

**OTHER MEASURE-
MENT TOPICS
INTRODUCED IN
PRIMARY GRADES:
CAPACITY, WEIGHT,
TIME**

What Do *You* Say?

1. What type of activities are appropriate for introducing capacity measurement to young children? Weight measurement?

2. What basic units—English and metric—should children be familiar with?

3. In the study of capacity or weight, what units are appropriate for measuring very small amounts? Large amounts?

4. What type of scales are more appropriate for use with very yound children—balance or spring-type scales?

5. How are experiences for capacity and weight measurement, begun in primary grades, extended for intermediate and upper-grade students?

6. How can teachers help develop children's awareness of the *need* to be able to tell time?

7. What other factors contribute to a child's awareness of time, apart from the ability to tell time from a clock?

8. How does regarding the clock as two spinners help children learn to tell time?

9. What is the advantage of focusing *only* on time "*after* the hour" (e.g., 4:25, 4:50) in early time-telling sessions with minutes?

Capacity

Children's experiences measuring capacity can be initiated in the early primary grades. The topic is a relatively easy one to teach when teachers provide opportunities for children to measure the capacity of many containers. The three-phase instructional sequence about to be outlined helps to develop the topic with children.

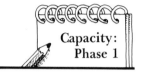

Capacity:
Phase 1

Developing the Concept of "Capacity"

As with length, the basic concept of capacity is largely rooted in the development of appropriate language. Using expressions such as *filling, contains, holds,* and *full* in the context of measurement activities will help children understand capacity. In addition to vocabulary development, it is important that teachers structure early activities so children are brought to realize that changing the shape of a container may or may not change its capacity.

Activity File:
Concept of Capacity

Activity 1: More Or Less

Give children pairs of containers. Have them pour water from one to the other to decide which holds more water.

Activity 2: Square And Round

Have children compare round containers with square containers to see which will contain more water.

Activity 3: Line Up

Have children pour water from container to container to decide which holds more water. Line up five or six containers in order from smallest to greatest capacity.

Capacity:
Phase 2

Becoming Familiar with Standard Units for Capacity

If children are to be successful in understanding the relationships between measures (such as 1000 milliliters = 1 liter and 2 cups = 1 pint) they must do more than just memorize a table of equivalent measures.

It is necessary that they become familiar with the units involved, developing appropriate mental imagery for each unit. The child who has the mental image of a milliliter as a *very small* amount, the size of a centimeter cube, is likely to choose that unit as an appropriate measure for small liquid quantities. The activities are designed to help children develop a useful set of mental images for the standard English and metric units as well as to learn the relationships between those units.

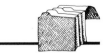

**Activity File:
Units for Capacity**

Activity 1: Looks Like A Liter
   Have children fill jars and bottles to show a liter of water in as many different-shaped containers as possible.

Activity 2: Looks Like A □
   Use the above activity for any unit being studied: quart, cup, gallon, milliliter, and so on.

Activity 3: Match Me
   Give children cards with various capacities written on them: 2 cups, 1 pint, 500 milliliters, 1/2 liter, 4 quarts, 1 gallon, and so on. Children find their partners (holding cards with equal amounts).

Activity 4: Concentration
   Let children play concentration with pairs of matching-capacity cards.

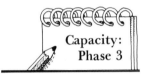

**Capacity:
Phase 3**

Developing Measuring Skills
   Real expertise with capacity measurement will develop only if children have frequent and varied experiences measuring capacity. The activities described below are examples of experiences that are appropriate for helping children develop important measuring skills.

Activity File

Activity 1: Making Measures
   Let children use milk jugs, assorted unmarked containers, and a permanent marker. They should pour measured amounts of water into the containers, then mark and label the water level to create a set of capacity measures.

Activity 2: Recipe
   Invite children to measure ingredients for afternoon refreshments (Kool-Aid, juice, and the like).

Activity 3: Measuring
   Measure water, measure sand, measure snow, measure rice, measure garden seed, measure dirt, measure often!

Weight
   The measurement of weight[*] can be accomplished in two ways, directly and indirectly. In order to measure weight directly some kind of balance scale is generally used. If the scale balances when an object is placed on one side of the scale against four one-pound

---

[*]The authors have chosen to ignore any controversy over the use of weight and mass. They have used the word *weight* here because it is more commonly used and understood.

weights on the other, then the object weighs four pounds (Figure 12.17).

If it takes more weight to balance one object than it takes to balance another object, then we know that the first object weighs more than the second.

The balance scale provides an effective means for comparing objects by weight and hence for establishing the notion of weight with young children. Two youngsters on a see-saw can quickly tell who is heavier.

In actual practice, however, nearly all weight measurements are done indirectly, using some sort of spring-scale. In this case what is really being measured is the pull of gravity on the object. The pull of gravity is stronger (therefore the "weight" greater) at sea level than it is on a high mountain. Of course, for all practical purposes, the apparent changes in weight as one moves from place to place are not enough to cause great concern.

Since, outside the classroom, most weights are calculated indirectly with spring-type scales, children also need experience weighing things this way. A carefully planned sequence of experiences for children would, over the years, include opportunities to use a variety of scales, from the simple balance to more finely calibrated spring scales. Many activities described for the instructional phases below suggest ways children can use scales in simpler experiments which develop or reinforce measurement concepts and skills.

Weight: Phase 1

Developing the Concept of Weight

What do we mean by *heavy* or *light*; by *heavier* or *lighter*? Attaching meaning to these words is part of the language development that underlies a child's understanding of weight. This development is best carried out informally, using familiar objects from the child's surroundings. The following activities suggest ways this can be done.

Figure 12.17

Activity File

Activity 1: Which Is Heavier
Instruct children to place objects in milk cartons and close the tops. Then have them pick two cartons and decide which is heavier. When they open the cartons to look at the objects, ask questions like "Are the big objects always heaviest?" "Does shape ever make any difference in weight?"

Activity 2: Line Up
Have children place objects in milk cartons and close the tops. Then ask them to line up the cartons from lightest to heaviest.

Activity 3: Ups And Downs
Using a see-saw, have children compare their weights.

Activity 4: Balancing
Let children use a balance scale to compare objects. "Which weighs more: ten paper clips or ten leaves? Math book or spelling book?" Etc.

**Weight:
Phase 2**

Developing Measuring Skills for Weight
In order to develop *useful* weight-measuring skills, children should have experiences that emphasize estimation and involve scales like those available to them outside the classroom. The following activities suggest ideas that may be used in meeting these goals for developing measuring skills with children.

Activity 1: Measure Me
Children should weigh themselves.

Activity 2: Line Up
Have children weigh objects and arrange them in order by weight. (Use small objects that would be weighed in grams or ounces; use larger objects that weigh several kilograms or pounds.)

Activity 3: Guesstimation
Gather what children think is five pounds of rocks. Then have them weigh the rocks to see how accurate the estimate was.

Activity 4: Sand And Water
Have children fill two identical containers—one with sand, the other with water. "Which do you think will weigh more?" Weigh to check.

Activity 5: Weighing
Have children weigh things: big things, small things, heavy things, light things, things at home, things in the classroom. Have them keep a record; make a poster.

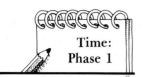

Developing the Concept of Time

Many people do not associate clock time with measurement; indeed, a clock *measures* the passage of time throughout each day. Here are suggestions for teaching time on the clock.

Appropriate first experiences are those that help children develop an awareness of time apart from the simple reading of a clock. The following ideas are most important in this regard.

For a given list of events (breakfast, reading, lunch, math, recess) there is an order in which they occur.

Activity 1: Shuffle

Paste pictures of various events (breakfast, riding bus to school, etc.) on cards and shuffle them. Children arrange the cards to show the order in which the events occur.

Activity 2: String Along

Hang picture cards on a string or clothesline to show the order in which they occur. Leave the line up for future reference. When daily schedule changes, change the picture string.

Some events are longer (take more time) than others.
Math takes longer than lining up for the bus.
It takes longer to watch *Sesame Street* than to get a drink.

Activity 1: Long And Short

Choose pairs of events. Have the children decide which takes longer.

Activity 2: Strips

Have the children cut paper strips different lengths to show "long" events and "short" events. Label or draw or paste pictures on the strips.

The distance (time) between events may vary in length.
There is more time between breakfast and lunch than between recess and math.

Activity: Hang Up

Hang cards with pictures or labels depicting events on a string or a clothesline. Let the distance between cards show the time between events.

How long is long?
How long an event seems depends on how much you are enjoying it.
"Time flies when you're having fun."

Activity 1: One Foot

Have the children stand on one foot for a minute.

Activity 2: Running
Have the children run in a circle for a minute.

Activity 3: Walking
How far can the child walk in a minute? Have him or her guess the distance first, then walk to see.

Activity 4: Eating
How long does it take to eat? What else takes as long?

Activity 5: Watching TV
How long is the child's favorite TV program?

**Time:
Phase 2**

Developing with Children Their Need to Read Clock Times
Personal need often points the way or inspires one to work to meet a goal. Too often children are required to attempt to read clock times before they experience the *need* to do so. As a result, they often have difficulty developing clock-reading skills.

Teachers (and parents, too) can set the stage for clock-reading sessions by developing with children the awareness of their need to tell time from a clock. Creating situations where it is to the child's *dis*advantage to be unable to read a clock is the key. Activities such as the following are suggested.

Activity 1: When's Recess?
Place children in charge of telling the class when it is time for recess; when it is time to eat; when it is time for the party; when it is time for other events they enjoy.

Activity 2: Time To Stop
Have a child work at a task until a given time. The child must know when to stop.

Activity 3: Favorite Program (A home activity)
Take time out to watch a favorite TV program if a child can tell when it is time to turn on the TV.

For each of these activities, it is the *child's* responsibility to keep track of the time. If the teacher or parent helps too much, there is no real *need* for the child to be able to tell time; and that is the point of the activity.

Of course, at first, some help may be necessary. For example, one could set a cardboard clock dial to show the time to be watched for (Figure 12.18). The amount of help that is appropriate will vary from child to child and must be left to the judgment of the teacher. However, it is important that enough responsibility be placed on children that they become aware that telling time is a useful skill.

**Time:
Phase 3**

Reading Clock Times

After a child has developed a feeling for time, clock reading skills can be introduced. For many, learning to read a clock will not be easy. Children must learn to deal with clocks having two (or three) hands that are different in size (and sometimes color), and move at different speeds. When a child looks at a clock, the hour and minute hands do not appear to move at all! How difficult, then, to relate time to the idea of hands moving at different speeds. Or, when children are told that the clock says 2:30, it is a difficult notion, since they can see the 2, but not the 30!

Recognizing that *telling time is relatively complex*, it is wise to try to eliminate as many of the confusing aspects of the situation as possible. One approach to simplifying the process of time-telling is to simplify the clock. Using a clock with only an hour hand allows the children—even as early as kindergarten—to "tell time" as soon as they have learned to read the numerals 1 to 12. This one-handed clock can be regarded as a spinner. The child reads time from the spinner dial by telling the number the clock hand is closest to. Using a one-handed clock, children learn to tell time almost immediately! (See Figure 12.19.)

Figure 12.18

Figure 12.19

Once children can confidently give the time in this way to the nearest hour, the teacher can then identify the hour hand on a two-handed clock and show how to tell time by focusing on the hour hand (Figure 12.20).

When children become comfortable with the hour hand and have developed a degree of competence using only this hand, the minute hand can be introduced as a means to greater precision. Children who have learned to read time on the hour by thinking of a clock as a spinner will find it easy to relate to a second spinner, *with a different set of numbers*, for minutes. "Telling time" can be thought of as reading those two spinners giving the hour and then the minutes. Gradually the association between the minutes and the twelve digits on the standard clock will be made (Figure 12.21).

If two-spinner experiences are followed by experiences with a geared clock, the child can be shown that as the hour hand moves from one hour to the next the minute hand goes all the way around. If the hour hand moves halfway to the next hour, the minute hand moves halfway around.

A specific point of difficulty for children is the concept of "before the hour" (Figure 12.22).

To simplify the learning task and to make times shown on clocks with hands consistent with digital clocks (which do not give the minutes before the hour), it is suggested that, in the beginning, children learn only to give the minutes *after* the hour (Figure 12.23).

Figure 12.20

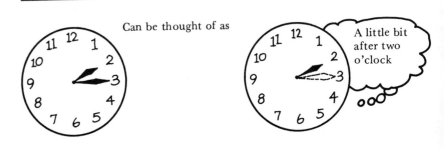

Figure 12.21

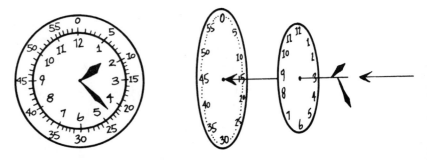

Figure 12.22

Figure 12.23

Once the child has learned to tell time in this way and is both confident and competent, learning to give the minutes before the hour becomes much easier and is often looked on by children as an interesting exercise.

**Activity File:
Telling Time**

Activity 1: Spinning Moves
Make a spinner out of a clock face, using only the hour hand. Have the children play board games where the spinner tells the child how far to move.

Activity 2: Spinning Time
Have the child tell you what time it is ("What is the hour?") frequently during the day.

Activity 3: Double Spinner
Using a geared clock, set the hands and ask a child to give the hour. Ask another child to give the minutes. Write the hour and minutes in the form (hour):(minutes).

Activity 4: Match Me
Pass out cards to the children like the ones in Figure 12.24.
Have the children find their partners.

Figure 12.24

Your Time: Activities, Exercises, and Investigations

1. Prepare, as your instructor directs, answers to the major questions on p. 430.

2. Outline a *sequence of activities* that could be used to teach primary children important concepts or skills related to each of the following topics. For each sequence, specify the concepts or skills you are considering.
   (a) Measurement of capacity
   (b) Measurement of weight
   (c) Time

3. Choose *one* sequence, and show how the activities of that sequence can be extended or adapted to meet learning objectives on the topic for intermediate or upper-grade students. Be sure to specify the new set of learning goals you are considering.

4. Likely or unlikely? Give your answer to each of the following, then measure or otherwise experiment to check.
   (a) The pencil weighed 600 grams.
   (b) The teaspoon held 5 ml of vanilla.
   (c) The man weighed 175 kg.
   (d) The cookie recipe (for 48 cookies) called for 2 l of milk.

5. Name something that:
   (a) Weighs about a kilogram
   (b) Weighs about a pound
   (c) Holds about a liter

6. Find answers to the following questions. Describe the procedure you used in determining each answer.
   (a) How much does one kernel of popcorn weigh?
   (b) How many grains of rice are in one pound? One kilogram?
   (c) Try the traditional hiccough cure. Hold your breath as you take ten swallows of water. How many milliliters of water did you drink?

MEASUREMENT TOPICS FOR INTERMEDIATE AND UPPER GRADES: AREA, VOLUME, ANGLE MEASURE

What Do *You* Say?

1. What type of activities are appropriate for introducing area measurement to children? Volume measurement?

2. How can basic area formulas for:
   • rectangles • triangles • parallelograms • trapezoids • circles
   be developed with children?

3. How might volume formulas for the prism and cylinder be intuitively developed with children?

4. How are volume formulas for a prism and pyramid related? For a cylinder and cone?

5. How can one use a folded paper for measuring simple angles and for introducing the protractor?

Area

The study of area begins in the upper primary or lower intermediate grades with the *measurement of area* and gradually turns to the development and use of *area formulas*. The two phases of instruction devoted to early measurement of area are now described.

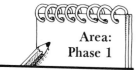

Area: Phase 1

Developing the Concept of Area and the Notion of Unit

The concept of area can be developed in much the same way as one develops the concept of length. Comparison of areas and the use of appropriate comparison vocabulary are the primary means by which one begins the development of the concept of area. However, the teacher should expect concepts to be more difficult for the child to grasp than the concept of length.

The size of an area depends on the interaction of several factors—length of the figure, width of the figure, and shape of the figure. For example, the longer of two figures may have the smaller area. It is for this reason that one should, very early in the study of area, begin to develop the notion of unit. A unit of area is a piece of or an amount of area. In order to determine the size of an area, one must determine *how many units will cover* the area.

**Activity File:
Unit of Area**

Activity 1: Cut And Paste
Give each child two odd-shaped regions. Have them cut pieces from the two regions that can be compared. Decide which total region was bigger (Figure 12.25).

Activity 2: Covering
Give children two odd-shaped regions to compare. Have them "measure" the regions by covering them with plastic counting disks. Which is larger? (See Figure 12.26.)

Activity 3: Squaring Off
Have children measure regions by laying squared paper over the regions, tracing the regions onto the squared paper, and counting squares.

Figure 12.25   *(a)*

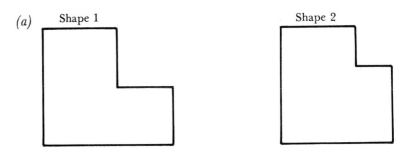

(b) Part C of figure 1 is larger than part D of figure 2, so we know that figure 1 is larger than figure 2.

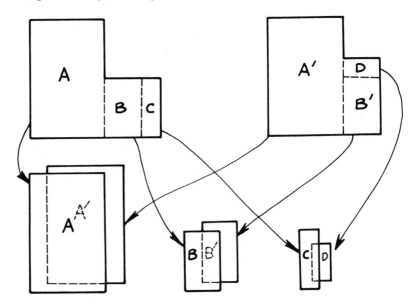

Figure 12.26

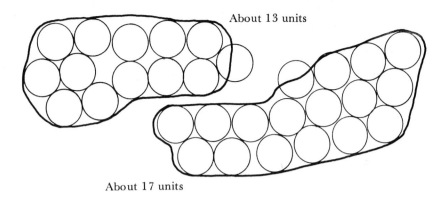

About 13 units

About 17 units

Figure 12.27

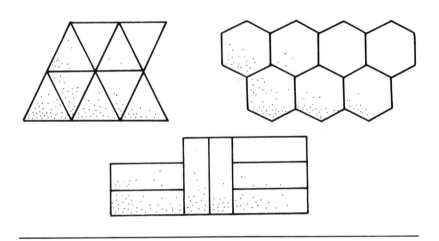

**Area:
Phase 2**

Developing Standard Units for Measuring Area

It can easily be seen that the circle is not satisfactory for determining precise area measurements. When one uses the circle there is space between the circles that is not being measured. So, it becomes obvious that shapes that will completely cover the region are more appropriate. Some examples of such shapes are triangles, hexagons, and rectangles, as shown in Figure 12.27.

Of course, the shape that becomes most useful for measuring area is the square. If the square unit has a side one inch long, then it is called a square inch (sq. in.). If the square unit has a side one meter long, then it is called a square meter (m^2). Other standard units of area are:

square foot (sq. ft.)
square yard (sq. yd.)
square centimeter (cm^2)
hectare (ha)
acre (a.)

**Activity File:
Standard Units of Area**

Activity 1: Squares
    Cut out squares that are 1 cm on a side. Choose a surface. Have a child cover the surface with the squares. How many cm²?

Activity 2: Tiles
    Cut out several cardboard squares 1 ft. on a side. Choose a surface. Have the child cover the surface with the cardboard tiles. How many sq. ft.?

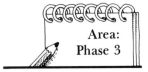

**Area:
Phase 3**

Developing Computation Formulas for Area
    One of the most interesting aspects of using area is that, in actual practice, area is almost never measured. Instead, some combination of lengths are measured and then a computation formula is used to compute the area. The next major phase of the study of area, then, consists of the development and use of area computation formulas for commonly encountered, special shapes.
    The easiest, and perhaps the most common, shape is the rectangle. The intent is to cover the region with square units. Begin by filling in one row of squares as in Figure 12.28. It can easily be shown that the number of squares in the first row equals the length of the rectangle (Figure 12.29).

Figure 12.28

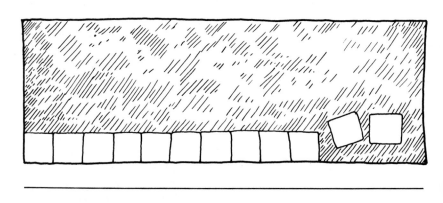

Figure 12.29

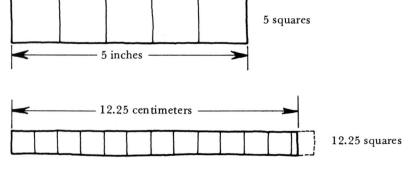

Figure 12.30

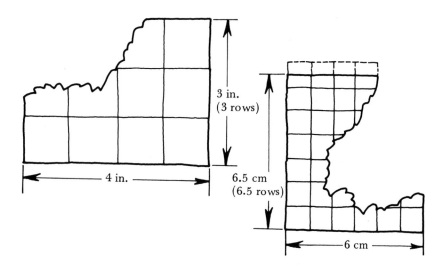

3 in.
(3 rows)

4 in.

6.5 cm
(6.5 rows)

6 cm

It is equally easy to show that the number of rows is the same as the width of the rectangle (Figure 12.30).

It follows that the number of squares it takes to cover a rectangle is equal to

(the number in the first row) x (the number of rows) or
(the length) x (the width)

The area formula for rectangles is, then,

$A = l \times w$

Of course, once children have been introduced to the terms *base* and *altitude*, an alternative formula, which is equivalent to the one given above, can be developed.

$A = bh$ (base of the rectangle times its height)

This latter formula for the area of a rectangle becomes the basis, in most school mathematics programs, for developing a computation formula for the area of a *parallelogram*. Remember that parts of a region can be rearranged without changing its total area. Start with *any* parallelogram—for example, one like that in Figure 12.31a. Cut a square corner (Figure 12.31b). Move the piece that was cut off to the opposite side (Figure 12.31c). Connect the two parts and the new shape is a rectangle (Figure 12.31d). The area of this rectangle is $A = bh$, where b is the base of the original parallelogram and h is its height. So, by multiplying the base of the parallelogram times its height, we find the area of a rectangle that happens to have the same area as the parallelogram.

Area of a parallelogram: $A = bh$

Figure 12.31

(a)

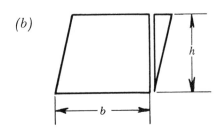

h (height)

b (base)

(b)

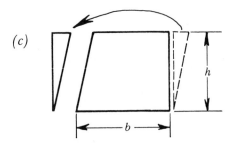

h

b

(c)

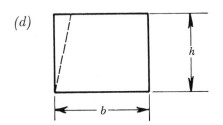

h

b

(d)

h

b

A similar technique is used to develop a formula for the area of a triangle. Start with *any* triangle (Figure 12.32a). Cut out a second identical (congruent) triangle (Figure 12.32b). If this second triangle is turned upside down (Figure 12.32c), and connected to the original triangle (Figure 12.32d), a parallelogram is formed whose area can be found (Figure 12.32e).

The area of the newly formed parallelogram is $A = bh$, where b is the base of the original triangle and h is its height. Of course, since the triangle area was doubled in forming the parallelogram, the area of the *original triangle* must be half that of the parallelogram. We multiply by one-half to find the triangle area computation formula.

Area of a triangle: $A = \dfrac{1}{2} bh$

Figure 12.32

(a)

(b)

(c)

(d)

(e)

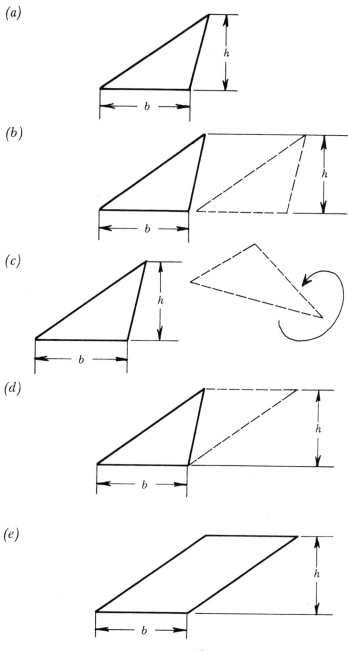

Figure 12.33

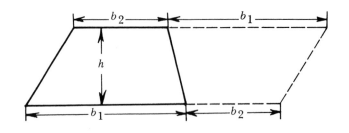

Figure 12.34   (a)

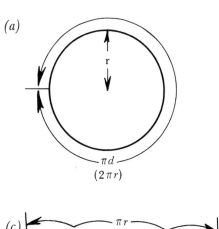

(b)

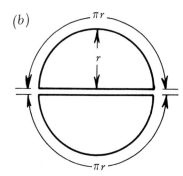

(c)

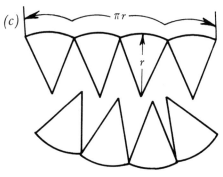

(d)

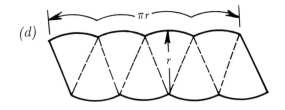

(e)

The technique for developing the area computation formula for a trapezoid is virtually identical. Double the figure to get a parallelogram and find half of the area of that parallelogram (Figure 12.33).

The area of the parallelogram is $A = (b_1 + b_2)h$, where b_1 and b_2 are the two bases of the trapezoid (the parallel sides) and h is the height of the trapezoid. Multiplying by one-half, we find the area of the original trapezoid.

Area of a trapezoid: $A = \dfrac{1}{2}(b_1 + b_2)h$

Procedures for discovering a computation formula for the area of a circle are not as straightforward. The same basic procedure, however, can be used to rearrange the shape into a figure for which the area can be found. Start with *any* circle (Figure 12.34a). Cut it in half (Figure 12.34b). Cut each half into four wedge-shaped pieces, as shown in Figure 12.34c). Fit the pieces together to get this shape, which is a little bit like a parallelogram (Figure 12.34d). If the pieces were cut small enough, the resulting shape would be so nearly a parallelogram that it would be difficult to tell the difference (Figure 12.34e). The base of this parallelogram would be πr, half the circumference of the original circle. Its height would be r. Its area, then, would be

$$A = (\pi r)(r) = \pi r^2$$

and this is the same as the area of the original circle.

Area of a circle: $A = \pi r^2$

Case Study: Greg

Greg is a fifth-grade child with no apparent disabilities who has been studying area computation formulas. He seemed to be doing well on the unit until he was asked by his teacher to complete a math lab activity.

Greg was given a card with a picture of a triangle and was told to measure the triangle and then compute the area. Using great care, he measured the sides of the triangle and recorded his findings (Figure 12.35).

Figure 12.35

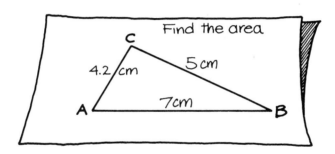

Greg then wrote the correct area formula $(A = \frac{1}{2}bh)$, but substituted 7 for b in the formula and 4.2 for h. Of course, his computation produced a wrong answer.

$$A = \frac{1}{2}bh$$

$$A = \frac{1}{2} \cdot 7 \cdot 4.2$$

$$A = 14.7$$

Greg's error surprised his teacher since he had completed all previous exercises correctly. However, on reviewing the examples to which Greg had been exposed, she discovered that up to this point either the base and height of the triangle had been identified or one of the "bottom" angles in the picture had been a right angle as in the examples shown in Figure 12.36.

Greg had been following an incorrect rule, which so far had produced correct answers. "The bottom side is the base and the shorter of the 'up and down' sides is the height."

Greg's teacher decided that she needed to do two things with him to correct his problem.

1. Discuss with him the relationship between the base and height of a figure—namely, that they are always perpendicular (Figure 12.37).

2. Work through a wide variety of exercises. The exercises should include both right triangles and nonright triangles. She would have Greg choose any side (not always the bottom side) to be the base and then identify the height to go with that base.

Figure 12.36

Figure 12.37

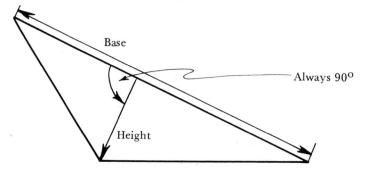

**Activity File:
Developing Area
Formulas**

Activity 1: Cut Up
   Give the children shapes cut from art paper. Have them cut up the shapes and arrange the pieces into rectangular shapes.

Activity 2: Tipsy Parallelograms
   Draw a parallelogram on squared paper. Have the child draw two different parallelograms on the paper with the base the same length and the height the same as in Figure 12.38. Compare the areas.

Activity 3: Posters
   Have children make posters to show how to:

1. Cut and rearrange parallelograms to get rectangles

2. Double a triangle to get a parallelogram

3. Double a trapezoid to get a parallelogram

4. Cut and rearrange a circle to get a parallelogram

Figure 12.38

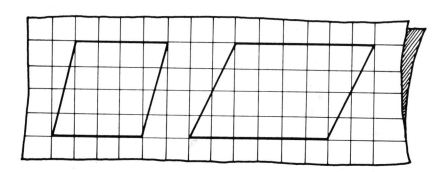

Volume
   While informal volume activities may be presented earlier, this study is ordinarily carried out in the upper-intermediate or middle-school grades. One question that commonly comes up is "What's the difference between capacity and volume?" The answer in theory is nothing; however, in practice, the term *capacity* is usually used to refer to how full a container is or to refer to a measurement of liquid volume. Volume is used in situations when we are interested in giving the number of units a solid figure can be cut into. We ordinarily express volume in cubic units such as cm^3 or cu. in.
   As with area, the study of volume falls into two major parts—the measurement of volume, and the development and use of volume computation formulas. Ideas follow for developing the topic with children.

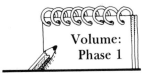

Volume:
Phase 1

Developing the Concepts of Volume and Unit of Volume

Comparison of solid figures is no simple matter to the child. With three given solid figures, one may be longer, a second one wider, a third taller. Comparing volumes involves consideration of all these factors along with the shape of the figure.

If two figures have the same shape (are similar), the comparison is easy. By comparing the lengths, the widths, the heights, or any single dimension, one can tell which figure is greater (Figure 12.39). If, however, the shapes are different, as in Figure 12.40, direct comparison of volume becomes very difficult. In order to compare two different-shaped solid figures, it is usually necessary to reduce them to more comparable shapes, either wholly or in parts. For example, consider two solid figures made of pliable clay like those pictured in Figure 12.41. By changing the shape of each to a sphere, a child can make the comparison easily. Or, suppose a child had two clay figures like those in Figure 12.42a. If the figures could be cut into cubes (all the same size), it would be possible to count the number of cubes contained in each figure and then make the comparison easily (Figure 12.42b).

Figure 12.39

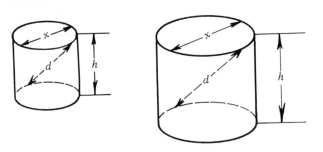

Figure 12.40

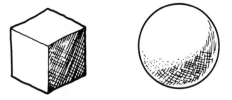

Figure 12.41

In the process of cutting into cubes the small cubes are used as units for measuring volume. Notice that a number is assigned to each shape—the number of cubes. The sizes of the figures are then compared by comparing these numbers. Of course, the shape of the unit need not be a cube, but for most solid figures, the cube seems to be the most convenient shape for measuring volume as well as the easiest shape to use as a unit of volume.

Once the shape of the volume measurement unit is established, an easy way to think of volume is to consider the question, "How many units are needed to *fill* the shape?" Children can generally get an intuitive grasp of the concept if they actually fill boxes with cubes. This is especially true if you can construct boxes with interesting shapes for them to use (Figure 12.43).

Figure 12.42   (a)

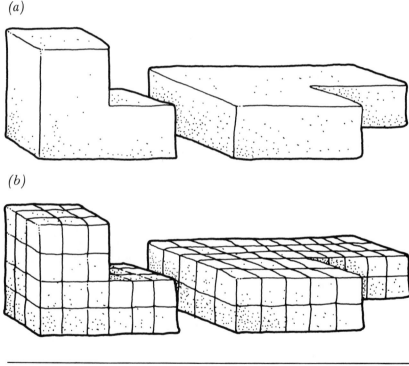

(b)

Figure 12.43

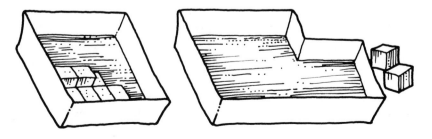

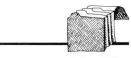

Activity 1: Having A Ball
Form odd-shaped solid figures with clay. Have children place them in order from smallest to largest by looking and guessing. Children can then roll the shapes into balls and compare the size of the balls to see if they were ordered properly.

Activity 2: Dip And Drip
Let children compare shapes by submerging them in water to see which figure displaces more water.

Activity 3: Sticking Together
One child constructs several shapes by gluing cubes together, keeping track of the number of cubes used for each shape. A second child arranges the shapes from smallest to largest, then checks with the first child for accuracy.

**Volume:
Phase 2**

Developing Standard Units for Measuring Volume

Children should have experiences with cubic units from both the English and the metric systems. If the unit cube has edges one inch in length it is called a cubic inch (cu. in.). If the edge is one centimeter long the unit is called a cubic centimeter (cm^3). Other standard units for measuring volume are:

cubic foot (cu. ft.)
cubic yard (cu. yd.)
cubic meter (m^3)

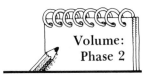

Activity 1: Shape Maker
Have children "copy" a given solid figure by gluing together inch-cubes (cubes that are 1 in. on a side).

Activity 2: Boxing
Construct a box that is 10 cm long, 10 cm wide, and 10 cm deep. Have the children fill the box with centimeter cubes to find its volume.

**Volume:
Phase 3**

Developing Computation Formulas for Volume

In real life, volume, like area, is seldom measured. Rather, some combination of lengths is measured and used in a computation formula to compute the volume. The majority of the time spent in the study of volume is spent in the development and use of those volume formulas.

There are two major groups of solid figures for which we develop volume formulas. They are the prism-like figures and the pyramid-like figures. The first group, the prism-like figures, include both prisms and cylinders (Figure 12.44).

To determine how many standard-sized cubes are needed to fill a figure, *two questions must be answered.* First, *How many cubes are needed for the first layer?* The number of cubes in the first layer is, of course, the same as the area of the base of the figure (Figure 12.45). Since the figure is the same size and shape from bottom to top (base to base) every layer of cubes will be the same. Now the second question can be addressed: *How many of these layers are needed?* If the unit cubes are cm^3, and if the height of the figure is 6 cm, then it will take six layers of cubes to fill the figure. The number of layers needed is the same as the height of the figure, as illustrated in Figure 12.46. The total number of unit cubes needed to fill a prism-like figure is

(The number of cubes in the first layer) x (The number of layers)
or (The area of the base of the figure) x (The height of the figure)

Figure 12.44

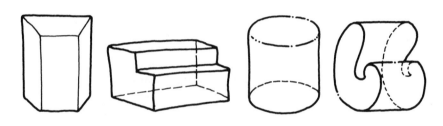

Figure 12.45

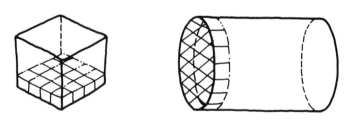

Figure 12.46

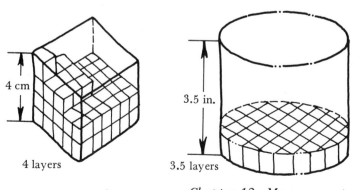

4 cm

4 layers

3.5 in.

3.5 layers

If we let B represent the area of the base and h the height, then the formula for the volume of a prism-like figure is:

Volume of a Prism or Cylinder: $V = Bh$

The second major group of solid figures for which volume formulas are developed are the pyramid-like figures which include both pyramids and cones (Figure 12.47).

Children can be asked to examine a pair of solids—one a prism, the other a pyramid—that have the same height and congruent bases. An example is shown in Figure 12.48a. If one were to fill the pyramid with water and then pour that water into the prism, the prism would be one-third full (Figure 12.48b).

The same result would be obtained regardless of the shape of the base of the prism. Comparable results are obtained when cones and cylinders (same height, congruent bases) are used (Figure 12.49).

The volume of the pyramid (or cone) is one-third that of the prism (or cylinder).

(Volume of pyramid-like figure) = 1/3 (Volume of prism-like figure with same base and same height

It follows, then, that to find the volume of any pyramid-like figure (pyramids and cones) we multiply 1/3 times the area of the base times the height.

Volume of a pyramid or cone: $V = 1/3\ Bh$

Figure 12.47

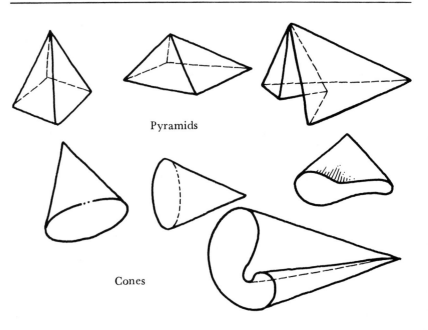

Pyramids

Cones

Figure 12.48   (a)   (b)

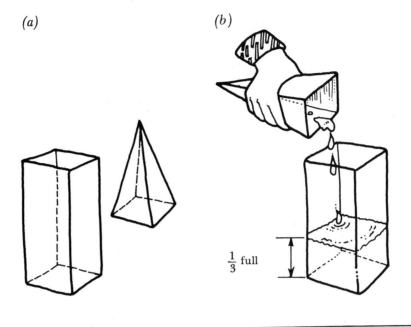

$\frac{1}{3}$ full

Figure 12.49

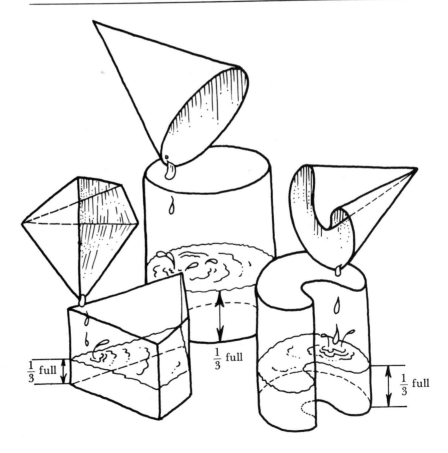

$\frac{1}{3}$ full $\frac{1}{3}$ full $\frac{1}{3}$ full

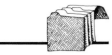

**Activity File:
Developing and Using
Volume Formulas**

Activity 1: Boxing
   Have children make a box that will contain sixty inch-cubes.

Activity 2: Stacks
   Have children make a stack of rods from a set of graduated centimeter rods. Challenge them to figure out how many centimeter cubes it would take to make the same size stack (Figure 12.50).

Activity 3: Using Formulas
   Give children opportunities to measure boxes or blocks and compute their volume.

Figure 12.50

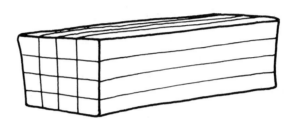

Angle Measurement
   Most school mathematics texts present angle measurement in the upper or middle-school grades. Two major phases of instruction on the topic are described next. In practice, the second is relatively easy for teachers to carry out if the first has been adequately done.

**Angles:
Phase 1**

Developing the Concept of Angle Measurement
   There are a number of ways to define and think about angles. Of these, two seem to be most useful to the elementary child.

Definition 1: An angle is the union of two rays with a common endpoint (Figure 12.51).

Two rays: $\overrightarrow{BA}$ and $\overrightarrow{BC}$
Common endpoint: B
Resulting angle: $\angle$ ABC

Figure 12.51

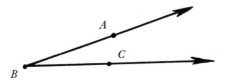

Figure 12.52

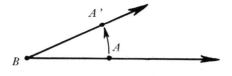

Figure 12.53

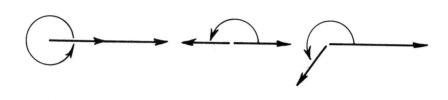

Figure 12.54

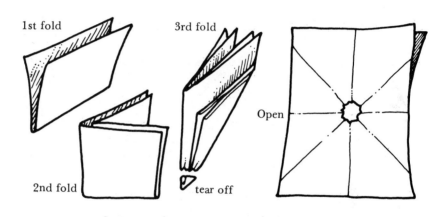

1st fold

3rd fold

2nd fold

tear off

Open

Definition 2: An angle is the amount of rotation of a ray about its endpoint. One side of the angle is the initial position. The other side is the final position (Figure 12.52).

Initial position: $\overrightarrow{BA}$
Final position: $\overrightarrow{BA'}$
Resulting rotation equal to: $\angle ABA'$

Definition 1 has particular value to the child in that it helps eliminate the tendency to think that lengthening the sides will make a bigger angle; of course, definition 2 has that same advantage since it is expressed in terms of rays other than line segments.

Using the second definition, children can be guided to an intuitive grasp of notions like "one full turn," "one-half turn," or "two-thirds of a turn." (See Figure 12.53.)

By folding a sheet of paper, the child can develop an effective tool for measuring eighths of a turn (Figure 12.54). Now, to measure an angle, extend the sides and place the paper over the angle

(Figure 12.55). Precision is gained with an additional fold, since it is then possible to measure angles to the nearest sixteenth of a turn (Figure 12.56).

To gain greater precision, more folds are needed in the paper. However, there is a limit to the number of folds that can be made in a sheet of paper, and folding accuracy usually leaves much to be desired. Also, like other kinds of measurement, there are commonly agreed-on standard units for measuring angles. They are the *degree* (1/360 of a turn), the *minute* (1/21600 of a turn), the *second* (1/1296000 of a turn), and the *radian* (which is normally not used until high school). For elementary school purposes, the degree yields sufficiently precise measurements and protractors marked in degrees are inexpensive enough that every child can usually be provided one. Measuring an angle with a protractor is illustrated in Figure 12.57; the angle measures 35 degrees.

Figure 12.55   Extend the sides so they can be seen beyond the edges of the paper. The angle measures about 1/8 of a turn.

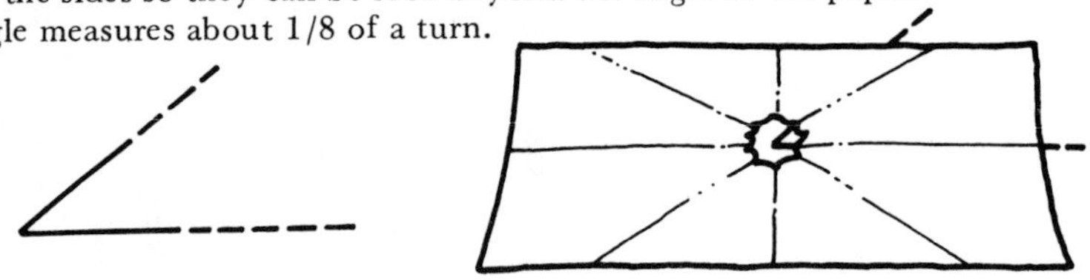

Figure 12.56

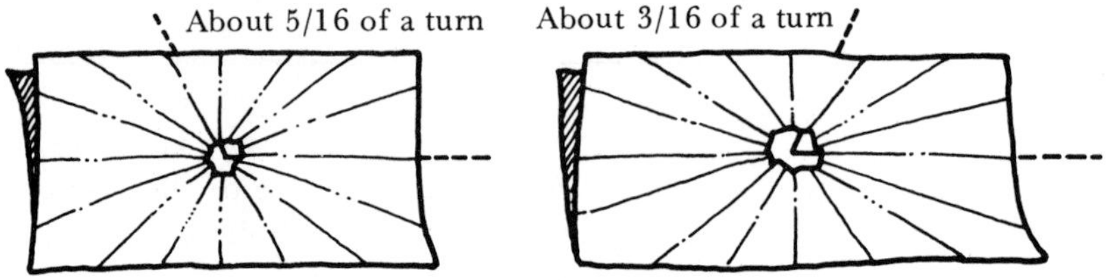

About 5/16 of a turn   About 3/16 of a turn

Figure 12.57

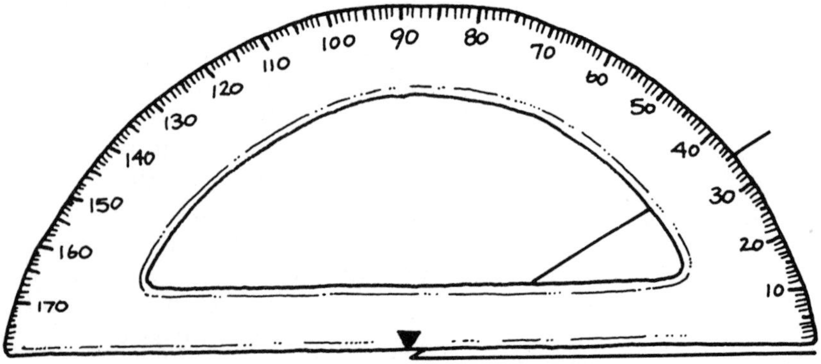

**Activity File:
Angles and Angle
Measurement**

Activity 1: Stringer
    Fasten both ends of a string to the ground or floor with the string stretched out tight. Fasten one end of a second string at one end of the first (Figure 12.58a). Move the second string to show half a turn, 3/4 of a turn, etc. (Figure 12.58b).

Activity 2: Take A Turn
    Have the child point with one extended arm, then rotate to a new position to show 1/4 turn, 1/8 turn, 3/8 turn, and so on.

Activity 3: Folding
    Have children fold paper as described earlier to measure an angle.

Activity 4: Compare
    Show two angles (by drawing, by pointing, using string, or the like). Have children decide which angle has the most "turn" (is bigger).

Figure 12.58   *(a)*

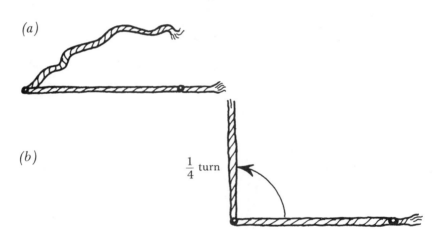

(b)

$\frac{1}{4}$ turn

**Angles:
Phase 2**

Developing Measuring Skill
    Children need to become familiar with protractors and their use. This is best accomplished by having them measure many angles. Care should be taken that the angles are shown in various positions— *not* all with one side on a horizontal line. Pictured in Figure 12.59 are three types of protractors that might be available to the child. (The American Printing House for the Blind carries braille protractors if these are needed.) Children should use a protractor to carry out angle measurements like those suggested in the following activities.

Figure 12.59

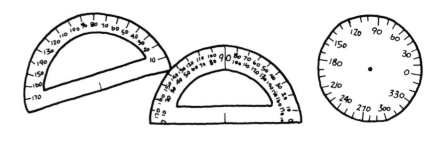

Figure 12.60

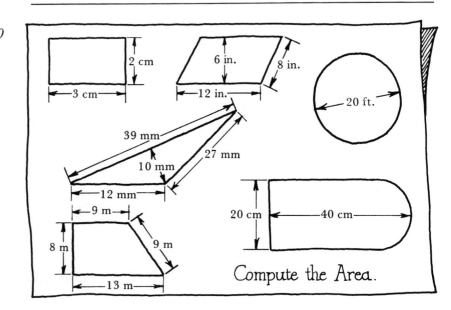

Compute the Area.

Figure 12.61

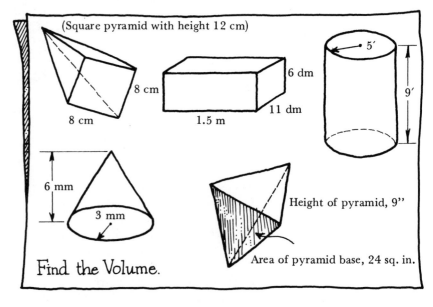

(Square pyramid with height 12 cm)

Height of pyramid, 9"

Area of pyramid base, 24 sq. in.

Find the Volume.

Activity 1: Draw And Measure

One child draws an angle; a second child measures it; the first child checks it.

Activity 2: Cut And Measure

Cut out of tagboard a variety of polygon shapes. Measure the angles at all the corners.

Activity 3: Measure

Measure many angles: on paper, in the room, outside, at home, wherever they are found.

Your Time: Activities, Exercises, and Investigations

1. Prepare, as your instructor directs, answers to the major questions on p. 441.

2. Find the area of the ceiling of the room.

3. Choose any can and find its surface area.

4. If the perimeter of a circle is 6.28 cm, find its area.

5. Compute the area of each of the figures in Figure 12.60.

6. Create a *sequence of related activities and written follow-up exercises* which lead a child through the development of an area formula into practical applications of the formula. Be sure to specify the basic figure you are considering in your instructional sequence.

7. How many centimeter cubes would it take to fill this room? Describe your procedure for answering this question.

8. Find the volume of each of the figures in Figure 12.61.

9. Assume children know the prism-cylinder formula. Outline a *sequence of related activities* that could be used to introduce children to the relationship in
   (a) The prism-pyramid formulas, *or*
   (b) The cylinder-cone formulas (which includes, as well, practical applications of the new formula developed in the sequence)

10. Use the paper "angle measurer" described in the preceding section to:
    (a) Draw a 45° angle (angle that is 1/8 of a turn)
    (b) Draw a triangle with *two* 45° angles. What do you find when you measure the lengths of the triangle sides?

11. How would you instruct a child to fold the angle measurer to obtain a 30° angle? A 60° angle?

12. Can you use the angle measurer to help in drawing:
    (a) A right triangle with one 30° angle?
    (b) A right triangle with one 60° angle?
    Compare the two triangles. What conclusions can you draw?

13 GEOMETRY: GUIDE TO PERCEPTUAL MOTOR LEARNING

Because of the tangible, activity-oriented nature of school geometry today, this area of mathematics is an exciting and highly motivating one for children. For special needs students, geometry
assumes a particularly important role, one rich in its potential to
reinforce and build perceptual-motor skills. Too often mathematics is limited to computational basics, and this potential is
neglected.

Many special needs children have perceptual handicaps that
hamper their success in mathematics. These imperceptions occur
in two major areas: seeing and hearing. Marianne Frostig (1972)
notes in particular the high correlation of *visual* perception problems to growth and achievement in mathematics. Because an
understanding of visually perceived relationships is essential to so
many phases of mathematics learning, Frostig feels that visual perceptual difficulties present even greater problems in mathematics
than in the reading curriculum.

Children with visual imperceptions, but with normal vision,
do not interpret the things they see as most people do. Inversions,
rotations, and other changes of symbols, signs, and words can
occur. Some children, for example, confuse the x and + signs;
others cannot distinguish between the =, ÷, and − signs. The
digits 6 and 9, as well as 3, 5, and 8, may also be confused. Consistent number reversals, such as 42 for 24, are quite common.
Fractions are often disoriented so that the numerator is not always
on the top; 1/2 may be interpreted as 112. A square may be perceived as four unrelated lines rather than one compact whole, or
may otherwise be distorted in size or disoriented in space. The floating, merging, or dislocation of letter, symbol, and shape parts is a
common phenomenon in the child's world of visual imperception.

Such disturbances also influence a child's success with number
sequencing, map and graph reading, perception of space and size
relationships with concrete materials and diagrams, and proper
spacing and alignment of numerals for computation. *Separate*

from the questions of learning mathematical concepts is the need for awareness of the importance of physical position, juxtaposition, and relatedness, in order to be accurate in computation.

Children with visual perception problems may appear inattentive and disorganized, when in reality they are just unable to control the shift in their focus of attention from one stimulus to another on a mathematics worksheet or textbook page. Often they cannot distinguish foreground from background in an illustration. They may be unable to find or keep their place on a page, or within a complicated computational problem. A major difficulty for many is that they have no conception of the relationship between their own bodies and the space around them. Consequently, they may attach little meaning to directional terms such as *up, down, between, before,* or *next to.* They may have poor visual-motor coordination that causes them to be awkward in their general movement patterns.

For many children, visual perceptual and visual-motor deficiencies interfere with progress in mathematics. Direct remedies involving the use of grids, masks, and color cues to delineate and focus, to maintain placement, position, and attention, are necessary. Such techniques, mentioned in several Trouble Shooting sections throughout this book, help prevent unfortunate errors not associated with *conceptual* understandings in mathematics. In addition to these approaches, however, long-range efforts to reinforce and build perceptions are necessary. A major premise of this chapter is that perceptual awareness can be enriched through carefully structured geometric activities.

This chapter examines the major thrusts of both three- and two-dimensional elementary school geometry. The approach taken is a hands-on, *sense*-ible one, with first steps taken in the child's environment of familiar, concrete objects. A child's study of three-dimensional shapes and their naming and description naturally divides itself into two phases: (1) first explorations, and (2) a closer look at solid shapes and their properties. As space size is reduced from three to two dimensions, we discuss the two-phase development of many plane geometry programs: (1) shape identification, and (2) basic properties of two-dimensional shapes.

Throughout a child's study of geometry, important characteristics of shapes gradually take on more precise meaning:

1. Straightness

2. Parallelism

3. Perpendicularity

4. Similarity

5. Congruence

6. Symmetry

These properties, basic to classifying and naming shapes, are useful and interesting topics of study in themselves.

In addition to the mathematical thrust, we emphasize a dual focus on motor activity and on describing what is seen or felt in many activities of the chapter. *We view this focus as a necessary aid to the perceptual-motor development that can be fostered through geometry.* When children act on objects physically, their mental images or perceptions become stronger. Carefully planned geometric activities such as those suggested can gradually lead children toward higher levels of geometric understanding and toward greater perceptual-motor strength.

What Do *You* Say?

1. Which shapes should be introduced first: three-dimensional or two-dimensional? Give reasons to support your thinking.

2. What three-dimensional shapes are commonly dealt with in the elementary school curriculum today?

3. How can important geometric properties such as congruence, parallelism, perpendicularity, similarity, and symmetry be highlighted in the study of three-dimensional shapes?

4. What role does body movement play in the simultaneous development of geometric concepts and perceptual-motor skills?

5. What types of geometric activities for three-dimensional shapes can be used to reinforce or build perceptual-motor skills?

6. What techniques can be used to increase transfer from the perception of three-dimensional objects to the perception of two-dimensional shapes?

FIRST EXPLORATIONS

Naming and Describing Solid Shapes: Phase 1

Moving, feeling, seeing, describing—this is the heart of a geometry program which brings children to *sense* and experience shape. A child's world is rich in geometry, so it is only natural that first explorations or space should involve familiar three-dimensional shapes. Both at home and at school a young child plays with blocks, boxes, cones, balls, and cans. Many school programs begin geometry informally by having children examine common shapes from their environment. Children manipulate and study them, use simple language to name and describe them, and gradually learn to identify each basic shape.

Sorting activities with "junk box" objects pave the way for more precise classifications. At first fewer and simpler solids are analyzed, and exact mathematical names are not required. Many of the "look-alike" sorting rules observed or created by children for the more familiar objects can be applied later to the solids themselves. Observations commonly suggested by children are presented in Figure 13.1.

Mathematically, many important notions emerge from these informal descriptions. As children group and describe solids on the basis of general characteristics, those properties or characteristics take on greater meaning. This activity also promotes a closer examination and comparison of individual shapes. Notions of straight or flat versus curved, same size and shape, parallelism and perpendicularity are grounded intuitively in these simple experiences.

Figure 13.1

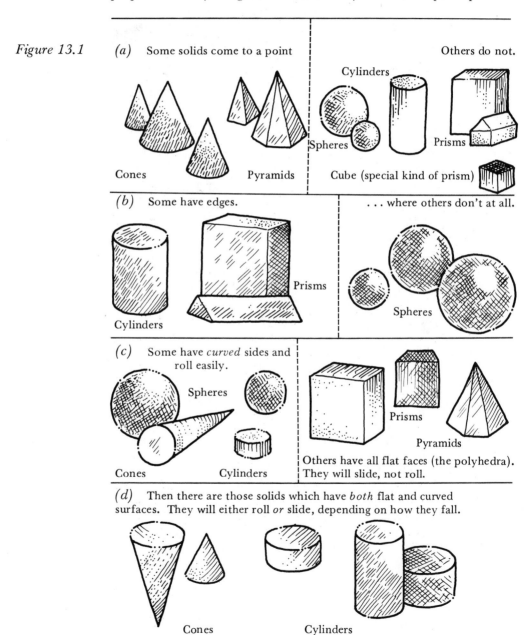

(a)   Some solids come to a point   Others do not.

Cones   Pyramids   Cylinders
               Spheres   Prisms
                                 Cube (special kind of prism)

(b)   Some have edges.   . . . where others don't at all.

Cylinders   Prisms   Spheres

(c)   Some have *curved* sides and roll easily.

Spheres
Cones   Cylinders   Prisms   Pyramids

Others have all flat faces (the polyhedra). They will slide, not roll.

(d)   Then there are those solids which have *both* flat and curved surfaces. They will either roll *or* slide, depending on how they fall.

Cones   Cylinders

In this early phase, a primary emphasis is that of vocabulary development. Flat and curved; large and small; round and straight; tall and short; and other descriptors can be taught as needed to help children who have difficulty discriminating or talking about visual differences in shapes. Often children's thoughts become more clearly organized when they verbalize what is experienced. At times it may be helpful to have children close their eyes, feel, and then describe a solid to better create a mental image of its general shape. Color and tactile cues can also be used. For example, one could cover all *flat* faces with yellow paper; or use felt to cover pyramid faces which *come to a point* (meet at the apex). Techniques such as these at the gross visual-motor level can be adapted later to help children compensate for visual imperceptions in more refined mathematical tasks. Several of the Case Study and Trouble Shooting sections of other chapters offer specific suggestions in this regard.

Movement is both fun for young children and necessary to their development. Learning control of movement at the gross motor level can be coordinated with informal geometry experiences that focus on naming and describing solid shapes. Such experiences can also build or reinforce two other important concepts:

1. *Laterality*, by which children "feel" the difference between the two sides of their bodies

2. *Directionality*, by which children project this feeling outward to recognize left and right of objects and other related concepts such as up, down, near, between, and so on.

Geometry activities such as those below are designed for young primary children. The emphasis on body movement and active learning, illustrated in the activities, can be initiated early and continued throughout the elementary school geometry program.

**Activity File:
Naming and Describing
Solids: 1**

Activity 1: Body Space (For large or small groups)

Focus: Vocabulary to name or describe solid shapes; vocabulary to describe spatial relationships.

Give each child a cube (or other solid shape) and have the children space themselves so that, when standing still with arms outstretched, they cannot touch. Have the children identify all the space their bodies can reach from this spot—in front, behind, to either side. This is their body space.

Have children feel and describe the cube:

- Flat faces

- Straight edges

- Six faces

- Eight corners

- Squares for faces

Have children use the cube to dramatize the vocabulary of body to object and other spatial relationships. Direct children to move the cube:

- Right to left and back again (across the body midline)

- Up high, down low

- On, over, around

- Across, under, near, far

- Beside, between

Example: "Set the cube down by your *right* toe. Now tiptoe *around* it. Now hold it high, *above* your head." Have children learn movements, large and small, by imitating what you do. Repeat similar directions using other shapes.

Activity 2: Spot It   (For individuals or small groups)
Focus: Vocabulary to name or describe solid shapes; vocabulary to describe spatial relationships.
Materials: Two mats such as that illustrated in Figure 13.2. Place three-dimensional shapes on one of the mats, leaving two or three spots empty.

Activity 3: Pattern It   (For individuals or small groups)
Focus: Vocabulary to name or describe solid shapes. Start a pattern using solids and challenge children to extend it   (Figure 13.3). Encourage children to describe the pattern. That in the

Figure 13.2

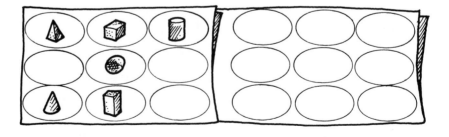

Figure 13.3   *Example:*

preceding example might be described as: "cube, ball, cube, ball," or "flat, curved, flat, curved . . ." (referring to surfaces of the solids).

Variation: To reinforce or build visual memory skills, have children study the pattern, then turn their backs on it while reproducing it.

A CLOSER LOOK: SOLID SHAPES AND THEIR PROPERTIES

Naming and Describing Solid Shapes: Phase 2

As children mature, so does the language used to identify and describe shapes, as well as the sophistication of the properties explored. The ball figure becomes a *sphere*; the box a *prism*; the tin can a *cylinder*; and *vertices* replace corners in a child's description of shapes. Primary teachers who have used the proper names for more common solids, at least part of the time, have informally paved the way for this language development. Solids which are commonly examined in the elementary-school geometry program are pictured in Figure 13.4

Children also learn to be more precise in classifying and describing solids. Not all cubes are *congruent*. They are *similar*, however. Prisms have bases which lie in *parallel* planes. Those bases are congruent to each other. Students learn which solids have height *perpendicular* to the base. In some classrooms children slice clay models of solids to better study patterns of *symmetry* or the shape of resulting sections. Mathematically, children grow in their understanding and use of important geometric concepts. The following activities are typical of those that can be used to nurture this growth.

Figure 13.4   Prisms (Have two faces which are congruent and parallel to each other)

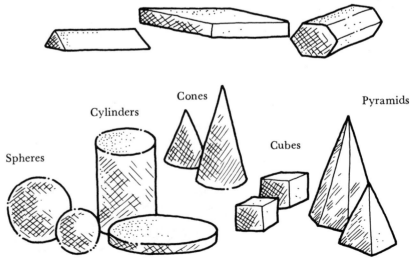

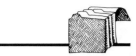

**Activity File:
Naming and Describing
Solids: 2**

Activity 1: Pattern It (For individuals or small groups—an advanced variation of the preceding Activity 3)

Focus: Vocabulary to name or describe solid shapes. Start a pattern using solids and challenge children to describe and extend it (Figure 13.5).

Note: The next three activities require children to recognize and name three-dimensional shapes, regardless of perceptual variations in size, position, or color.

Activity 2: You Name It (For small or large groups)

Focus: Identify a solid shape from verbal description of its basic properties.

Give each child one solid shape. Decide on a solid and describe its general characteristics. Anyone who has the shape stands, shows, and names the shape.

Activity 3: Odd Ball (For small or large groups)

Focus: Identify a solid which is basically different from others in a given group.

Display several solids and invite children to examine them to find the Odd Ball—the one shape that doesn't belong. For instance, a pyramid could be displayed along with prisms of different sizes and colors. Children should be required each time to give the reason for their selection.

Activity 4: How Alike, How Different (For small or large groups)

Focus: Select from a larger group two solids which are alike/ different in some way.

Place several solids along the chalk tray. Include, if possible, size and color variations of each shape. Select one solid and call on children individually to find and describe another which is like

Figure 13.5   *Example:*

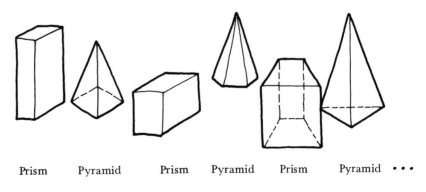

Prism   Pyramid   Prism   Pyramid   Prism   Pyramid   • • •

Chapter 13   Geometry: Guide to Perceptual Motor Learning   471

(or different from) it in some way. The challenge is to get children to classify solids in many different ways. Sample dialogues follow. The first focuses on choosing solids that are like each other in at least one way. In the second, the children in the photograph were asked to find, name, and describe solids which were different from that selected in some way.

Teacher selects: a Cylinder

Child 1: "I choose the ball. It is curved. So is your cylinder."
Child 2: "This cone has a curved edge. So does the cylinder."
Child 3: "Your cylinder has two bases. So does my prism."

Teacher selects: a Cone

Child 1: "The height of your cone is perpendicular to its base. That's not true for my cone."
Child 2: "Your cone has only one base; this cylinder has two."

Activity 5: I Spy! (For small or large groups)
   Focus: Recognize a solid shape in its environmental surroundings. Have children match solids to examples inside and outside of school. Field trips are ideal for this purpose. Children should be encouraged to use proper names when identifying the solids (e.g.,

sphere rather than "ball"). Challenge them to describe as many properties of each shape as possible. Things children might see on a field trip or neighborhood walk are pictured in Figure 13.6.

Activity 6: Choose   (For small or large groups)

Focus: Discriminate among shapes by correctly selecting a given solid from a larger group.

Ask children to find one shape from a box which is mixed with several others. Color and tactile cues, such as those suggested in an earlier section, may be used as needed. Children should be encouraged to identify properties characteristic of the shape they locate.

Figure 13.6

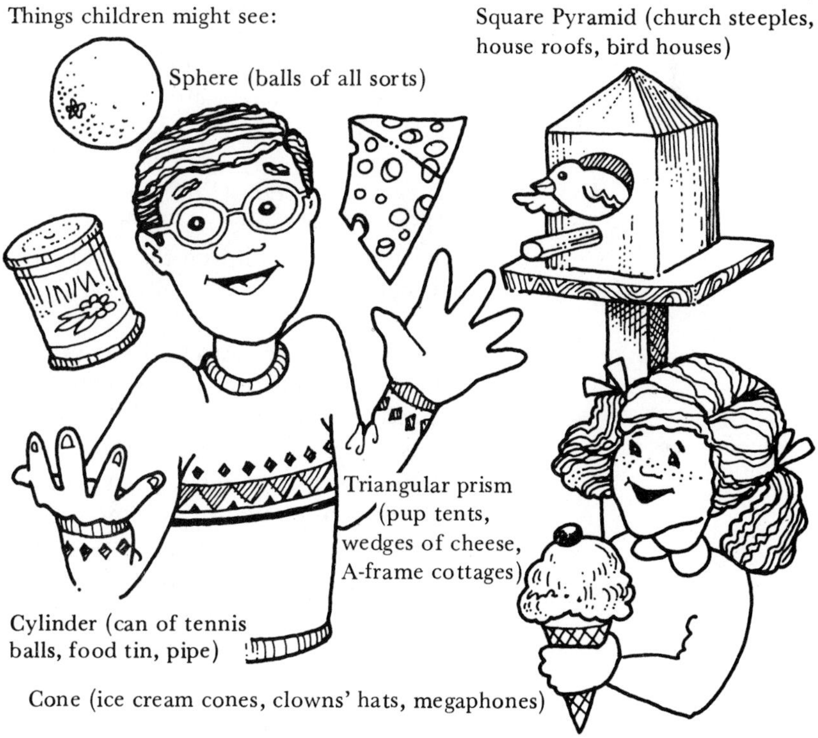

Things children might see:

Square Pyramid (church steeples, house roofs, bird houses)

Sphere (balls of all sorts)

Triangular prism (pup tents, wedges of cheese, A-frame cottages)

Cylinder (can of tennis balls, food tin, pipe)

Cone (ice cream cones, clowns' hats, megaphones)

GEOMETRY OF TWO DIMENSIONS

Trouble Shooting

Two separate issues are discussed in this section. The first concerns children who have difficulty recognizing pictures of three-dimensional shapes, even when they are of familiar objects. The second deals with the transition from perception of solids to perception of the shapes of the *faces* of the solids, which marks the beginning of plane or two-dimensional geometry in many school programs today.

Relating Solids to Two-Dimensional Pictures

Some children have difficulty relating pictures of objects to the objects themselves. Several psychologists claim that recognizing an object from its two-dimensional drawing must be totally learned (Konkle, 1974). Activities like the following are useful in developing this skill. They help children achieve the ability to translate the visual image of a three-dimensional object to a two-dimensional plane, and vice versa. Teachers make a valuable contribution toward building perceptual skills when they include them in their geometry program.

Activity File:
Recognizing Pictures
of Solids

Activity 1: In Its Place (An activity for one and all)
Tape or tack a drawing of each three-dimensional shape to its storage space in a box or on a shelf in the classroom. Instruct children to put each solid "in its place," on or near its picture, as an activity in itself and at the end of each session with space figures.

Activity 2: Match (For small or large groups)
Distribute shape cards such as those described in the preceding activity and ask children to find a block like that pictured on their card.

Faces of Solids. As children study characteristics of solids which make them "like" or "different from" other solids, they build a framework of both two- and three-dimensional concepts. A child who is given the opportunity to manipulate and explore with three-dimensional shapes will be in a position to relate more readily to differences in plane or two-dimensional shapes. While it is true that perception of three-dimensional objects does not necessarily carry over to perception of two-dimensional shapes, transfer can be increased by having children participate in activities such as those that follow.

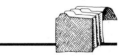

Activity File:
Identifying Faces
of Solids

Activity 1: Body Shapes (For individuals or groups)
It is not enough for perceptually or motor-impaired children to perceive shapes only with their eyes. Their whole bodies must be used to explore and experience. Have children curl their bodies round like the edge of a cylinder; or use arms, legs, or square shoulders to illustrate the corners they see on square and rectangular surfaces. Children will find they can make many body shapes using straight, curved, and angular combinations. While this general technique is a good one to use with most children, it is necessary for some.
Have children use movement patterns to trace the shapes of

solid surfaces. They can make these at different levels, high and low. They can make all sizes, large and small. They can walk, skip, or hop along the outline of basic shapes which appear as faces of solids. Encourage children to talk about shapes they make or trace.

Activity 2: Rope Games (For small groups)

Have children curve a rope to outline shapes they see on solids. They could also be directed to tie a knot in the rope for each vertex (corner) they count on a surface. By posting a child at each knot, "people shapes" can be formed.

Activity 3: Trace! (Group activity)

Have children fit the faces of solids into traced layouts or have them trace the faces onto colored paper. Encourage children to compare each other's drawings and search for likes and differences in shapes. Children can later make their own drawings and match them to shapes they trace.

Activity 4: Sort (For individuals or small groups)

Have children cut out the tracings they have made in Activity 3 and sort them into piles. Children can group all shapes with square corners, or those that have line symmetry. Many other rules can be generated for the Sort.

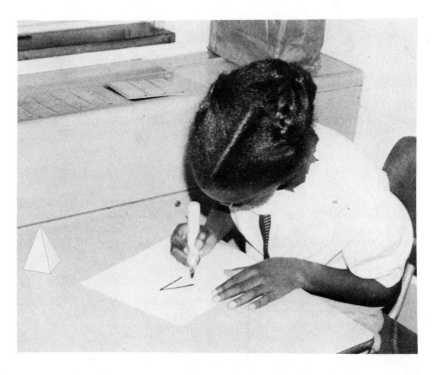

1. Prepare, as your instructor directs, answers to the major questions posed on page 466.

2. Which solids have:
   (a) Edges but no vertices?   (c) Only one vertex?
   (b) No edges?   (d) Only curved edges?

3. Prisms, like pyramids, are named by the shapes of their bases. Prisms have two bases. They are in parallel planes and are congruent to each other. A procedure for drawing prisms is illustrated in Figure 13.7. Hidden edges are shown with dotted segments.
   (a) Draw a rectangular prism (shoebox shape, with rectangles as bases).
   (b) Draw an octagonal prism (octagons, eight-sided figures, as bases).
   Every pyramid has one base (Figure 13.8). Line segments join every vertex of the base to a vertex not in the plane of the base (the apex). Sketch the following pyramids. Start with the base. Then draw the apex and the line segments joining it to the base.
   (a) Hexagonal pyramid (hexagon, a six-sided polygon as base).
   (b) Triangular pyramid (triangle, a three-sided polygon as base).

4. *Likes and differences.* Describe two ways that:
   (a) Pyramids and prisms differ
   (b) Cones and cylinders are alike
   (c) Cones and pyramids differ
   (d) Cones and pyramids are alike
   (e) Cones and cylinders differ
   (f) Pyramids and prisms are alike

5. *Name it.*
   (a) It has six faces, twelve edges, and eight vertices. The faces are rectangular regions. Every vertex is the intersection of three edges. There are three sets of four congruent edges.
   (b) All the faces are square regions of the same size. Every vertex joins exactly three edges.
   (c) Eight triangular faces and one octagon together form the surfaces of this polyhedron. In all, there are nine vertices.
   (d) There are only two flat surfaces, and these are circular.

6. | Triangular Prism | Octahedron | Square Pyramid |

   These statements refer only to the three solids named above. Answer true or false for each statement.
   (a) The triangular prism and square pyramid have the same number of faces.
   (b) The triangular prism and the square pyramid have the same number of edges.
   (c) The octahedron has the most faces.
   (d) The octahedron has the most edges.
   (e) The octahedron has the most vertices.
   (f) Each of the three solids is a polyhedron.
   (g) All three solids are regular polyhedra.

7. *Generating solids.* One can rotate or slide polygons to generate solids (Figure 13.9).
   (a) What solid is generated by sliding the rectangle along the straw and perpendicular to it? As in Figure 13.10, start at Point C and stop at Point D.
   (b) What solid is generated by rotating the triangle one full turn about the straw? (See Figure 13.11.)

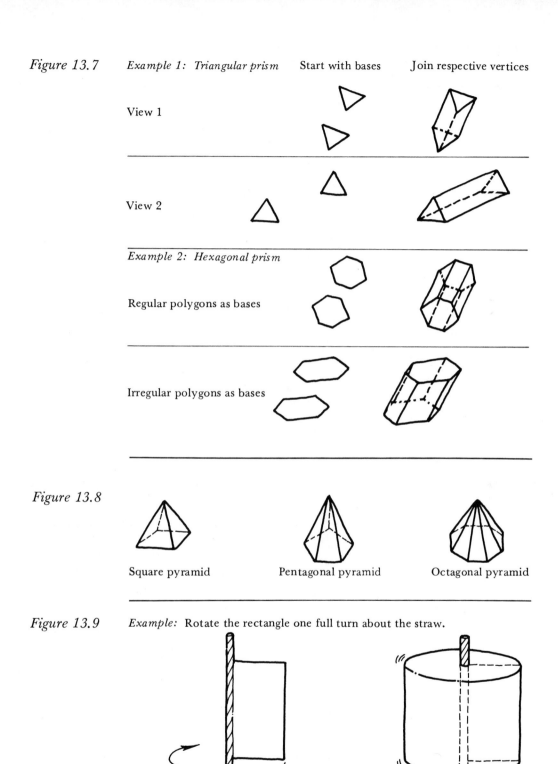

Figure 13.7   Example 1: Triangular prism   Start with bases   Join respective vertices

View 1

View 2

Example 2: Hexagonal prism

Regular polygons as bases

Irregular polygons as bases

Figure 13.8

Square pyramid   Pentagonal pyramid   Octagonal pyramid

Figure 13.9   Example: Rotate the rectangle one full turn about the straw.

The solid resulting from this sweeping motion: a cylinder

Chapter 13   Geometry: Guide to Perceptual Motor Learning   477

Figure 13.10

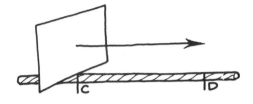

Figure 13.11

Figure 13.12   (a)

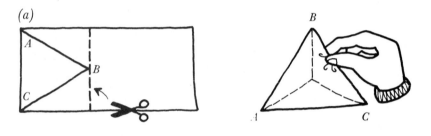

Figure 13.13

POLYHEDRON (All surfaces of the solid are FLAT)	NUMBER OF		
	Vertices	Faces	Edges
Cube	8	6	12
Triangular Prism	6	5	9
Triangular Pyramid	4	4	6

Figure 13.14

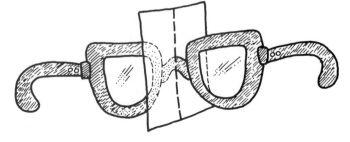

8. *Making a tetrahedron.* Seal the flap of an ordinary envelope. On one side of the envelope draw an equilateral triangle (similar to triangle ABC shown in Figure 13.12a), using the end edge as one side. Cut the envelope along the dotted line and pinch the end of the envelope to form a tetrahedron with ABC as one face (Figure 13.12b). How many other faces does the tetrahedron have? What kind of pyramid is the tetrahedron? (*Note:* Use this tetrahedron for problem 10 below.)

9. The table of Figure 13.13 lists the number of vertices, faces, and edges of several solid figures. Complete the table. Select any two polyhedra you like to fill the missing entries. Leonhard Euler, an eighteenth-century mathematician, proved a special relationship between the vertices, faces, and edges of polyhedra. Can you write a formula that states this relationship?

10. *Planes of symmetry.* Some solids can be "cut in half" so that the *two halves are mirror images of each other.* The plane along which the cut is made is called a *plane of symmetry.*
    (a) Many objects in our environment have at least one plane of symmetry: a tennis table, a pair of gloves, and eyeglasses, among others (Figure 13.14). Children with visual perception difficulties may have difficulty recognizing symmetry in figures. Name three objects commonly found in elementary school classrooms that have at least one plane of symmetry. Suggest ways you might help children perceive the symmetry of each. Ideas discussed in this chapter should help you. If possible, try your ideas with children.
    (b) The rectangular prism (Figure 13.15)
        (1) How many vertical planes of symmetry does it have?
        (2) How many horizontal planes of symmetry does it have?
        (3) What is its total number of planes of symmetry?
    (c) What is the total number of planes of symmetry for the solids pictured in Figure 13.16?
    (d) Examine a cube (Figure 13.17).
        (1) How many vertical planes of symmetry does it have?
        (2) How many diagonal planes of symmetry do you find?
        (3) How many horizontal planes of symmetry does it have?
            Note: A cube has nine planes of symmetry in all. Did you find them?
    (e) Use the tetrahedron from problem 6.
        (1) How many planes of symmetry are there for each edge of the tetrahedron?
        (2) Are there any planes of symmetry not passing through an edge?
        (3) How many planes of symmetry does a tetrahedron have?
    (f) Draw a picture of a solid that has no planes of symmetry.

11. Of all the solid figures mentioned in this chapter, which is the most common in our environment? Name the solid and give reasons to explain the frequency of its occurrence.

12. Effective programs of remediation often involve capitalizing on a student's strengths to deal with or help compensate for areas of weakness. This chapter discusses several ideas for capitalizing on strengths (e.g., auditory, haptic) for children with visual perception difficulties. Summarize these ideas and show, by example, how the techniques can be used to help an intermediate level child recognize and name basic three-dimensional shapes.

13. Geometry activities can be structured to reinforce and build eye-hand coordination and other motor capabilities. Children who succeed in activities requiring gross motor movement may gradually be given tasks requiring finer motor involvement. Design a *sequence of activities* requiring children to focus on likenesses or differences in given solids; let it begin with gross motor demands and gradually require finer motor movements.

14. *Think*, then tell what you think. Does a solid that has more faces and edges than another also have more vertices? Give an example, or reasons to support your thinking.

15. What is the value for regular class students of a geometry program such as that presented in this chapter?

Figure 13.15

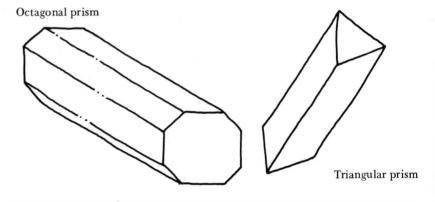

Figure 13.16 Octagonal prism

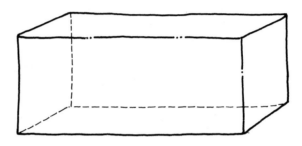

Triangular prism

Figure 13.17

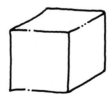

GEOMETRY OF TWO DIMENSIONS

What Do *You* Say?

1. What polygons are commonly studied in elementary-school geometry programs?

2. What basic properties of circles and polygons are studied by most school children?

3. What basic sequence guides teachers in introducing two-dimensional shapes to children?

4. What visual-tactual approaches help children recognize and name shapes?

5. What multipurpose teaching aids are useful for helping children learn about two-dimensional shapes and their properties?

6. What types of activities are appropriate for helping children learn about particular shape properties? Can you give examples for younger children? For older children?

Naming and Describing Plane Figures

Circles and squares, triangles and rectangles—children meet these and other geometric figures in their exploration of two-dimensional shapes. Recognizing shapes and naming them comes first (Figure 13.18). Simple sort activities are gradually extended. Children learn that polygons have three or more sides. The straight-line segments forming the sides meet only at the corners or vertices of a polygon.

Classifying polygons by the number of sides is introduced early (Figure 13.19). Children learn that even polygons with the same number of sides have many sizes and shapes (Figure 13.20).

Figure 13.18

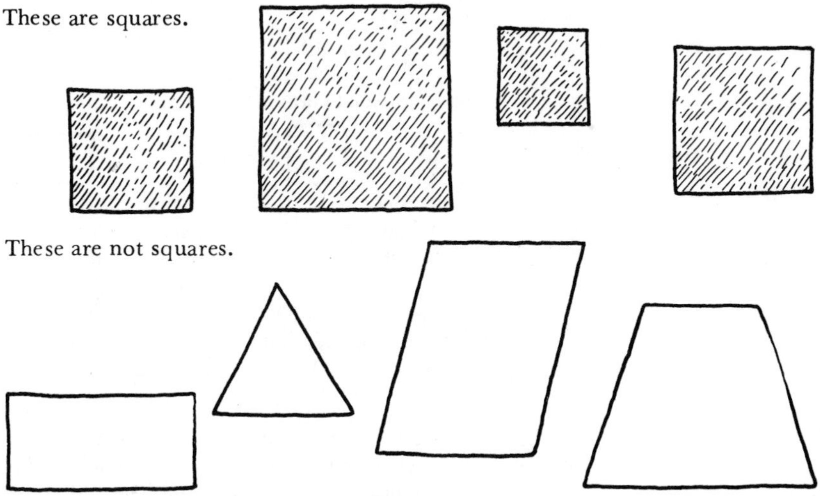

These are squares.

These are not squares.

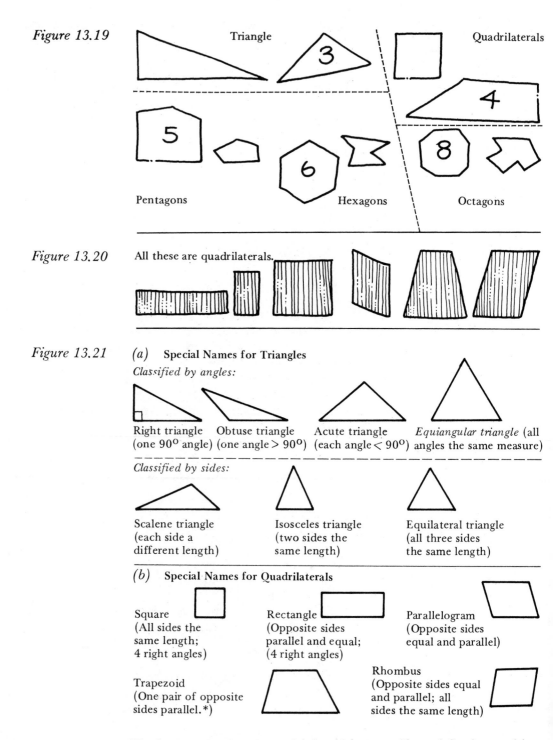

Figure 13.19

Triangle

Quadrilaterals

3

4

5

8

Pentagons

Hexagons

6

Octagons

Figure 13.20   All these are quadrilaterals.

Figure 13.21   (a)   **Special Names for Triangles**
Classified by angles:

Right triangle   Obtuse triangle   Acute triangle   *Equiangular triangle* (all
(one 90° angle) (one angle > 90°)   (each angle < 90°)   angles the same measure)

Classified by sides:

Scalene triangle   Isosceles triangle   Equilateral triangle
(each side a   (two sides the   (all three sides
different length)   same length)   the same length)

(b)   **Special Names for Quadrilaterals**

Square   Rectangle   Parallelogram
(All sides the   (Opposite sides   (Opposite sides
same length;   parallel and equal;   equal and parallel)
4 right angles)   (4 right angles)

Trapezoid   Rhombus
(One pair of opposite   (Opposite sides equal
sides parallel.*)   and parallel; all
                                                                    sides the same length)

*Teachers may encounter materials in which trapezoids are defined as quadri-
laterals with *at least* one pair of parallel sides. In this case all parallelograms,
rectangles, squares, and rhombii would be special cases of trapezoids.

In the upper grades children more carefully analyze the properties of shapes. They learn very precise ways of classifying polygons (Figure 13.21).

By the time children are in the upper grades, most have also dealt with the *five basic properties of two-dimensional figures* (Figure 13.22). Most children have also plotted points in a rectangular

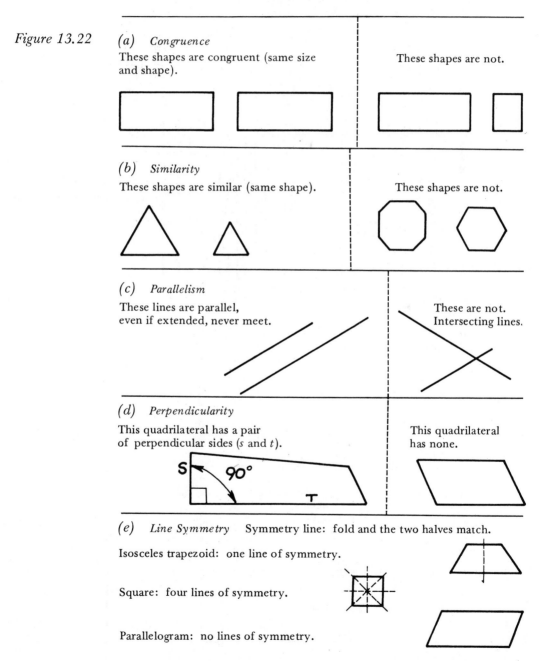

Figure 13.22

(a) *Congruence*
These shapes are congruent (same size and shape).

These shapes are not.

(b) *Similarity*
These shapes are similar (same shape).

These shapes are not.

(c) *Parallelism*
These lines are parallel, even if extended, never meet.

These are not. Intersecting lines.

(d) *Perpendicularity*
This quadrilateral has a pair of perpendicular sides (s and t).

This quadrilateral has none.

(e) *Line Symmetry*   Symmetry line: fold and the two halves match.

Isosceles trapezoid: one line of symmetry.

Square: four lines of symmetry.

Parallelogram: no lines of symmetry.

coordinate system (Figure 13.23), or even drawn pictures by drawing lines between points they have plotted.

Note: Plotting coordinates into pictures is great fun for students, third grade and up. Picture plotting also provides meaningful practice in visual-motor coordination skills. Excellent materials for this purpose are available to teachers through school supply companies. Refer to the list given in Appendix B.

Many children have explored sliding, flipping, or turning figures (Figure 13.24).

Figure 13.23

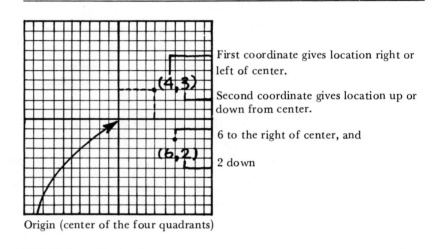

First coordinate gives location right or left of center.

Second coordinate gives location up or down from center.

6 to the right of center, and

2 down

Origin (center of the four quadrants)

Figure 13.24

Before   After

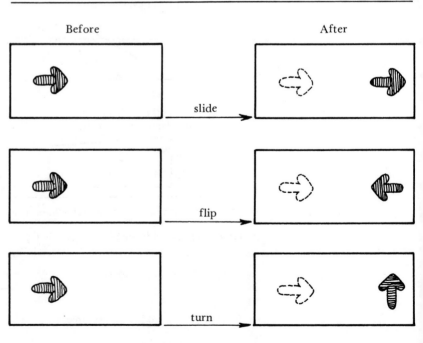

slide

flip

turn

Multipurpose Geometry Aids

Active exploration is the key whereby children best learn about shapes and their properties. Most teachers gradually build up a collection of homemade and commercial teaching aids for this purpose. In many cases given geometric materials can be used to illustrate a number of different shapes and their characteristics. Here are three examples of materials with this flexibility.

Example 1: Pattern Pieces.

Some teachers make their own sets of pattern pieces (Figure 13.25). Laminated construction paper or heavier tagboard sets are popular. If monies are available, wooden blocks can be purchased from school supply companies. The pieces can be used:

1. To reinforce shape recognition and naming. For example, give each child a set of pieces and play Simon Says: "Simon says to hold the square high." "Simon says to put the green triangle on top of the circle."

2. To help children focus on the five basic properties of shapes.

   - Which pieces have sides that are parallel? Find one piece. Trace around the parallel sides with a green crayon.

   - Which pieces have sides that are perpendicular? Find one piece. Trace around one pair of perpendicular lines with a red crayon.

   - Find the triangle piece. Place a mirror on the triangle as shown by the dotted line of Figure 13.26. Do you see the whole shape when you look into the mirror? The dotted line is a symmetry line for the triangle. How many other symmetry lines does the triangle have?

Figure 13.25

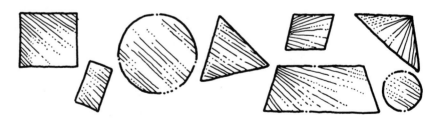

Figure 13.26   (a)   (b)   (c)

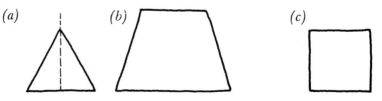

- Find all pieces congruent to the trapezoid (Figure 13.26b).

- Use only the square (Figure 13.26c). Make a larger square. How many pieces are needed? Can you add more pieces to make an even larger square? Use another pattern piece. If you put several like pieces together, can you form a larger pattern having that shape? (Informal introduction to similarity)

Example 2: Plexiglass.

Red or deep orange translucent plexiglass can serve many uses in a geometry lesson. Plexiglass reflects as a mirror would. In addition, you can peer through the plexiglass and actually trace the image seen. Or, given a figure with line symmetry, one can locate that line by moving the plexiglass until the reflected image on the glass coincides with the portion of the figure behind it. The mat of Figure 13.27 will be used to illustrate several exercises using colored plexiglass. The dotted line is the symmetry line on which the plexiglass should be placed.

- Identify the shape. Use your plexiglass to trace the image of the rectangle. Write *A* inside your tracing.

- Find two similar figures. Use your plexiglass to trace them. Write *B* inside each.

- Trace the images of two congruent shapes. Write *C* inside each.

- Use a green crayon. Trace the reflected image of a pair of parallel lines.

- Use a blue crayon. Trace two lines perpendicular to each other.

- Trace the image of the square. Now use your plexiglass to draw in all the symmetry lines.

Figure 13.27

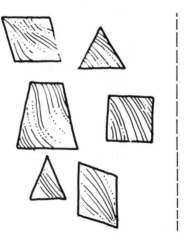

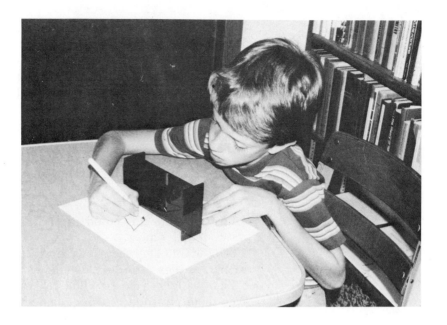

Note: Red plexiglass pieces, called *Miras*, can be purchased commercially from school supply companies. See the address list of Appendix B.

Example 3: The Classroom.
Every classroom is a geometry teaching aid all its own. Look around you.

- Which shape is most common?

- Point to a pair of parallel lines; to a pair of perpendicular lines.

- Can you find two congruent shapes in the room? Two similar shapes?

- Find a shape you think has line symmetry. Can you trace the shape and paper fold to show the symmetry line?

These and other questions help children focus on the richness of geometry in their environment.

SHAPE IDENTIFICATION

Naming and Describing Plane Figures: Phase 1

For its basic activities of shape identification and discrimination, some special educators (e.g., Ebersole, 1968) view geometry as a program critical to success in reading. Distinguishing a square from a rectangle which is not a square, for example, is prerequisite to finer discriminations required in recognizing letters (distinguishing *b* from *p*, for instance). Similarly, focusing on likes and differences in geometric figures relates to identifying like parts of words (e.g., p*an*, r*an*).

Geometric experiences like those that follow help children recognize and name two-dimensional shapes. The general approach taken here and in most elementary school textbooks today is one of active exploration. Visual-kinesthetic approaches, with emphasis on orally describing what is seen or felt, should be used as necessary with children having visual imperceptions. Students might, for example, be encouraged to:

- Close their eyes, touch, then describe what is felt
- Finger trace, then describe or name what is felt or seen
- Box or circle a figure to stop it from "floating" when perceived.

Color or texture cues could also be used to emphasize certain likenesses or differences.

A basic four-step sequence, illustrated in Figure 13.28, guides teachers as they introduce two-dimensional shapes to children.

First, with eyes closed, children can feel and describe a shape given them. Introduce correct names of shapes at this time. Children look at the shape and find a matching shape. They may enjoy finding a partner in the room who has a shape just like theirs. The children can place the shapes on top of one another to check that they are the same. In the beginning congruent shapes can be colored or textured alike to emphasize the desired match.

Tracing is a next step in helping children recognize and name shapes. Templates or dot-to-dot patterns may be provided for this purpose. It is sometimes helpful to supply a matching shape that can be traced. Children can finger-trace the shape, then complete the dotted pattern with crayon or pencil. A more challenging trace task is illustrated in Figure 13.29.

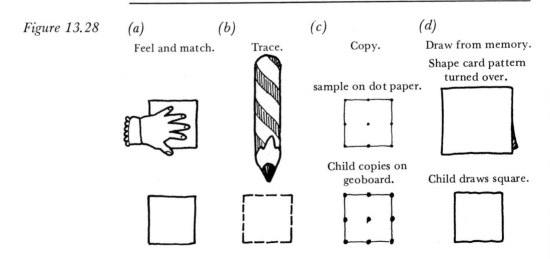

Figure 13.28   (a)   (b)   (c)   (d)

Feel and match.   Trace.   Copy.   Draw from memory.

                                               Shape card pattern
                                               turned over.

                               sample on dot paper.

                               Child copies on
                               geoboard.   Child draws square.

Figure 13.29   Use a green crayon. Trace the triangle.

A step above tracing is copying. High-contrast backgrounds in the sample cards sometimes help. In special cases it may be necessary to highlight each vertex with color or tactile cues. These cues help children focus on important aspects of a shape.

The harder of the four steps is drawing the shape from memory. At first shape cards may be shown as prompts, then turned over or removed. A more challenging task is to draw the shape correctly upon request, without the shape card prompt. The abilities involved in tracing and reproducing a geometric shape or pattern, simple as they seem, are perceptual prerequisites to those needed for copying a row of written problems onto paper from a chalkboard or a mathematics textbook page.

As children work with and describe shapes, they learn the basic characteristics of each. Precise terminology is not required of very young or developmentally immature children. Gradually, however, most children learn to understand and use terms like the following to describe given plane figures:

- Open/closed curve
- Inside/outside
- Straight sides
- Parallel sides
- Perpendicular sides
- Line symmetry
- Congruent/similar to

Activities like the following reinforce shape-identification skills. They can be adapted to the level and ability of the students involved.

**Activity File:
Shape Identification**

Activity 1: Drop
Provide several blocks of each basic shape. Children drop the shapes through matching holes cut in box lids (Figure 13.30).

Activity 2: Pantomime
Pantomime to initiate a sort (Figure 13.31).

Activity 3: Sherlock Holmes

Give each child a tagboard magnifying glass.
The child hunts down many examples of the shape in the glass.
Get children to name the shapes they find.

Variation (for older students): Provide a scavenger check list of shapes for students to find.

Figure 13.30

Figure 13.31

Figure 13.32

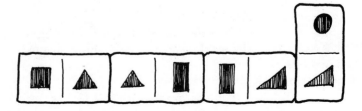

Activity 4: Match
Make dominoes and glue shapes on each domino half (Figure 13.32). See how many shape matches can be made. Name the shape each time to score a point.

Variation (for older students): Place shapes on half the dominoes and identifying names (e.g., obtuse triangle, isosceles trapezoid) on the other half.

Activity 5: Look Alikes
See if children can follow oral directions in locating shapes and placing them on a mat. Use directional terms, such as *on top of, above, below, right, left,* and so on.

Example. "Find the square. Place it in the upper left corner. Now find a scalene triangle. Put it to the right of the square."

It is interesting to frame questions so that, if properly followed, a picture results. Have children compare pictures. (They *should* look alike!)

BASIC PROPERTIES OF TWO-DIMENSIONAL SHAPES

Naming and Describing Plane Figures: Phase 2

Five important properties of shapes are repeatedly examined throughout the elementary-school geometry curriculum:

1. Symmetry

2. Congruence

3. Parallelism

4. Perpendicularity

5. Similarity

Of these properties, the concept of *symmetry* is perhaps the most intuitive. Children are usually aware of the symmetry of a valentine or a butterfly, and they enjoy paperfolding or inkblot activities in which they create their own symmetric designs.

Young children often determine line symmetry by inspection. If they have normal visual perception skills, they'll notice how a shape "balances" on either side of a central axis. To check what the eye may not perceive, a fold test or a mirror test may also be used.

The fold test.
Trace the figure on a piece of paper. See if the paper can be folded so the two halves match exactly. The fold line is the line of symmetry.

The mirror test.
Place a mirror or piece of colored plexiglass on the figure so that half the figure, together with the reflection of that half, form the

entire figure. The ling along which the mirror is placed is the line of symmetry.

Activities for line symmetry using the fold or mirror test appear early in many primary geometry programs.

The concept of congruence is also readily noted by most children. For example, it is basic to the motion of a child making a puzzle piece match. Congruency can be determined by placing two figures on top of each other to see if they match in size and shape. When figures cannot be moved, tracings can be used. Two figures are congruent when a tracing of one fits exactly on top of the other. Like symmetry, the idea of congruence is introduced early in the elementary geometry curriculum.

The terms *parallelism* and *perpendicularity* both refer to special relationships between pairs of lines. Parallel lines are *everywhere equidistant*. For any given pair of lines, no matter where one could choose to measure, the distance between them is always the same (see Figure 13.33). Even when extended, parallel lines never meet. Train tracks are an environmental example of parallels.

Figure 13.33

Figure 13.34

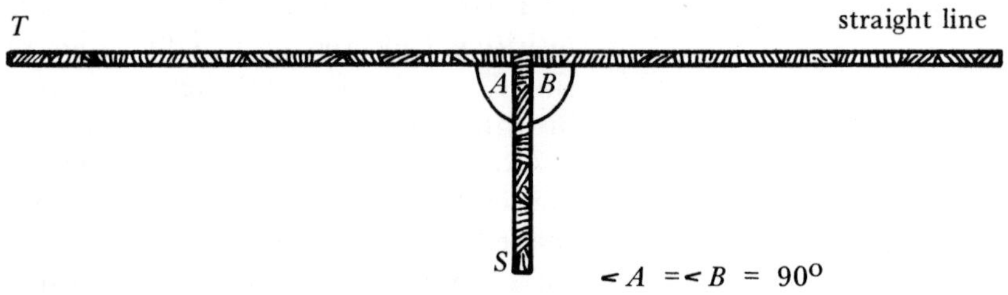

straight line

$< A = < B = 90°$

Figure 13.35

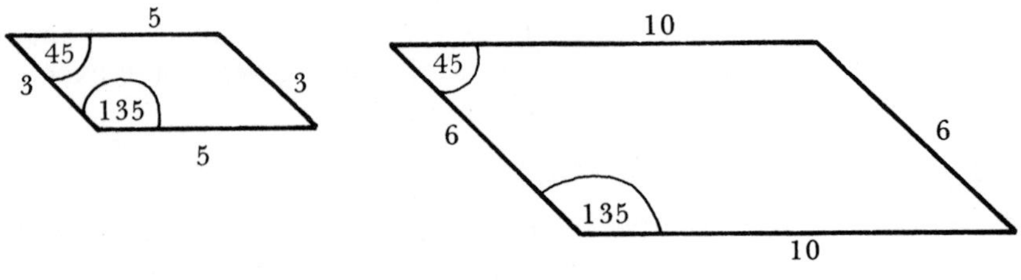

Some mathematicians define perpendicular lines as lines that meet to form congruent adjacent angles. When this happens, as in Figure 13.34, the exterior sides of the angles form a straight line and we say line *s* is perpendicular to line *t*. More simply, we can observe that lines which meet at right angles are perpendicular to each other. Look at this page and note the square corner (right angle) formed by the edges of the page. The two edges that meet to form this corner are perpendicular to each other. Examples of perpendicularity meet us wherever we turn. Children are often surprised how many they can discover in a very short time.

The formal treatment of similarity usually appears later in the geometry program because more subtle perceptions are involved. Similar figures have the same shape, but can differ in size. Corresponding angles of similar figures are equal and corresponding sides are proportional. In Figure 13.35, the ratio of corresponding sides is 2 to 1. Many textbooks ask children to enlarge or reduce pictures as a practical application of similarity. One approach to this exercise is described in the following sample activities. We include ideas for introducing or reinforcing other basic shape properties discussed in this section.

Activity File: Properties of 2-D Shapes

Activity 1: Fold And Cut
Focus: Early work with line symmetry.
Materials: Paper, pencil, and scissors.
Fold a paper in half and sketch half a valentine, Christmas tree, or snowflake on one half (Figure 13.36). Then cut and unfold to see the whole shape. Multiple folds can be used to make more intricate snowflake designs.
Note: The use of paint or inkblot patterns produced by folding provides another interesting opportunity to introduce symmetry.

Activity 2: Pattern Twins
Focus: More advanced work with line symmetry.
Materials: Pattern pieces or Cuisenaire rods; mirror; deck of pattern cards like that of Figure 13.37.

Figure 13.36

opens to   opens to   opens to

Figure 13.37

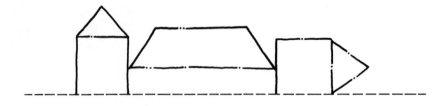

Figure 13.38   Tangram pieces

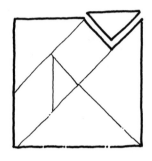

Figure 13.39

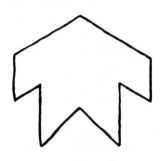

Figure 13.40   Sample Cue Sheet

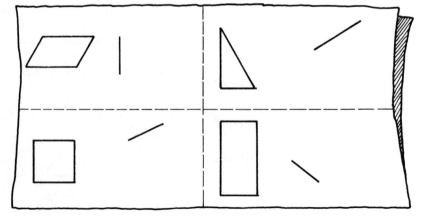

Lay pattern pieces above the line of the pattern card to complete the design given. Lay additional pieces below the line so the bottom half is symmetrical to the top. Use a mirror to check.

Variation: Move one or more pieces of a given design to make it symmetrical.

Activity 3: Tangrams

Focus: Early work with congruence.

Tell the story of Tan, the son of a Chinese nobleman. "Tan's father gave him a beautiful red tile for his birthday. One day as Tan was playing with the tile, it dropped and broke. Tan tried hard to put the tile together again, but couldn't. Tan found some pieces of the tile that were the *same size and shape*." Give children sets of tangram pieces (Figure 13.38) and ask them to find those that are the same size and shape. Have the children place the pieces on top of each other to check. Some children may need to close their eyes and feel how they are the same.

Tell how Tan made puzzles with his pieces. Give them a tangram puzzle such as that in Figure 13.39, and ask them to use their pieces to make one exactly the same size and shape as Tan's.

Note: Several interesting books on tangrams are available to teachers. One example is that by Read, 1965.

Activity 4: Complete A Shape

Focus: More advanced work with congruence.

Materials: Box of pattern pieces and cue sheets as shown in Figure 13.40.

Children find the pattern piece for the shape on the left and use it to draw a congruent shape. The given cue line must be used as one side of the drawing.

Activity 5: Draw, Touch, And Feel

Focus: Early work with parallelism.

Materials: Ruler, pencil, and paper.

Draw train tracks on paper. Use a ruler to help. Just draw a horizontal line along two sides of the ruler edge. Use the ruler to check that the two lines are the same distance apart at each end. "The lines you draw are parallel."

Show children how to hold the ruler so that *vertical and oblique parallels* can also be drawn. Ask them to find and touch parallel sides or edges of objects around the room. Get them to explain why the sides are parallel.

Activity 6: You See It And You Don't

Focus: More advanced work with parallelism.

Materials: Photos or pictures from magazines which show parallel lines.

Get children to identify pairs of lines they know are parallel in the pictures. Trace the lines, then extend them. Do they meet? Now measure them in at least two places. Are they the same distance apart? Discuss the idea of perspective—how the proximity and position of the eyes in relation to an object can affect the object's appearance. This makes lines appear to meet, as in a picture, when in reality, they do not. Sometimes it is important to focus on what you know, not on what you see.

Activity 7: Square Corners
  Focus: Early work with perpendiculars.
  Materials: Two pencils, one short, one long.

1. Place the pencils on a desk and move them together to make a capital T. How many square corners are formed?

2. Use a piece of notepaper and show how the square corner of the notepaper just fits into the corners of the T shape (Figure 13.41). If children are ready, point out that sides which meet to make a square corner really form a *right angle*. Ask them to bend their right arm to make a square corner (right angle) at the elbow. Ask children to tell the number of square corners formed by the sides of these objects:

Table Filing card Book Picture frame Window pane Door
Chalkboard Desk top

Activity 8: Experiment!
  Focus: More advanced work with perpendiculars.
  Materials: Plexiglass, pencil, and paper.
  Fold the paper in half and draw a simple clown figure on one half, as in Figure 13.42. Give the clown three buttons on its suit and a tassle on the top of its hat. Now place the plexiglass on the fold line and draw the reflection image of the clown. Examine the clown and its image. Draw lines to connect four or five sets of matching points. Describe the relation of these lines to the line of symmetry. Experiment with other figures. *Is the symmetry line always the perpendicular bisector of lines joining matching points?*

Activity 9: Shadow Play
  Focus: Early work with similarity.
  Materials: Overhead projector and pattern pieces.
  Place pattern pieces, one at a time, on the overhead glass. Get children to compare the projected image to the pattern-piece shape. "Both are the same shape, but they differ in size." If children are ready, introduce the term *similar*.

Multipurpose Activities
  Many activities can be varied at will to emphasize either shape recognition and naming or basic properties of shapes. The following

two activities are examples that focus on this flexibility within basic activities.

Activity 1: See It? (For individuals or groups)

Asking children to identify and extend a pattern is an interesting way to focus on geometric figures or their properties. Patterns can be very simple (Figure 13.43). They can be more sophisticated (Figure 13.44). Patterning tasks help strengthen visual memory and visual sequencing skills, and these become important as children see how letters relate spatially to form a word, and how numbers relate to each other (before, after, between, greater, less) on a ruler or number line.

Figure 13.41

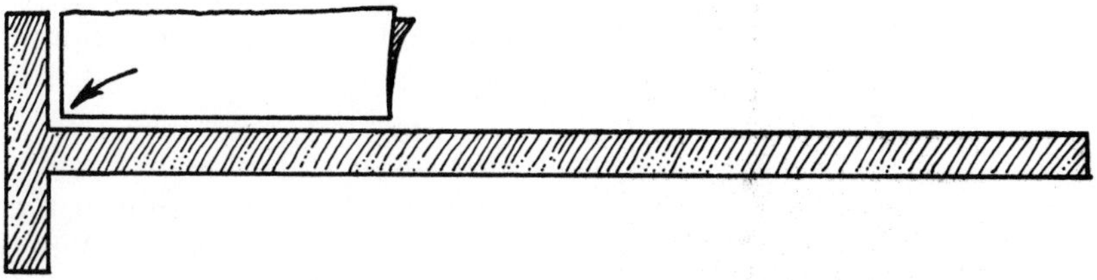

Figure 13.42

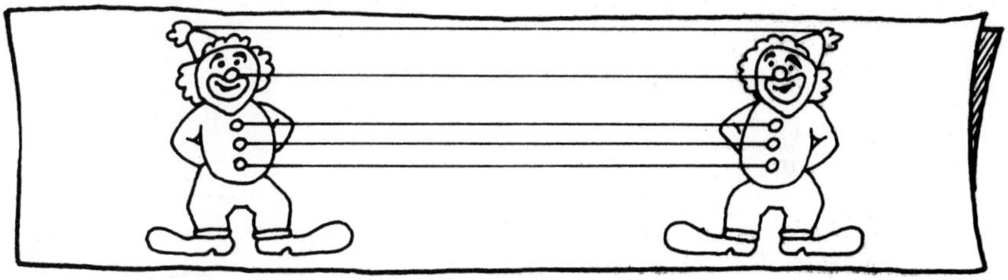

Figure 13.43

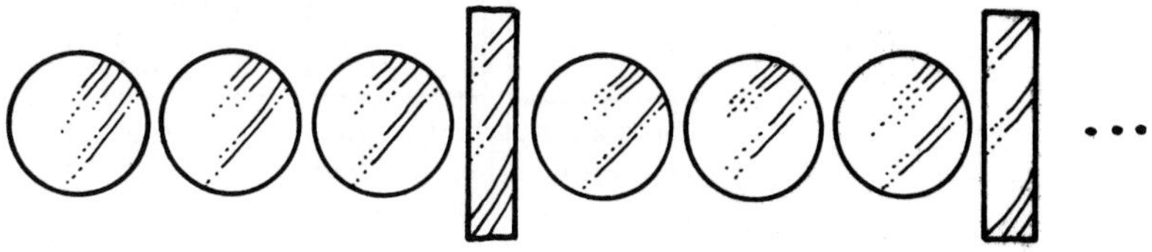

Figure 13.44

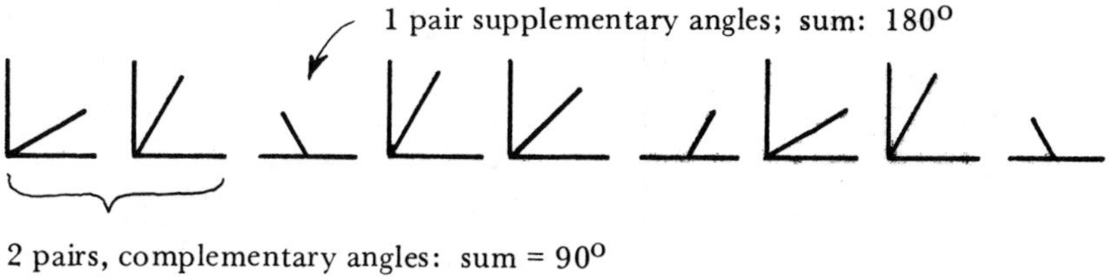

1 pair supplementary angles; sum: 180°

2 pairs, complementary angles: sum = 90°

Activity 2: Find It (For two or three players)

Materials: Gameboard like that shown in Figure 13.45; a marker for each player; geoboard and rubber bands; die marked 1, 2, 3; card deck with problems on one side, answers on the other.

Procedure:

1. Mix the cards and place them in the envelope on the playing board.
2. Players roll the die. High point goes first. Place markers on *Start.*
3. In turn, draw a card and follow its directions.
4. Turn the card over to check. If correct, roll the die to see how far to move.

Winner: First to reach *Finish.*
(Card deck can be changed from time to time to emphasize the current topic of study.)

Special note on pentominoes.

Pentominoes are an extension of the idea of dominoes. A domino is formed by joining *two* squares. When *five* squares are joined in all possible ways, twelve special shapes called pentominoes are formed. Pentomino activity booklets, available from many of the publishers/school-supply companies listed in Appendix B, give excellent ideas for the middle school curriculum for exploring basic properties of shapes. Most of these activities are also good for fostering visual perception or visual-motor coordination skills.

Figure 13.45

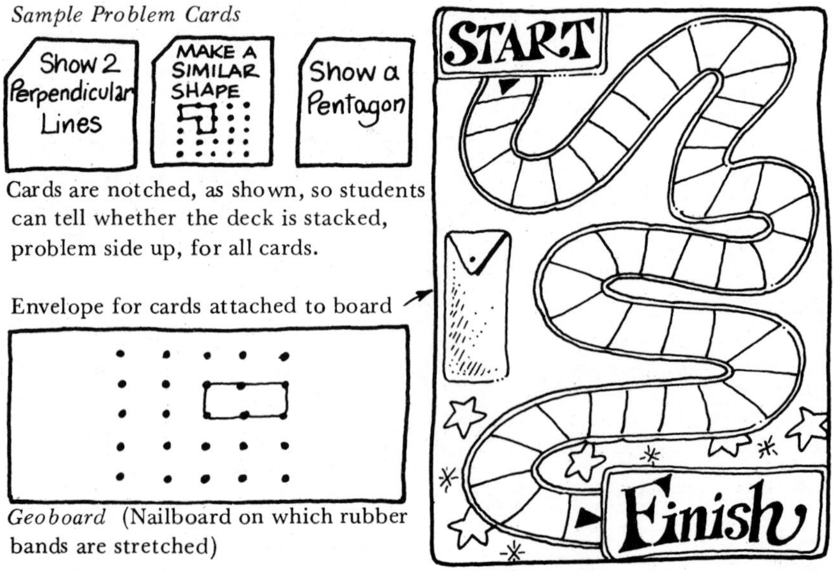
Sample Problem Cards
Show 2 Perpendicular Lines
MAKE A SIMILAR SHAPE
Show a Pentagon
Cards are notched, as shown, so students can tell whether the deck is stacked, problem side up, for all cards.
Envelope for cards attached to board
Geoboard (Nailboard on which rubber bands are stretched)
START
Finish

Your Time: Activities, Exercises, and Investigations

1. Prepare, as your instructor directs, answers to the major questions on p. 481 of this chapter.

2. Several multipurpose geometry teaching aids were described in this chapter. Choose one other aid you think fits this category and show how it can be used to help children learn about two-dimensional shapes and their properties. Glancing through the geometry sections of elementary school textbooks may give you an idea.

3. Create an activity and show how, with a little adaptation, it can be used to emphasize either shape identification or one of the five basic properties of two-dimensional shapes. The multipurpose activities of this chapter (pp. 485–487) may give you an idea. The Math Lab Activities of Appendix C might also be consulted. The teacher's manuals for school geometry units is another good reference.

4.

Two-dimensional shapes: major concepts and skills	
Shape recognition	Symmetry
Congruence	Parallelism
Similarity	Perpendicularity

For each category above, describe an activity which reinforces the concept or skill.
(a) Make your activity different from any found in this chapter.
(b) Tell the ability level of the child for whom your activity is intended.
(c) Describe how the activity satisfies one of the geometry goals listed in Appendix A for a child at this level.

As you complete each of the following exercises, consider the role similar questions or activities might play in a school geometry program.

5. Describe two ways rectangles and parallelograms are alike. Name one way in which they differ.

6. A regular polygon is one that has (a) all sides of equal length, and (b) all angles of equal measure. Examine each shape in Figure 13.46 and tell whether it is a regular polygon. Be able to tell why you think so for each case.

7. What regular polygons are formed by:
(a) Using a paper strip and tying a knot (Figure 13.47)?
(b) Tracing an equilateral triangle and folding each vertex into the center?
(c) Drawing a circle with a compass, making the opening of the compass equal to the radius of the circle, marking off this length as many times as it will fit around the circle, and then connecting the marks?

8. In Figure 13.22, the square was shown to have four lines of symmetry.
(a) Sketch each of the other regular polygons in the table of Figure 13.48 and determine the lines of symmetry for each.
(b) What pattern does your completed table suggest?
(c) How many symmetry lines does a circle have?

Figure 13.46

Figure 13.47

Figure 13.48

Regular polygon	Number of symmetry lines
Equilateral triangle	
Square	4
Regular pentagon	
Regular hexagon	

Figure 13.49

	Rectangle	Rhombus	Square	Trapezoid	Parallelogram
The diagonals bisect each other.	yes				
The diagonals are perpendicular to each other.		yes			
All sides are congruent.			yes		
All angles are congruent.					
Opposite sides are parallel.					
Opposite angles are parallel.					
Opposite sides are congruent.					
Has at least one line of symmetry.					
Has at least one right angle.					

Figure 13.50

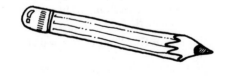

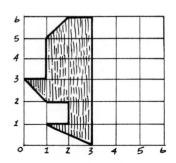

9. The seven pieces of the tangram puzzle are shown in Figure 13.38.
   (a) Which tangram pieces are congruent to each other? Similar?
   (b) It is possible to form several squares using different combinations of tangram puzzle pieces. Squares can be made from 1, 2, 3, 4, 5, and 6 puzzle pieces. *How many of these squares can you form?*

10. Draw four different quadrilaterals on your paper. Find the midpoint of each, then connect the midpoints in order. What do you notice about your results? Must this always happen when you connect the midpoints of quadrilaterals?

11. Copy and complete the table of Figure 13.49.

12. Which of these points would not be graphed if you were showing the flip image of the shaded region in Figure 13.50?
    (a) (6,5)   (c) (5,5)
    (b) (4,6)   (d) (5,1)

13. When guy wires are used to prop a telephone pole, what shape do they make with the ground? What shape is formed by the brace in each corner of a card table? What shape appears most frequently in bridges and high-voltage wire towers? Use paper fasteners and cardboard strips to form a triangle and a square. Which shape is rigid? Which shape needs an extra diagonal strip to make it hold its shape? Experiment with other shapes. Is the triangle the only rigid shape? Why would engineers and carpenters need to know the principles you've just explored?

14. Use graph paper to plot the following points.
    (a) (2,3)   (f) (7,34)   (k) (2,1)
    (b) (2,4)   (g) (9,4)   (l) (1,3)
    (c) (5,4)   (h) (9,3)   (m) (2,3)
    (d) (5,6)   (i) (13,3)
    (e) (7,6)   (j) (11,1)
    Connect the points in order: *a* to *b*, *b* to *c*, and so on. What picture is formed? Create or find other point plotting directions that give an interesting picture.

References

Ebersole, M., N. C. Kephart, and J. B. Ebersole. *Steps to Achievement for the Slow Learner.* Columbus, Ohio: Charles E. Merrill, 1968, pp. 122–33.

Frostig, M. *Pictures and Patterns.* Teacher's Guides. Chicago: Follett, 1972.

Konkle, G. S. *Shapes and Perceptions: An Intuitive Approach to Geometry.* Boston: Prindle, Weber and Schmidt, 1974.

Read, R. C. *Tangrams, 330 Puzzles.* New York: Dover, 1965.

APPENDIX A A TYPICAL SET OF K-8 MATHEMATICS OBJECTIVES

The objectives reflecting a typical K–8 program in mathematics are broken into four strands. The first strand is Number and Numeration. This strand contains the objectives that are related to the development of conceptual understanding in the area of number and number use. The second strand is that of Operations. This strand contains the objectives which are related to algorithmic processes. The third strand is Geometry and Measurement. This strand concentrates on the development and application of concepts which deal with spatial information in mathematics. The fourth strand is that of Problem Solving. The objectives in this strand are focused on applications of the appropriate level of mathematics which are germane to the school curriculum. The objectives are presented by grade levels with the four strands separated within each grade level.

Kindergarten

Number and Numeration
At the end of kindergarten the child will:

1. Recognize likenesses in common objects.

2. Recognize differences in common objects.

3. Sort objects by common properties.

4. Compare groups by matching.

5. Correctly use comparison terminology: as many as, not as many as, more than, less than. . . .

6. Order groups by cardinality.

7. Correctly use order terminology: before, after, between, first, next, last. . . .

8. Associate a number with a group of six or fewer objects at sight (objects not arranged linearly).

9. Rote count to ten.

10. Associate a number with a group of ten or fewer objects by counting.

11. Write the numerals 0 through 9.

12. Recognize 1/2 and 1/3 of a region.

13. Associate 1/2 and 1/3 of a region with the correct fractional numerals.

Operations
At the end of kindergarten the child will:

1. Sort objects on the basis of likenesses and differences as perceived by the child.

2. Sort a collection of objects on the basis of color (red, yellow, blue, green, orange, white, black. . .).

3. Sort a collection of objects on the basis of size (large, small, thick, thin, big, little. . .).

4. Sort a collection of objects on the basis of texture (rough, smooth, sticky, sharp, wet, dry. . .).

5. Sort a collection of objects on the basis of use (tool, eating utensil, car. . .).

6. Use the appropriate language for the classes formed in objectives 1–5.

7. Compare objects on the amount of a particular property they possess (i.e., length, height, size, mass, volume. . .) and label the one having more of the property and the one having less of the property.

8. Compare three objects on the basis of the amount of a given property they possess and rank them from the smallest to the largest.

9. Follow directions involving positional language (up-down; left-right; above-below; high-low; top-bottom. . .).

10. Compare the relative size of objects shown in pictures and indicate which has more of a given attribute.

11. Explain the use of common measuring devices such as calendars and clocks.

12. Name the days of the week.

Problem Solving
At the end of kindergarten the child will:

1. Recognize and name the penny, nickel, and dime.

2. Complete a simple verbal picture pattern by indicating the next objects to appear in the pattern.

3. Select the next objects to appear in a visual pattern by circling the appropriate objects.

4. Solve simple picture problems involving sequences of events.

5. Model simple number stories involving sums and differences of four or less.

Grade 1

Number and Numeration
At the end of the first grade the child will:

1. Compare groups by matching.

2. Correctly use comparison terminology: as many as, equal to, not as many as, not equal to, more than, greater than, fewer than, less than.

3. Order groups by cardinality.

4. Correctly use order terminology: before, after, between, next, first, second, third, . . . , ninth, tenth.

5. Associate a number with a group of eight or fewer objects at sight.

6. Rote count to 100.

7. Count on to 100 from any given number less than 100.

8. Associate a number with a group of less than 100 objects by counting.

9. Associate bundles of ten objects with the numbers 10, 20, 30, . . . , 90.

10. Find the number in a group of 99 or fewer objects by forming groups of ten and telling how many tens and how many left over.

11. Write the numerals 0–100.

12. Correctly associate regions with the unit fractions 1/2, 1/3, . . . , 1/10.

13. Write unit fractions for shaded regions having one out of fewer than ten subregions shaded.

Operations

At the end of the first grade the child will:

1. Give the number that is one more than any given number 0 to 99.

2. Give the number that is one less than any given number 0 to 99.

3. Combine two groups and tell how many in the new group.

4. Associate the combining of groups with addition and the plus sign.

5. Write addition sentences (3 + 2 = 5) to go with the combining of groups.

6. Model an addition sequence by combining appropriate groups of objects.

7. Associate the vertical form of addition sentences with the combining of objects and the associated horizontal form.

8. Work both horizontal and vertical addition examples by combining groups and counting.

9. Organize and see relationships among basic addition facts with sums of ten or less.

10. Commit to memory basic addition facts with sums of ten or less.

11. Add three addends having a sum of ten or less.

12. Understand that addends can be rearranged without changing the sum.

13. *Informally* use the associative principle of addition, i.e.,
    $5 + 8 = 5 + (5 + 3) = (5 + 5) + 3 = 10 + 3 = 13$.

14. Use already memorized facts to figure out harder basic addition facts, i.e., $8 + 7 = 8 + 6 + 1$.

15. Add two-digit numbers where no regrouping is required.

16. Take away objects and tell how many are left in a prescribed setting.

17. Compare groups of objects and tell how many more.

18. Associate take away with subtraction and the minus sign.

19. Associate comparison with subtraction and the minus sign.

20. Write subtraction sentences in both the horizontal and vertical form to go with both take away and comparison situations.

21. Work both horizontal and vertical forms of subtraction examples by taking away objects and counting.

22. Organize and see relationships among the basic subtraction facts with minuends of ten or less.

23. Commit to memory basic subtraction facts with minuends of ten or less.

24. Use already memorized facts to figure out harder basic subtraction facts.

25. Subtract two-digit numbers where regrouping is not required.

Geometry and Measurement
At the end of the first grade the child will:

1. Compare a collection of objects on the basis of a given attribute (color, shape, size, function. . .).

2. Order a collection of objects by size or amount of a given attribute they possess.

3. Identify examples of a curve, a segment, a simple closed curve, a circle, a triangle, a rectangle, and a square.

4. Identify examples of a cube, a pyramid, a cylinder, and a sphere.

5. Relate geometric shapes studied to objects in the environment.

6. Recognize and identify similar shapes.

7. Measure a given object with a given arbitrary unit (paper clip, eraser, centicube. . .).

8. Measure a given object using a collection of standard units (cm or inch).

9. Develop own ruler using adding machine tape and cm or inch cubes for units.

10. Measure line segments and identify those having the same length.

11. Sort a collection of shapes on the basis of similarity and then identify those having the same shape and size (congruent figures).

12. Examine a collection of geometric figures and identify those having line symmetry.

Problem Solving
At the end of the first grade the child will:

1. Tell time to the nearest hour.

2. Name the days of the week and months of the year.

3. Recognize and give the value for the penny, nickel, dime, quarter, half-dollar, and dollar coins.

4. Count the value of a collection of coins up to a total of 50 cents.

5. Complete a picture sequence by selecting or drawing the next objects in the pattern.

6. Solve and interpret picture drawings illustrating the models for addition and subtraction.

7. Make charts illustrating graphical relations (i.e., bar graphs, showing eye colors, bar graphs showing heights, or some similar graphical relationships).

8. Write equations for simple story problems presented in picture or word form.

9. Choose the correct equation for a given number story from a list of alternatives.

10. Solving simple open sentences of the form $a + x = b$ and $a - x = b$.

Grade 2   *Number and Numeration*
At the end of grade two the child will:

1. Name numbers to 99 in tens and ones

2. Tell how many in a group of 99 or fewer objects by tens and ones.

3. Recognize that 10 tens is one hundred.

4. Name numbers to 999 in hundreds, tens, and ones.

5. Tell how many in a group of 999 or fewer objects by hundreds, tens, and ones.

6. Correctly use ordinals: first to thirty-first.

7. Count on from any given number 0–998.

8. Rote count to 100 by ones, twos, fives, tens, and twenty-fives.

9. Give the number that is one more than or one less than any given number from 1 to 999.

10. Compare numbers to 1000 using *greater than, less than,* or *equal to.*

11. Identify 1/2, 1/3, . . ., 1/10 of a region by shading the appropriate amount of the region.

12. Recognize and use decimals representing tenths.

13. Write a decimal (one-place) or a fraction (denominator less than ten) which represents a given shaded region of a standard figure.

Operations
At the end of grade two the child will:

1. Give from memory the addition facts.

2. Give from memory the subtraction facts.

3. Use memorized addition and subtraction facts to figure out harder addition and subtraction situations for two-digit problems involving no regrouping.

4. Relate addition and subtraction facts such as $2 + 3 = 5$, $3 + 2 = 5$, $5 - 2 = 3$ and $5 - 3 = 2$.

5. Recognize the addends can be arranged in any order without affecting the sum.

6. Informally use the associative property in grouping addends.

7. Write numbers in expanded forms showing the number of hundreds, tens, and ones in each.

8. Add with regrouping 2 two-digit numbers.

9. Subtract 2 two-digit numbers with regrouping involved.

10. Associate the combining of equivalent groups with multiplication and the times sign.

11. Figure out easy multiplication facts by combining and counting.

12. Figure out easy multiplication facts by repeated addition.

Geometry and Measurement
At the end of grade two the child will:

1. Measure a given segment using either cm or inches.

2. Give an estimate of the length of a specified segment in either cm or inches.

3. Measure and give the perimeter of a polygon in cm or inches.

4. Compare the length of a pair of segments using the concepts of *greater than* or *less than* with respect to length.

5. Identify open and closed curves, simple closed curves, circles, triangles, rectangles, and squares.

6. Explain how a square unit can be used to measure the area of a polygonal region.

7. Measure the area of a polygonal region using cm^2 or in^2 as units.

8. Calculate the areas of a square or rectangle knowing the measures of the sides of the figures.

9. Compare geometric solids on the basis of size.

10. Identify rectangular solids and cubes.

11. Explain the concept of volume in terms of filling and packing.

12. Give the volume of a simple rectangular solid by filling it with given cubic units.

13. Recognize the liter as a unit of capacity.

14. Judge whether the capacity of a given container is greater than or less than that of a liter.

Problem Solving
At the end of grade two the child will:

1. Tell time to the minute using a regular two-hand clock.

2. Name and give the values of all coins and the one- and five-dollar bills.

3. Count a collection of change having a value less than or equal to a dollar.

4. Select and write number sentences describing verbal problems through the level studied in the regular classroom instruction.

5. Estimate the size of a given sum or product.

6. Round numbers to the nearest 10, 100, or 1000.

7. Complete a given picture sequence by selecting or supplying the next figure.

8. Complete a given number pattern sequence by supplying or selecting the next three numbers in the pattern.

9. Construct a simple bar graph for a given set of information.

10. Give an interpretation for a given bar graph.

Grade 3   *Number and Numeration*
At the end of grade 3 the child will:

1. Read and write numerals to 999,999.

2. Order numbers to 999,999.

3. Model numbers from 0 to 999.

4. Compare numbers using the appropriate signs $<$, $=$, or $>$.

5. Relate regions to the appropriate fractions.

6. Relate regions to the appropriate decimals.

7. Relate part of a group to a fraction.

8. Write fractions equivalent to given fractions.

9. Read and write mixed numerals.

10. Compare fractions using models.

11. Compare fractions by selecting the appropriate sign $<$, $=$, $>$.

12. Model decimals through hundredths.

Operations
At the end of grade three the child will:

1. Give from memory all basic addition and subtraction facts.

2. Relate addition and subtraction.

3. Model whole-number addition.

4. Add any number of whole numbers.

5. Model whole-number subtraction.

6. Subtract any whole number from any larger one.

7. Model multiplication facts by combining equal-sized groups and counting.

8. Figure out multiplication facts by successive additions.

9. Recall from memory the first half of the multiplication facts according to the method used for teaching the facts.

10. Mentally figure out harder facts using the teaching strategies for fact mastery used in the classrooms.

11. Recall from memory multiplication facts with explanations for their associated products for the first half of the multiplication table.

12. Relate division to the separation of a group into equal-sized subgroups.

13. Figure out easy division facts by successive subtractions.

14. Figure out easy division situations by separating and counting.

15. Solve basic division situations by manipulating models.

16. Solve basic division problems with or without remainders.

17. Recall from memory all basic multiplication facts.

18. Multiply easy two-digit numbers by one-digit numbers with no regrouping.

19. Find fractional parts of quantities.

20. Find decimal parts of quantities.

21. Solve two-decimal-place addition and subtraction problems.

Geometry and Measurement
At the end of grade three the child will:

1. Handle linear measurement situations involving cm, in., m, and yd.

2. Conceptually handle measurements involving km and miles.

3. Estimate distances in gross terms for km and miles and with some accuracy for measures equal to or less than a m or a yd.

4. Measure the perimeter of a given polygon in cm, in., m, and yd.

5. Identify the inside and outside of a simple closed curve.

6. Identify by naming the radius, center, and diameter of a given circle.

7. Identify shapes as being similar (having the same shape) or congruent (having the same shape and size).

8. Measure the area of a given polygonal region in terms of cm^2 or in^2.

9. Find the surface area of a rectangular solid given the dimensions of its edges.

10. Identify the parts of an angle (vertex and sides).

11. Identify and use the following units of capacity (cup, pint, quart, gallon, and liter).

12. Find the mass of a given object in pounds and kilograms.

13. Describe the temperature of an object in either degrees Celsius of degrees Fahrenheit.

Problem Solving
At the end of grade three the child will:

1. Make and read a bar graph.

2. Make and read a picture graph.

3. Make and read a line graph (connected ordered pairs).

4. Use coordinates to locate a point on a coordinate system.

5. Count money to $10.

6. Make change to $1 from any value less than $1.

7. Tell time to the minute from a regular clock.

8. Complete picture or number sequence patterns by selecting or supplying the next elements in the pattern.

9. Select or write number sentences describing verbal problems through the level studied in the classroom portion of operations strand.

Grade 4   *Number and Numeration*
At the end of grade four the student will:

1. Read and write numbers to 1,000,000.

2. Order numbers to 1,000,000.

3. Compare numbers to 1,000,000.

4. Model numbers to 9999.

5. Use common fractions to represent parts of regions or sets.

6. Use decimals to name regions or parts of sets.

7. Compare fractions with respect to greater than, less than, or equal.

8. Compare decimals with respect to greater than, less than, or equal.

9. Give fractions equivalent to a given fraction.

10. Use mixed numbers to represent a region.

11. Change mixed numbers into fractions.

12. Change fractions to mixed numbers when possible.

13. Use decimals to label points on a number line.

14. Relate decimals to fractions.

15. Identify prime numbers less than 40.

Operations
At the end of grade four the student will:

1. Add any number of whole numbers.

2. Find the difference between any two whole numbers whose difference is non-negative.

3. Recall from memory all basic multiplication and division facts.

4. Relate multiplication and division.

5. Give some multiples of any given whole number.

6. Identify whole numbers as odd or even as a result of dividing by two.

7. Identify the factors of whole numbers less than 50.

8. Give prime factorization of whole numbers less than 50.

9. Find the common factors of some pairs of whole numbers less than 50.

10. Model whole-number multiplication.

11. Multiply any 2-digit or 3-digit whole number by a one-digit number.

12. Multiply any 2-digit or 3-digit whole number by 10 or 100.

13. Multiply any 2-digit or 3-digit number by a multiple of 10 through 90.

14. Multiply any 2-digit or 3-digit number by a 2-digit number.

15. Model whole-number division.

16. Check division with remainder problems by multiplying and adding.

17. Divide any whole number by a one-digit number.

18. Divide any whole number by a multiple of 10 up to 90.

19. Estimate with reasonable accuracy sums, differences, products, and quotients of whole numbers.

20. Add and subtract fractions with like denominators.

21. Add and subtract fractions with related denominators.

22. Add and subtract decimals with three or less decimal places.

23. Add and subtract mixed numbers with common denominators.

24. Add and subtract fractions with unlike denominators.

25. Find the product of two decimal numbers having two or fewer decimal places.

26. Find the product of two fractional numbers.

27. Find fractional parts of quantities.

Geometry and Measurement
At the end of grade four the student will:

1. Display the competencies with linear measurement and area measurement from the lower grades.

2. Measure the volume of a given container in liters, milliliters, cups, pints, quarts, and gallons.

3. Estimate the capacity of a given container in terms of liters or quarts.

4. Measure the mass of a given object in kilograms or pounds.

5. Determine the volume of a given rectangular solid by filling it with cubic units and counting them.

6. Identify and describe a pentagon, a hexagon, and an octagon.

7. Identify and draw examples of acute, right, or obtuse angles.

8. Identify shapes that are similar and those that are also congruent.

9. Identify which transformation (slide, twist, or flip) takes a given figure into a second of two congruent figures.

Problem Solving
At the end of grade four the student will:

1. Record sums of money up to $100.

2. Make change up to $10.

3. Record time to the second.

4. Select or write the number sentence describing any verbal problem studied in the operations strand of the fourth-grade level.

5. Work appropriate word problems involving the concept of speed $(d = rt)$ or averages.

6. Construct and interpret bar, picture, and line graphs.

7. Graph points (x,y) on a rectangular coordinate grid.

8. Complete number pattern sequences.

9. Determine the rule generating number pattern sequences given in the form of input—unknown rule—output. (function machine games)

Grade 5   *Number and Numeration*
At the end of grade five the student will:

1. Complete all the reading and writing of numeral competencies of the prior grades.

2. Read and write Roman numerals.

3. Identify and give in writing all primes less than 100.

4. Simplify fractions to lowest terms.

5. Use mixed numbers to label points on a number line.

6. Change fractions to decimals.

7. Give standard equivalences for fractions in terms of decimals.

Operations
At the end of grade five the student will:

1. Recall all basic facts of addition, multiplication, subtraction, and division.

2. Find the answers to addition, subtraction, multiplication, and division problems involving whole numbers, fractions, and decimals.

3. Estimate the sums, differences, products, and quotients for problems involving the addition, subtraction, multiplication, and division of whole numbers, fractions, and decimals.

4. List factors of numbers up to 100.

5. List common factors of pairs of numbers less than 100.

6. List multiples of any number.

7. List common multiples of any pair of numbers less than 100.

8. Find the least common multiple of a pair of whole numbers less than 100.

Geometry and Measurement
At the end of grade five the student will:

1. Find the areas of triangles and rectangles by use of the formulas $A = (1/2)bh$ and $A = 1 \times w$.

2. Find the volume of a rectangular solid by using $V = 1 \times w \times h$.

3. Relate the metric and English systems for temperature and mass by listing the units and showing their applications to everyday problems.

4. Show the relationship between the liter and 1000 cm^3 and between the ml and cm^3.

5. Identify and draw examples of segments, rays, lines, perpendicular and parallel lines and segments.

6. Identify and describe right, acute, and obtuse angles.

7. Measure angles using a protractor.

8. Identify and describe congruent segments, angles, and polygons.

9. Describe the angle sum for a triangle, square, and rectangle.

10. Find the lines of symmetry of a given geometric figure.

11. Make a scale drawing (enlargement or reduction) of a given geometric shape.

12. Recognize and describe the following solid geometric regions: rectangular prism, rectangular solid, triangular prism, triangular pyramid, square pyramid, cylinder, cone, sphere.

Problem Solving
At the end of grade five the student will:

1. Handle money situations, including making change, through $100.

2. Draw and interpret graphs (bar, picture, line, circle, or coordinate).

3. Construct a specified scale drawing given the ratio and appropriate drawing tools.

4. Handle application of ratio and average as they apply to standard grade level word problems.

5. Perform estimation (extrapolation) when given a situation involving change governed by a fixed ratio and asked what will happen.

6. Select or write the equation that goes with any of the standard operations studied at the grade level as required by word problems.

7. Determine the rule generating a given number pattern sequence (function machine games).

Grade 6   *Number and Numeration*
At the end of grade six the student will:

1. Read and write whole numbers of twelve digits or less.

2. Read and write decimals through ten-thousandths.

3. Order decimals and whole numbers within the bounds of items 1 and 2.

4. List prime numbers less than 100.

5. Give fractions as the ratio of two quantities.

6. Simplify fractions.

7. Change fractions to decimals and vice versa.

8. Change fractions to mixed numbers and vice versa.

9. Give reciprocals of nonzero fractional numbers.

10. Use integers to show distance and direction on the number line.

11. Give the opposite of any integer.

Operations
At the end of grade six the student will:

1. Handle the measurement activities in the metric and English systems studied at lower grade levels.

2. Convert from one metric unit of measure to another metric unit of the same form of measure; i.e., 35 cm = ___ m.

3. Find the areas of compound figures by breaking them up into smaller regions that are then summed for the final area.

4. Use formulas for the areas of triangles, squares, rectangles, and circles.

5. Estimate the areas of regions given their important dimensions.

6. Describe the relationships between similar figures and congruent figures.

7. Describe and identify examples of parallel lines, perpendicular lines, skew lines, perpendicular bisectors of segments, angle bisectors, and medians of a triangle.

8. Describe the effects of each of the following transformations on a plane figure: slide, flip, and turn.

9. Illustrate the following solids: prisms, parallelepipeds, pyramids, cylinders, spheres, and cones.

10. Draw nets for the construction of given solid regions.

11. Draw the cross-section of a given solid in a given direction.

Problem Solving
At the end of grade six the student will:

1. Use money skills to $1000.

2. Apply money skills in estimating situations.

3. Apply money skills in money management (budget) and banking situations (savings, checking).

4. Handle the construction and interpretation of bar, picture, circle, line, and point graphs.

5. Make a scale drawing using raw data or a given geometric region as a model.

6. Set up and solve application problems appropriate to the grade level involving ratio, average, percent, discount, and scale.

7. State the probability of a given event, given the sample space for the event.

8. List the outcomes for a given probability experiment.

9. Tell whether two events are equally likely or not.

10. Write and solve simple equations involved with verbal problems illustrating the use of grade level operations applied to everyday situations.

Grade 7

Number and Numeration
At the end of grade seven the student will:

1. Read and write whole numbers, fractions, decimals, and integers.

2. Correctly represent numbers in each system on a number line.

3. Use exponents to represent repeated multiplication.

4. Use scientific notation to represent large and small numbers.

5. Relate fractions, decimals, percents.

6. Identify numbers as prime or composite.

7. Use the prime factorization of a pair of numbers to find their lcm and gcd.

8. Identify additive and multiplicative inverses for fractional numbers.

9. Give the ratio of one number to another.

10. Express ratios as percents and vice versa.

Operations
At the end of grade seven the student will:

1. Perform all basic operation with whole numbers, fractions, decimals, and integers.

2. Recognize and use the commutative, associative, and distributive properties in computational settings.

3. Simplify complex fractional situations.

4. Estimate answers to computational problems using scientific notation.

5. Solve proportions using the cross product method.

Geometry and Measurement
At the end of grade seven the student will:

1. Perform the metric and English measurement activities found in the objectives for grades K–6.

2. Classify triangles according to right, scalene, acute, isosceles, obtuse, equilateral.

3. Measure angles with a protractor.

4. Construct congruent segments, angles, and quadrilaterals with a straightedge and compass.

5. Find the areas and perimeters of triangles, quadrilaterals, and circles.

6. Identify the edges, vertices, and faces of a polyhedron.

7. Determine the surface area of a given prism.

8. Determine the volume of a given prism.

9. Determine the surface area of a given cylinder.

10. Determine the volume of a given cylinder.

11. Draw cross-sections and nets for given polyhedra and cylinders.

Problem Solving
At the end of grade seven the student will:

1. Round numbers to the nearest 10, 100, 1000, 1/10, 1/100, or 1/1000.

2. Evaluate given algebraic expressions for specific values of the variables involved.

3. Analyze tabular data using graphs, charts, and other forms of representation.

4. Read and use flow charts describing how to perform a given operation.

5. Find ratios equivalent to a given ratio.

6. Solve proportions where one part of one of the ratios given is missing.

7. Evaluate percent problems using the equivalent ratio model.

8. Recognize the percent as being a composite of amount, base, and percent.

9. Apply ratio and percent ideas to solving speed, unit pricing, and other practical applied problems.

10. List possible outcomes in a simple probability experiment.

11. Determine possible outcomes to an experiment using trees, lists, and counting methods.

12. Find the mean, median, mode, and range for a given set of data.

Grade 8

Number and Numeration
At the end of grade eight the student will:

1. Correctly use positive and negative exponents.

2. Express decimal numbers using scientific notation.

3. Find square roots.

4. Identify various decimals as rational numbers, whole numbers, irrational numbers.

5. Use the radical sign to write irrational roots of non-square integers.

6. Recognize pi as the ratio of the circumference to the diameter of a circle.

7. Identify and relate the additive and multiplicative inverses of a given rational number.

8. Identify and relate the additive and multiplicative identities of numbers in the rational number system.

Operations
At the end of grade eight the student will:

1. Perform the basic computational algorithms with whole numbers, integers, fractions, decimals, and mixed numbers.

2. Recognize and use the various properties of the operations in carrying out the computations.

3. Solve percentage problems and ratio problems using proportions.

4. Simplify complex fractions.

5. Estimate the answers to computational procedures using a variety of methods.

6. Carry out multiplication and division of powers of numbers using rules of exponents.

Geometry and Measurement
At the end of grade eight the student will:

1. Provide the appropriate areas, perimeters, or volumes associated with rectangles, triangles, quadrilaterals, pyramids, cubes,

parallelepipeds, cones, circles, spheres, and cylinders and prisms.

2. Relate metric measurements in the metric and English systems using linear, capacity, mass, temperature, and volume situations.

3. Extend the metric measurement capabilities to include the density of an object.

4. Give the number of significant digits in a given measurement.

5. Tell which of two measurements is more precise.

6. Give the relative accuracy of a given measurement.

7. Apply the Pythagorean Theorem to solve for missing lengths in a right triangle or a situation involving right triangles.

8. Solve for the missing lengths in a pair of similar triangles using the ratio of lengths of corresponding sides of the triangles involved.

9. Solve for missing parts in a triangle using trigonometric ratios (sin, cos, and tan).

Problem Solving
At the end of grade eight the student will:

1. Write and solve linear equations related to verbal problems.

2. Write and solve linear inequalities related to verbal problems.

3. Substitute for variables in given formulas and solve for the value of the desired variable.

4. Graph using ordered pairs in all four quadrants.

5. Construct and use both line and curved graphs on a coordinate axis system.

6. Fit or select an equation for a given line graph on a coordinate system.

7. Determine the square root of a given number to two decimal places.

8. Solve problems involving equivalent ratios.

9. Solve problems where a missing part of the proportion must be determined.

10. Use proportions in setting up and solving trigonometric ratios using sin, cos, and tan.

11. Use proportions in solving percent problems with percents greater than 100 or less than one.

12. Use the statistical concepts of mean, median, mode, and range.

13. Determine the number of possibilities for a given event by counting, making a tree, or by multiplying.

14. Determine the probability of an event given the sample space.

15. Determine whether or not two events are dependent or independent.

16. Determine the probability of the joint occurrence of two independent events by multiplying.

17. Determine the probability of one event or another's happening by use of the addition property.

APPENDIX B   SOURCES OF MATHEMATICS TEACHING AIDS AND TESTS

Teaching Aids

Addison-Wesley Publishing
Company
2725 Sand Hill Road
Menlo Park, CA 94025

American Printing House
for the Blind
1839 Frankfort Avenue
Louisville, KY 40206

Concept Company, Inc.
P. O. Box 273
Belmont, MA 02178

Creative Playthings
Edinburgh Road
Cranbury, NJ 08512

Creative Publications
P. O. Box 328
Palo Alto, CA 94303

Cuisinaire Company
of America, Inc.
12 Church Street
New Rochelle, NY 10805

Developmental Learning
Materials
7440 Natchez Avenue
Niles, IL 60648

Dick Blick
P. O. Box 1267
Galesburg, IL 61401

Educational Teaching Aids
159 West Kinzie Street
Chicago, IL 60610

Ideal School Supply Company
11000 South Lavergne Street
Oak Lawn, IL 60453

The Judy Company
310 North Second Street
Minneapolis, MN 55401

National Council of Teachers
of Mathematics
1906 Association Drive
Reston, VA 22091

Selective Educational
Equipment, Inc. (SEE)
3 Bridge Street
Newton, MA 02195

**Information
About Tests**

Arithmetic Readiness Inventory
Charles E. Merrill Publishing Company
1300 Alum Creek Drive
Columbus, OH 43216

Illinois Test of Psycholinguistic Abilities
University of Illinois Press
Urbana, IL 61801

KeyMath Diagnostic Arithmetic Test
American Guidance Service, Inc.
Publisher's Building
Circle Pines, MN 55014

Kraner Preschool Mathematics Inventory
Learning Concepts
2501 North Lamar
Austin, TX 78705

Level K Mathematics Test of Basic Experiences
CTB/McGraw Hill
Del Monte Research Park
Monterey, CA 93940

Peabody Individual Achievement Test
American Guidance Services, Inc.
Publisher's Building
Circle Pines, MN 55014

Stanford Diagnostic Mathematics Tests
Harcourt Brace Jovanovich, Inc.
Test Department
757 Third Avenue
New York, NY 10017

Stanford-Binet Intelligence Scales
Houghton Mifflin
110 Tremont Street
Boston, MA 02107

Wechsler Intelligence Scale for Children
Psychological Corporation
304 East 45th Street
New York, NY 10017

APPENDIX C MATH LAB ACTIVITIES FOR THE SPECIAL NEEDS CHILD

Number and Numeration

Activity 1: Number Train
  Activity Type: Diagnostic/evaluative; Developmental; Practice.
  Group Size: Individual; Small group.
  Objective: To place numerals in a correct sequence.
  Materials: Deck of cards labeled 0–10.
  Procedure: Cards are shuffled. Place deck face down. Top card is placed face up on table.
  In turn, each child draws a card from the deck and places it in correct sequence in relation to other card(s) on the table. As play continues, each player is responsible for seeing that all cards on the table are in the proper order or sequence.
  Variation: Deck can be expanded to 20–30.

Activity 2: Skip Counting
  Activity Type: Diagnostic/evaluative; Developmental.
  Group Size: Individual; Small group; Class.
  Objective: To practice sequences of numbers.
  Materials: Ten strips of posterboard with ten numbers on each strip, arranged according to a patterned sequence (see sample strips in Figure C.1); pocket made from posterboard or cloth.
  Procedure: Insert the strip in a pocket and show only two numbers. Show the first two numbers of sequence. Encourage children to guess the next number, then the next, etc. The strips may be

Figure C.1

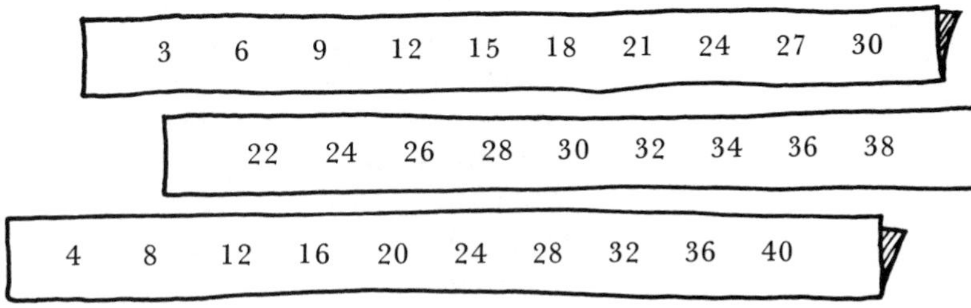

| 3 | 6 | 9 | 12 | 15 | 18 | 21 | 24 | 27 | 30 |

| 22 | 24 | 26 | 28 | 30 | 32 | 34 | 36 | 38 |

| 4 | 8 | 12 | 16 | 20 | 24 | 28 | 32 | 36 | 40 |

made so that a sequence reads backwards as in the following example: 50, 45, 40, 35, 30, 25, 20, 15, 10, 5.

Activity 3: How Many?
    Activity Type: Diagnostic/evaluative; Developmental.
    Group Size: Individual; Small group.
    Objective: To identify the number of elements in a group.
    Materials: Cards made from black construction paper with dots of glue to represent 0–9; include different representations for each number.
    Procedure: Ask a child to tell the number of dots on each card. If there are four or less dots on a card, challenge children to say it fast. Otherwise encourage children to *count on* from a smaller sight group.

Activity 4: Hidden Matching Sets
    Activity Type: Developmental.
    Group Size: Small group.
    Objective: To recognize equivalent sets by identifying them through matching.
    Materials: Large sheet of cardboard; set of 18 pairs of matching cards showing pictures of objects (1–9 pictures per card); attach a string to each card.
    Procedure: Cards are hidden under the large sheet of cardboard with the strings extending around the edge.

In turn, children pull two strings. If the set pictures match in number, players keep the cards. If not, the cards are replaced.

Play continues with each child pulling two strings until all cards are matched. The winner is the child with the most matching pairs of cards.

    Variation: Matches may be set-set; set-numeral; set-word; numeral-word.

Activity 5: Lace A Pattern
    Activity Type: Diagnostic/evaluative; Developmental.
    Group Size: Individual; Small group.
    Objective: To recognize pattern sequences.
    Materials: Lacing string; beads of assorted colors and shapes; cards with pictures of bead patterns.
    Procedure: A child looks at a bead pattern on a card and reproduces the pattern exactly in that order on a string. An example of a pattern card is shown in Figure C.2.
    Variation: Ask that the child reverse the pattern of the card on the lacing string.

Activity 6: Count Along Animals
    Activity Type: Practice.

Group Size: Small group.

Objective: To practice counting, 0–9.

Materials: Gameboard like that of Figure C.3; number cube 1, 2, 3, 1, 2, 3; numeral cards, 0–9; gameboard markers.

Procedure: Children take turns throwing the number cube. After each throw, players move their marker the number of spaces indicated on the number cube.

Whenever a child's marker lands on a space where an animal is pictured, the child gets to draw a card from a stack of numeral cards. If, for example, the child draws a "3" numeral card and the space marker is on a cat, the child meows three times.

Figure C.2   Example of a pattern card:

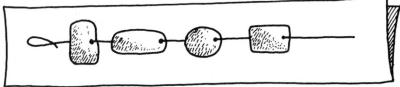

Figure C.3

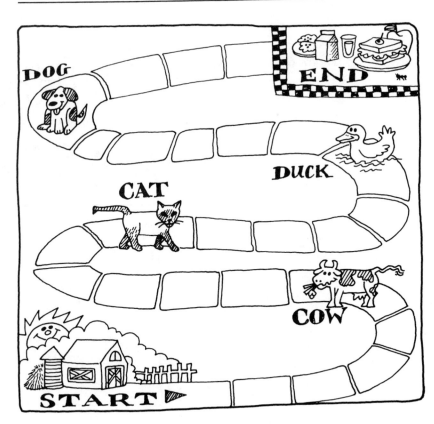

Activity 7: Counting Number Fence
    Activity Type: Practice.
    Group Size: Individual; Small group.
    Objective: To practice counting, 0–10.
    Materials: tongue depressors; masking tape.
    Procedure: Use the masking tape as shown in Figure C.4a to attach eleven tongue depressors to the chalkboard, a chart, or sturdy cardboard. Children can use this counting number fence to practice counting and to determine what number comes before, after, or between another number. Blank tongue depressors can be used to cover numbers, as in Figure C.4b.

Whole Number Addition

Activity 1: Match Dots To Facts
    Activity Type: Diagnostic/evaluative; Developmental.
    Group Size: Individual.
    Objective: To match domino dots to basic addition facts.
    Materials: Lady bugs and matching fact circles as shown in Figure C.5; the materials can be placed in a file folder as shown in Figure C.5.
    Procedure: A child matches the combination of dots to the number fact it illustrates.

Figure C.4

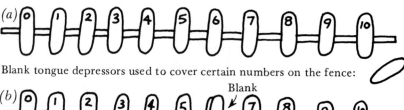

Blank tongue depressors used to cover certain numbers on the fence:

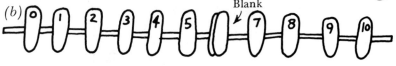

Figure C.5

Match the spots to the number facts.

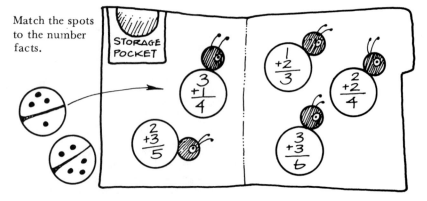

Activity 2: Family Tree Of Six

    Activity Type: Developmental; Practice.

    Group Size: Individual; Small group.

    Objective: To group chips to find the members of a given addition fact family.

    Materials: Laminated posterboard tree; cardboard chips; paper clips; marker.

    Procedure: For a family tree of six, children count out six chips. The teacher or other members of the group can check to see that there are exactly six chips. Children then try to find all pairs of numbers whose sum is 6. They can experiment by arranging the chips in various groupings. Children can see, for example, that a group of two and a group of four make six. This fact, 2 + 4 = 6, is then written on the trunk of the tree.

    Other groupings of chips are then made until all combinations of six are found. Then the six counters are placed on tree branches.

    Challenge children to find all combinations of a family tree, including commutatives such as 2 + 4 = 6 and 4 + 2 = 6.

    Variation: This activity can be carried out for all of the basic addition facts. Family Trees of 2, 3, 4, 5, 6, 7, . . . , 18 can be made.

Activity 3: Roll To 100

    Activity Type: Developmental; Practice.

    Group Size: Small group.

    Objective: To add one- and two-digit whole numbers.

    Materials: Base-ten blocks; place value chart for each child as shown in Figure C.6, 2 number cubes labeled 2, 4, 5, 6, 7, 8, and 1, 3, 6, 7, 8, 9.

    Procedure: Children, in turn, roll both number cubes, add the numbers, and get that number of base-ten blocks to put on their place value chart. (For example, if a 7 and 8 is rolled, the child gets 1 ten-block and 5 ones to put on the chart.) Appropriate trades of base-ten blocks are made each time. For example, if a child already has 15 on board and rolls a 9 and 8, 17 is added to 15. Now trades are made and the child puts 3 ten-blocks and 2 ones on the chart. The winner is the one who first reaches exactly 100.

Figure C.6

Activity 4: Over One Hundred
  Activity Type: Practice.
  Group Size: Small group.
  Objective: To practice addition of one- and two-digit whole numbers.
  Materials: Standard deck of playing cards.
  Procedure: Dealer gives each player three cards face down.

  First player places a card on the table and declares its value. (Ace = 1 or 11; 2 through 8 = face value; 9 = 0; 10 = 10; face cards = 10 each) That player then draws another card from the deck.

  The second player plays one card and adds its value to the value of all cards on the table. Play continues until one of the players goes over 100. That person is the winner.

Activity 5: Domino Addition
  Activity Type: Practice; Application.
  Group Size: Individual; Small group.
  Objective: To practice the basic addition facts.
  Materials: Dominoes; matching addition fact cards as shown in Figure C.7; make dominoes from black paper using glue to make white stripes and dots; mat can be placed inside file folder as shown in Figure C.7.
  Procedure: Children match the dominoes with the number facts they show.

Activity 6: Build An Addition Wall
  Activity Type: Practice; Application.
  Group Size: Small group; Class.
  Objective: To practice the basic addition fact families.
  Materials: Seven wall cards like those shown in Figure C.8a; "bricks" (cards) with pairs of addends such as 3 + 4, 5 + 3, 1 + 4, 5 + 1, etc.
  Procedure: Each player is given a wall card. Then the player "builds" the wall by placing bricks on the wall as shown in Figure C.8b. In the sample, the fact family of five is built.

  For example, the 7 wall card can have the following bricks placed on it: 7 + 0, 0 + 7, 1 + 6, 6 + 1, 5 + 2, 2 + 5, 3 + 4, 4 + 3.
  Variation: For children requiring more structure in activities, draw as many bricks on a wall card as there are facts for that number.

Activity 7: Aim For 999
  Activity Type: Practice; Application.
  Group Size: Small group; Class.
  Objective: To practice addition of 1-, 2-, and 3-digit whole numbers.

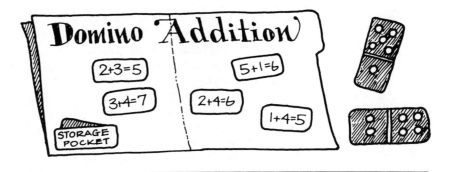

Figure C.7

Figure C.8   Wall Card   (a)

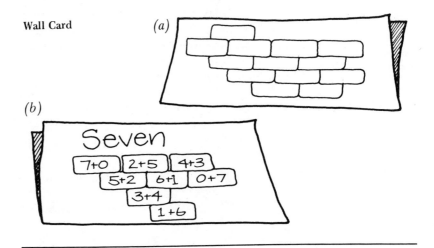

(b)

Figure C.9

Materials: Set of ten cards labeled 0–9, paper and pencil.
Procedure: Players draw the chart of Figure C.9 on their papers.
Cards are then shuffled and placed face down. The top card is
turned over. Each player records that digit under any column.
As each card is played, the children record that digit on their
chart. Once a digit is written, it cannot be changed.

When six cards have been played, children add the two
three-digit numerals. The winner is the child with the greatest
sum for that round. Play five rounds.

Activity 8: Kicking The Football

Activity Type: Practice; Application.

Group Size: Small group.

Objective: To use place value concepts in addition of whole numbers.

Materials: Football gameboard (Figure C.10); two number cubes, one white (for tens): 0, 1, 2, 2, 3, 3, and 1 red (for ones): 10, 12, 14, 16, 18, 8; markers in the shape of a football (different color for each player).

Procedure: Each child selects a football marker. First player throws both number cubes. If a child throws 2 tens and 16 ones, the player places the football on the 36-yard line.

Each child gets one turn. The object of the game is to kick the football the farthest. One point is scored each round by the player with the longest punt.

Any child who kicks the ball more than fifty yards always scores one point if that child can state the total number of yards kicked correctly. The first player to score ten points is the winner.

Whole Number Subtraction

Activity 1: Estimate Differences

Activity Type: Developmental.

Group Size: Small group.

Objective: To estimate the difference of two four-digit whole numbers.

Materials: Gameboard (see Figure C.11a); twenty problem cards with one subtraction (4-digit minus 4-digit) problem per card (see Figure C.11b), answers can be written on the backs of cards.

Procedure: The card deck is shuffled and spread out on the table, face down. In turn, players each take a card, estimate the difference and place the card on the correct space of the gameboard. The fastest player (if correct) gets two points. The winner is the player with the most points when all cards in the deck are used.

Activity 2: Subtract–0

Activity Type: Diagnostic/evaluative; Developmental.

Group Size: Small group.

Objective: To subtract with renaming/regrouping using base-ten blocks.

Materials: Two number cubes labeled: 3, 4, 5, 6, 7, 8; 4, 5, 6, 7, 8, 9; base-ten computer chart for each player (see Figure C.12).

Procedure: Each child begins with 1000, 1 block (B).

In turn, children toss the cubes and the product of the toss is subtracted from 1000. For example, if a 7 and 8 are tossed, then 56 is the product which must be subtracted from 1000. Appropriate trades are made with the base-ten blocks when necessary. The winner is the first to reach zero.

Figure C.10

Gameboard

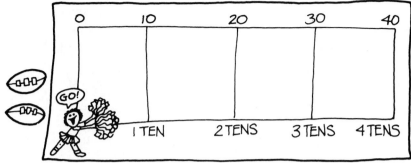

Figure C.11

(a) Sample cards:

(b) Gameboard

Estimates

100	400	700
200	500	800
300	600	900

Figure C.12

B	F	L	U
1	0	0	0

Activity 3: Get To Zero!

    Activity Type: Developmental; Practice.

    Group Size: Small group.

    Objective: To subtract from 500 using base-ten blocks and a place-value chart.

    Materials: Deck of ten cards labeled 0, 1, 2, . . . , 9; base-ten blocks; place-value chart for each player (sheet of paper with three columns labeled) as in Figure C.13.

    Procedure: Each child starts with five hundred-blocks (flats). Cards are shuffled and placed face down between players.

In turn, children draw two cards and form a two-digit number. For example, if a 3 and 7 are drawn, then 37 or 73 can be formed. Since the object is to get to zero by subtracting from 500, it would be to the child's advantage to form 73.

The player makes appropriate trades, if necessary. For example, one hundred-block can be traded for 9 ten-blocks and one set of ten unit-blocks.

The winner is the first to reach zero.

Activity 4: One Thousand One

    Activity Type: Practice.

    Group Size: Small group.

    Objective: To subtract 1-, 2-, or 3-digit numbers from 1001.

    Materials: Three number cubes with: 0, 1, 2, 3, 4, 5,; 3, 4, 5, 6, 7, 8; 4, 5, 6, 7, 8, 9.

    Procedure: To begin a round, children write 1001 on their papers. In turn, children roll number cubes, form a three-digit number from the cubes and subtract that number from 1001. The winner of the round is the player with the smallest difference. The winner of each round gets two tally points. Play ten rounds.

    Variation: Winner is the player with largest answer.

Whole Number Addition and Subtraction

Activity 1: Feeding Trixy The Dog

    Activity Type: Practice; Application.

    Group Size: Class.

    Objective: To practice the basic facts of addition and subtraction.

    Materials: One toy dog, Trixy; two bowls; 30 fact bone cards (see Figure C.14).

    Procedure: The children form two teams. Two bowls are placed near Trixy. The players on each team take turns trying to give the answer to a basic fact shown on a bone fact card. If the answer is correct, the child gets to place the fact bone in that team's bowl. The object of the game is to see which team can feed Trixy the most bones.

    Variation: Whole-number operations of addition, subtraction, multiplication, or division facts may be written on the bones.

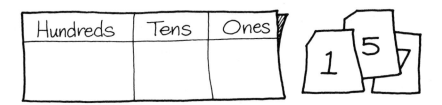

Figure C.13

Figure C.14   Examples

Activity 2: Three Tries To Make The Number
   Activity Type: Practice.
   Group Size: Small group.
   Objective: To practice combinations of whole number addition and subtraction facts.
   Materials: Three dice, paper, and pencil.
   Procedure: One player picks a number between 1 and 18. The next player rolls the dice and tries to make that number, using addition or subtraction with the three numbers shown on the dice. All three dice must be used and all addition, all subtraction, or a combination of addition and subtraction may be used in the same turn. The player can throw the dice up to three times to try to make the number called. For example, one player picks 7 and the next rolls 2, 4, and 5. This player may say $5 + 4 - 2 = 7$. If the player makes the number in one of the three tries, the player gets two points. Procedure continues until one person gets twenty points. That player is the winner.

Whole Number Multiplication

Activity 1: Multi-Fact Find
   Activity Type: Practice.
   Group Size: Small group.
   Objective: To practice basic multiplication facts.
   Materials: Laminated gameboard with factors 2–9 arranged in mixed order, as shown in Figure C.15; card deck with products written on them.
   Procedure: To begin, cards are shuffled. The top card is then turned over and played on an appropriate square.

Figure C.15

X	6	2	7	3	5	8	4	9
3								
4								
9								
6								
2								
8								
7								
5								

Players in turn must play a card adjacent to one already on the board.

Each time a child makes a play, two points are earned. If a card cannot be played adjacent to one on the board, it must be placed at the bottom of the deck.

The winner is the player with the highest score after ten rounds.

Variation: Addition or subtraction may also be used for such an activity.

Activity 2: Factor Feat

Activity Type: Practice.

Group Size: Small group; Class.

Objective: To recognize factors of a given number.

Materials: Laminated Factor Feat gameboard as shown in Figure C.16; markers.

Procedure: First player (team) makes a selection of any one number on the playing board. Initially this number is the player's score.

Second player (team) tries to list all factors of that number. These are added to the player's score. This player then chooses another number from the board and adds it to the total score for that round.

The game continues in alternating turns. Each number is crossed out as it is used. When all numbers have been used, the player with the highest score wins.

Activity 3: Product War
  Activity Type: Diagnostic/evaluative; Practice.
  Group Size: Small group.
  Objective: To practice the basic multiplication facts.
  Materials: Two decks of cards: each set has 44 cards labeled
0–10 (four of each number).
  Procedure: The two decks of cards are shuffled and placed
face down in two piles.
  Two players each draw one card from each deck and place them
face up in the playing area (see Figure C.17).
  The first player to say the correct product takes both cards. Play
continues in this manner.
  The winner is the player with the most cards after all cards are
used.

Activity 4: Multo-Bingo Grid
  Activity Type: Practice.
  Group Size: Small group; Class.
  Objective: To practice the basic multiplication facts through a
bingo game.

Figure C.16

2	3	4	5	6	7	8
9	10	11	12	13	14	15
16	17	18	19	20	21	22
23	24	25	26	27	28	29
30	31	32	33	34	35	36
37	38	39	40	41	42	43
44	45	46	47	48	49	50

Figure C.17   Sample Play

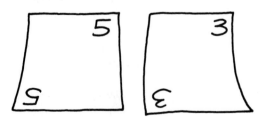

Materials: For each player, a 6 x 6 laminated grid on poster-board (like the one of Figure C.18); marker; twenty-five cards marked 18, 21, 24, 27, 30, 36, 42, 48, 54, 60, 42, 49, 56, 63, 70, 48, 56, 64, 72, 80, 54, 63, 72, 81, 90.

Procedure: Each player randomly marks the following numbers along left column of the grid: 10, 9, 8, 7, 6. Mark along the top row: 9, 8, 7, 6, 3. Place a multiplication sign in the upper left corner. Students now have their own individual multiplying grids.

The caller (teacher or other student) shuffles the twenty-five cards and then draws the top one and reads it. The players record the number in only *one* space on their grid. The caller continues to draw and call until someone says MULTO-BINGO (five in a row).

Variation: Addition.

Whole Number Division

Activity: Remainders Do It

    Activity Type: Practice.

    Group Size: Small group.

    Objective: To practice division with remainders where the divisor is a one-digit number.

    Materials: Number cube labeled 4, 5, 6, 7, 8, 9; gameboard as shown in Figure C.19; gameboard markers.

    Procedure: In turn, children throw the number cube and divide the first number on the gameboard by the number indicated on the cube. The remainder of the problem dictates the number of spaces the player moves.

First player to reach the *Finish* space is the winner.

Whole Number Computation

Activity: Operate

    Activity Type: Diagnostic/evaluative; Practice.

    Group Size: Individual; Small group; Class.

    Objective: To rename numbers in as many ways as possible using any of the four basic operations.

    Materials: Five sheets of laminated construction paper for each child; marker for each.

    Procedure: Leader or teacher calls out a number (1–15). Children try to write all possible ways of obtaining that number by using any of the four basic operations on whole numbers. These can be posted on the classroom bulletin board. An example for the number 6 is shown in Figure C.20.

Decimals

Activity 1: Building A Train Of Decimals

    Activity Type: Diagnostic/evaluative; Developmental.

    Group Size: Small group.

    Objective: To correctly order decimal cards in a sequence.

Figure C.18

Multi-Bingo Grid

×	6	3	8	7	9
6					
8	48				
9		27			81
7	42	21	56	49	63
10			80	70	

Figure C.19

Gameboard

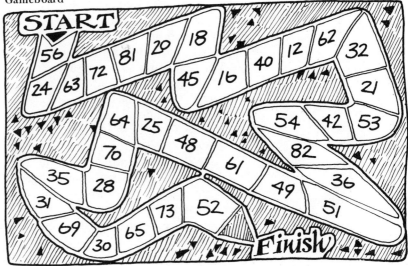

Figure C.20

Example: Number is 6

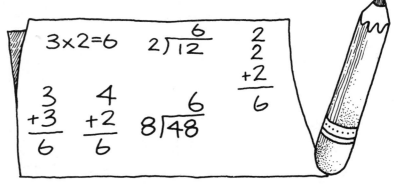

Materials: Set of twenty-five decimal cards that have decimal values less than one (see Figure C.21).

Procedure: Build a train of decimals by playing the cards in order from least to greatest.

Activity 2: Color 100%

Activity Type: Diagnostic/evaluative; Developmental; Practice; Application.

Group Size: Small group.

Objective: To relate decimal values with percent equivalents.

Materials: Two 10 x 10 square centimeter grids laminated on posterboard; two sets of six markers (each set a different color); two number cubes marked as follows: (a) 40%, .25, .1, 5%, .15, 30%; (b) .35, 45%, 10%, .01, 15%, .2.

Procedure: In turn, players roll one of the cubes. Then, using a different colored marker for each roll, players color squares on their grids to represent the decimal or percent that is rolled. If a number is rolled that is larger than the total of the uncolored parts of all the unit squares, the player loses turn.

Winner is first to completely color in 100% of the playing board.

Activity 3: Greater Than, Less Than, Equal To

Activity Type: Developmental; Practice.

Group Size: Small group.

Objective: To decide whether two decimals are equivalent, greater than, or less than the other.

Materials: Deck of thirty-six cards, labeled as shown in Figure C.22a; a laminated three-column answer sheet for each player (see Figure C.22b); grease pencils.

Procedure: Cards are shuffled and placed face down. Children work in pairs.

In turn, students draw and record the decimal in the proper column (first or third) on the answer sheet. Both children decide what sign, < or >, should be written in the middle column of the sheet. Repeat procedure ten times.

Figure C.21

.1	.15	.01	.06	.63
.2	.26	.02	.07	.78
.3	.35	.03	.08	.83
.4	.42	.04	.6	.92
.5	.51	.05	.7	.31

Decimal Train

Activity 4: Decimal War
 Activity Type: Practice.
 Group Size: Small group.
 Objective: To practice decimals recognition of the value of a decimal.
 Materials: Set of 52 cards with decimals written on them (see Figure C.23).
 Procedure: The deck is shuffled, split, and placed face down in two piles.

 Two players each draw a card from one of the piles and place it **face up** between them. The player who has the decimal card with **the greater** decimal value takes both cards. If players draw equivalent values, they must draw additional cards to break the tie. The **winner,** when the deck is used, is the player with the most cards.

Figure C.22

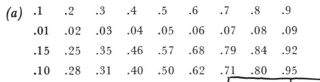

(a) .1 .2 .3 .4 .5 .6 .7 .8 .9

.01 .02 .03 .04 .05 .06 .07 .08 .09

.15 .25 .35 .46 .57 .68 .79 .84 .92

.10 .28 .31 .40 .50 .62 .71 .80 .95

(b)

Player 1	Sign	Player 2

3 column answer sheet for each player

Figure C.23

.5   .23   .02

Fractions

Activity 1: Draw It Or Say It
 Activity Type: Diagnostic/evaluative; Developmental.
 Group Size: Individual.
 Objective: To use the word *of* in shading multiplication of fraction problems in order to see the result.
 Materials: Ten sets of matching cards, laminated, like those of Figure C.24.
 Procedure: Mix the cards and challenge children to find matching pairs.

Variation: Have students draw a card and (a) write the multiplication sentence which is pictured, or (b) sketch the multiplication illustration to match the equation drawn.

Activity 2: Greater Than Or Less Than
   Activity Type: Developmental; Practice.
   Group Size: Small group.
   Objective: To determine whether one fraction is greater than or less than another fraction.
   Materials: Number cube labeled: 1/2, 1/3, 1/4, 1/5, 1/5, 1/8; set of fraction strips; laminated three-column answer sheet as shown in Figure C.25; grease pencils.
   Procedure: In turn, children roll the number cube and take equivalent fraction strip and write the fraction in the proper column (first or third) on the answer sheet. Then both players compare fraction strips to see what sign ($<$, $>$) should be recorded in the middle column of the sheet.
   Repeat procedure ten times.

Activity 3: Equifrac Card Game
   Activity Type: Practice.
   Group Size: Small group.
   Objective: To practice matching cards that have equivalent values.
   Materials: An Equifrac Card deck (see sample deck of 52 cards in Figure C.26).
   Procedure: Shuffle the Equifrac cards and deal the entire deck. Players match equivalent fractions from their hands and place the cards face up on the table. If a player has two, three, or four

---

Figure C.24   10 cards with pictures of shaded multiplication problems such as

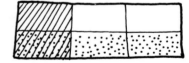

10 cards with sentences such as 1/2 of 1/3 = 1/6

Figure C.25

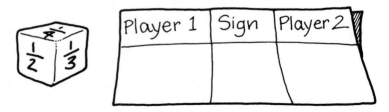

equivalent fractions, the player places all the cards on the table. The dealer begins play by calling for a card from *one* of the players. *Sample call:* "Give me 5/8." If the player has the card requested by the dealer, the dealer places the two cards on the table. The dealer continues to call for cards from this player. If the player does not have a card requested by the dealer, this ends the dealer's turn, and the child to the right of the dealer has a turn. Play continues in this manner until one player runs out of cards.

Measurement

Activity 1: Time Question Cards
  Activity Type: Diagnostic/evaluative; Developmental.
  Group Size: Individual; Small group.
  Objective: To tell what time a clock reads and to draw the time on a clock.
  Materials: Set of twenty cards that are laminated (using two types of time cards as shown in Figure C.27); markers.

Figure C.26

$\frac{5}{9}$	$\frac{10}{18}$	$\frac{15}{27}$	$\frac{20}{36}$	$\frac{1}{8}$	$\frac{2}{16}$	$\frac{3}{24}$	$\frac{4}{32}$
$\frac{1}{3}$	$\frac{2}{6}$	$\frac{3}{9}$	$\frac{4}{12}$	$\frac{5}{8}$	$\frac{10}{16}$	$\frac{15}{24}$	$\frac{20}{32}$
$\frac{2}{3}$	$\frac{4}{6}$	$\frac{6}{9}$	$\frac{8}{12}$	$\frac{7}{8}$	$\frac{14}{16}$	$\frac{21}{24}$	$\frac{28}{32}$
$\frac{1}{9}$	$\frac{2}{18}$	$\frac{3}{27}$	$\frac{4}{36}$	$\frac{1}{2}$	$\frac{2}{4}$	$\frac{4}{8}$	$\frac{3}{6}$
$\frac{1}{5}$	$\frac{2}{10}$	$\frac{3}{15}$	$\frac{4}{20}$	$\frac{1}{4}$	$\frac{2}{8}$	$\frac{3}{12}$	$\frac{4}{16}$
$\frac{2}{5}$	$\frac{4}{10}$	$\frac{6}{15}$	$\frac{8}{20}$	$\frac{3}{4}$	$\frac{6}{8}$	$\frac{9}{12}$	$\frac{12}{16}$
$\frac{1}{6}$	$\frac{2}{12}$	$\frac{3}{18}$	$\frac{4}{24}$				

Figure C.27

Procedure: Children draw and follow directions on the cards. Answers can be checked by another child or an answer key can be provided.

Activity 2: Time Dominoes
    Activity Type: Developmental.
    Group Size: Small group.
    Objective: To tell what time is represented on a clock and to read what time it is in words.
    Materials: Set of domino cards that represent time as shown in Figure C.28a.
    Procedure: Play Time Dominoes as a game of conventional dominoes. Children build a domino train by matching ends as shown in Figure C.28b.

Activity 3: Money Jigsaw Cards
    Activity Type: Developmental; Application.
    Group Size: Individual; Small group.
    Objective: To match the picture of coins to their appropriate values.
    Materials: Set of ten jigsaw cards as shown in Figure C.29.
    Procedure: Children match coin stamps to the correct money value on the jigsaw pieces.

Activity 4: Completing The Circle
    Activity Type: Developmental; Practice.
    Group Size: Small group.
    Objective: To add/subtract angle measurements and then draw in the angle in a circle.
    Materials: Three dice (see sample labeling of Figure C.30); a circle for each player (one radius drawn); pencils; protractors.
    Procedure: To determine who goes first, each player throws the three dice and adds or subtracts the numbers as indicated. The person with the largest angle measurement goes first.
    In turn, players roll the dice and add or subtract as indicated to determine the measure of the central angle to be drawn in their circle. Students use protractors to draw the angles (center of circle as vertex). Continue this procedure, drawing adjacent angles until the circle has been entirely completed. The final angle must fit exactly. The first player completing a circle is the winner.

Activity 5: Change For A Dollar
    Activity Type: Practice; Application.
    Group Size: Small group; Class.
    Objective: To practice in making change for a dollar in as many ways as possible.
    Materials: Paper; pencil; play or real money (coins).

Figure C.28   (a)

(b)

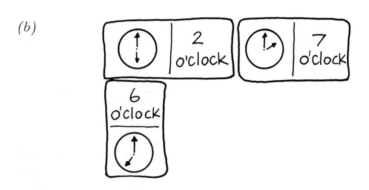

Figure C.29   (a)

(b)

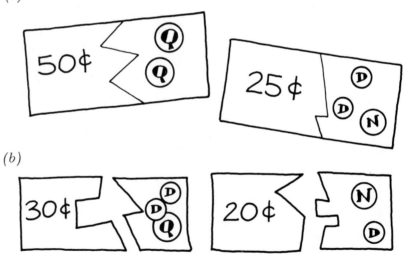

Procedure: Children form two teams. Each team lists as many different ways as they can for making change for a dollar within a given time limit. For example, two half-dollars, 100 pennies, ten dimes, etc.

When time is called, the winning team is the team with the longest list. Teams trade papers to check each other's work for correct responses. No combination may be repeated.

Activity 6: Make A Dollar
    Activity Type: Practice; Application.
    Group Size: Small group.
    Objective: To practice making change for a dollar.
    Materials: Three number cubes labeled as follows: 5¢, 10¢, 30¢, 6¢, 7¢, #; 1¢, 20¢, 25¢, 8¢, 9¢, #; 2¢, 3¢, 4¢, 15¢, 10¢, #.
    Procedure: A small group of children and a banker are needed. First player throws the three number cubes and asks the banker for the total amount indicated by the number cubes. For every # that appears, the player must give the banker 5 cents.

Players can use real or play money to help them make necessary trades. The first player to reach $1. is the winner.

Activity 7: Estimate
    Activity Type: Diagnostic/evaluative.
    Group Size: Individual; Small group.
    Objective: To estimate various lengths of objects in the room.
    Materials: Pencil; handout like that below.
    Procedure: Have children complete the chart. For each item they should first estimate, then measure.

Object	Estimate	Actual
Pencil		
Chalkboard		
Chalkboard eraser		
Fingernail		
Wrist		
Arm length		
Head size		
Bottom of shoe		
Height of door		
Width of paper		
_____		
_____		
_____		

Geometry

Activity 1: Attribute Grid

  Activity Type: Diagnostic/evaluative; Developmental; Practice.
  Group Size: Individual; Small group.
  Objective: To identify 1-, 2-, and 3-differences between attribute blocks on a grid.
  Materials: 5 x 5 gameboard grid like that of Figure C.31; set of attribute blocks (two sizes, four colors, four shapes).
  Procedure: In turn, players place an attribute block on the grid. *One point* is received if the block differs in *one* way (attribute) from a block on the board (i.e., if it is horizontally adjacent to a block). *Two points* are received if the block differs in *two ways* (attributes) from a block on the board (i.e., if it is vertically adjacent to a block). *Three points* are received if the block differs in *three ways* (attributes) from a block on the board (i.e., if it is diagonally adjacent to a block on the board). Winner is the person with the most points.

Activity 2: Forming Shapes

  Activity Type: Diagnostic/evaluative; Developmental; Practice.
  Group Size: Individual; Small group.
  Objective: To use geoboards to form different geometric shapes.
  Materials: Rubber bands; geoboards.
  Procedure: Leader names a geometric shape such as a square, rectangle, or triangle. Children put rubber bands around the nails on the geoboard to form the geometric shape.

Activity 3: Match The Shapes

  Activity Type: Diagnostic/evaluative; Developmental.
  Group Size: Individual.
  Objective: To match a shape piece with a picture already drawn of that shape.

Figure C.31

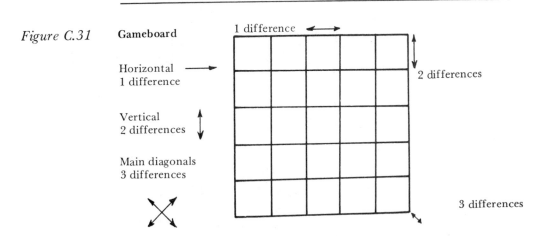

Gameboard

1 difference

Horizontal 1 difference

Vertical 2 differences

Main diagonals 3 differences

2 differences

3 differences

Materials: File folder, as shown in Figure C.32. (Cut the shapes out of construction paper, trace them on the folder, then store the shape pieces in the pocket.)

Procedure: Children match the shape pieces with the outlines.

Activity 4: Geo-Concentration
   Activity Type: Diagnostic/evaluative; Developmental.
   Group Size: Small group.
   Objective: To use new geometry vocabulary in determining when a match is made between a geometric word and its symbol.
   Materials: Deck of thirty cards like those shown in Figure C.33.
   Procedure: Shuffle the cards and place them face down in a 5 x 6 array. Children take turns turning over two cards. If the cards match, the child keeps those two cards. If the cards do not match, they must be replaced. The winner is the child with the most matching pairs of cards after all cards have been played.

Activity 5: Symmetry In The Alphabet
   Activity Type: Developmental.
   Group Size: Individual; Small group; Class.
   Objective: To identify which letters of the alphabet have symmetry and then to find the lines of symmetry.
   Materials: A handout with the letters of the alphabet written in block form.
   Procedure: Children determine which letters of the alphabet have:

- Line symmetry

- Point symmetry

- Both line and point symmetry

- Neither line nor point symmetry (the letters are not symmetric).

Figure C.32

Figure C.33

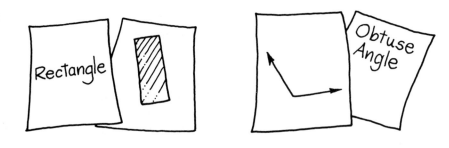

Figure C.34

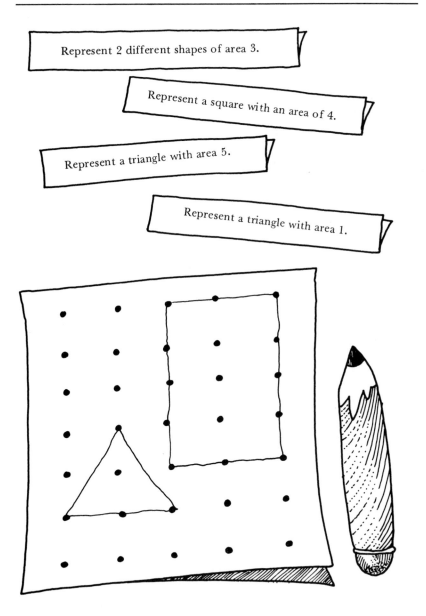

Activity 6: Area Of Shapes On Geoboard

Activity Type: Developmental.

Group Size: Individual; Small group.

Objective: To use a geoboard and identify shapes of different named areas.

Materials: Geoboard; dot paper; ten cards listing a problem on each card (see sample cards in Figure C.34).

Procedure: Have children use the geoboard to try to represent the shape and area indicated on a card. Then have them draw the shape onto dot paper.

INDEX

place value 171–184
plus 203
polyhedron 478
practice activities 64–65
precision 416–418, 460
prenumber experiences 118–120, 138–139
preoperational stage 82
prism 456–457, 477, 480
problem solving 4
processing skills 47
proportions 404–407
Public Law 94–142
pyramids 477

Q
quadrilaterals 482

R
ray 458
reactive teaching 67–68
reception skills 25
rectangles 444–445, 487–489
reinforcement 108–109
remediation 265
repeated addition 275
research on teaching 94–96
resource personnel 16
retardation, levels of 19, 50

S
schema 80
screening tests 25–26
semiconcrete representations 85
sensory handicapped 38–43
sensorimotor stage 81–82
seriation
    activities 140–146
    developmental sequence 139–140
    role in curriculum 138–139
sets 133–134
similarity 483, 493, 496
sight groups 153, 156
skills 4, 59–60, 77
slow learner 37–38
special needs 10–11, 17, 37–49
speech or language impaired 40–41
spiral curriculum 87
squares 481, 487–489

subtraction
    of decimals
        algorithms 341–342, 344
        error patterns 358–362
        models 342–344
    of fractions
        algorithms 382–390
        error patterns 402–403
        models 382–389
    of whole numbers
        activity file 254–256, 258–261
        algorithms 254–262
        basic concepts 221–227
        basic facts 221–233
        big ideas 199–200, 254
        comparison 224–225
        computation 245–262
        difficulties 263–265
        fact mastery 200–201, 227–230
        models 221–227
        regrouping 257–265
        relation to addition 202–203, 222–223, 231
        remedying problems 265
        special helps 266–269
        strategies 228–230
        take-away method 221–222
        teacher background 196–201
        trouble shooting 263–269
        zeros in minuend 261
supplementary angles 497
symbolic stage 85
symmetry
    line 465, 475, 483, 489, 491, 493–495
    plane 479

T
tangrams 495
teaching strategies 94–96
tests
    achievement 14
    Arithmetic Readiness Inventory 25
    Comprehension Tests of Basic Skills 21
    criterion-referenced 21
    KeyMath 23, 33

Level K Mathematics Test of Basic Experiences 25
ITPA 25–26
Iowa Test of Basic Skills 21
Metropolitan Achievement Test 21
norm-referenced test 21
Peabody Individual Achievement Test 23
Stanford Achievement Test 21, 30
Stanford-Binet Individual Intelligence Test 20
Stanford Diagnostic Mathematics Tests 23, 30
SRA Achievement Series 21
Weschler Intelligence Scale for Children 20
thinking strategies 210–212, 228–230, 278–281
time
    activities for 435–436, 439–440
    teaching of 435–440
trainable students 51
transitional activities 62, 114–115
trapezoid 448–449
treatment 71–74
triangles
    classification of 482
    measurement of 446–447

U
union of sets 134

V
visually impaired 41–43
volume
    activities for 63, 454, 458
    computational formulas for 454–457
    measurement of 451–458
    units of 452–454
weight
    activities for 434
    measurement of 432–434
    units for 434
whole numbers—see specific area

Z
zero, concept of 151, 156–157